COLOSSAL MAGNETORESISTANCE, CHARGE ORDERING and RELATED PROPERTIES of MANGANESE OXIDES

Editors

C. N. R. Rao

Jawaharlal Nehru Centre for Advanced Scientific Research

and

Indian Institute of Science, Bangalore

India

B. Raveau

ISMRA

University of Caen

France

World Scientific

Singapore • New Jersey • London • Hong Kong

Published by

World Scientific Publishing Co. Pte. Ltd.

5 Toh Tuck Link, Singapore 596224

USA office: 27 Warren Street, Suite 401-402, Hackensack, NJ 07601

UK office: 57 Shelton Street, Covent Garden, London WC2H 9HE

British Library Cataloguing-in-Publication Data
A catalogue record for this book is available from the British Library.

First published 1998
Reprinted 2004

COLOSSAL MAGNETORESISTANCE, CHARGE ORDERING AND RELATED PROPERTIES OF MANGANESE OXIDES

ISBN 981-02-3276-4

Printed in Singapore by Multiprint Services

PREFACE

Metal oxides constitute the most amazing class of materials with a wide range of properties. They exhibit a variety of phenomena such as ferroelectricity, ferromagnetism, superconductivity and so on. The area of metal oxides became specially important after the discovery of high-temperature superconductivity which created a great sensation. In the last four years, a new aspect of metal oxide has come to the fore. This has to do with colossal magnetoresistance exhibited by certain manganese oxides, in particular rare earth manganates of perovskite structure. Some of these oxides exhibit magnetoresistance to the tune of 100%. Since 1993, there has been intense activity in the area of colossal magnetoresistance in manganese oxides and related systems, part of the motivation being the potential technological applications. Studies of colossal magnetoresistance have led to many other new phenomena and properties such as charge-ordering and orbital ordering.

In view of the importance of colossal magnetoresistance, charge-ordering and related phenomena exhibited by manganese oxides to the physics and chemistry of solid materials, we considered it worthwhile to bring out a volume of contributed articles by leading experts. We believe that this small voloume consisting of eleven articles nearly represents the present status of a subject which is growing rapidly. We start with a review on the status and then include contributions covering the present status of the subject by a number of experts. We trust that this volume will be of great use to students, teachers and practitioners of solid state physics and chemistry as well as material science.

C.N.R. Rao

B. Raveau

CONTENTS

COLOSSAL MAGNETORESISTANCE, CHARGE ORDERING AND OTHER NOVEL PROPERTIES OF MANGANATES AND RELATED MATERIALS

C.N.R. RAO* and A.K. RAYCHAUDHURI†

**Solid State & Structural Chemistry Unit and †Department of Physics*
Indian Institute of Science, Bangalore 560012, India
and
Jawaharlal Nehru Centre for Advanced Scientific Research
Jakkur, Bangalore 560 064, India

Colossal magnetoresistance (CMR) and related properties of perovskite manganates of the general formula $Ln_{1-x}A_xMnO_3$ (Ln = rare-earth; A = divalent ion) are discussed in detail. The manganates are ferromagnetic at or above a certain value of x (or Mn^{4+} content) and become metallic at temperatures below the curie temperature, T_c. This behavior is commonly attributed to double-exchange. CMR is generally a maximum close to T_c or the insulator-metal (I-M) transition temperature, T_{im}. The T_c and %MR are markedly affected by the size of the A site cation, $\langle r_A \rangle$, thereby affording a useful electronic phase diagram when T_c or T_{im} is plotted against $\langle r_A \rangle$ or pressure. The commonalities and correlations found in the properties of manganates are examined along with certain unusual features in the electron-transport properties of these materials. Recent studies clearly indicate that double-exchange alone cannot explain the many fascinating features of the manganates. Some of the $Ln_{1-x}A_xMnO_3$ compositions exhibit charge-ordering and related effects. Charge ordering is crucially dependent on $\langle r_A \rangle$ or the e_g bandwidth and the charge-ordered insulating state transforms to a metallic ferromagnetic state on the application of a magnetic field, charge-ordering and double-exchange being competing interactions. The importance of Jahn-Teller interaction in determining the various properties of manganates is highlighted. CMR found in a few inorganic solids other than the perovskite manganates is pointed out.

1. Introduction

Magnetoresistance is the relative change in the electrical resistance or resistivity of a material produced on the application of a magnetic field. It is generally defined by,

$$MR = [\Delta\rho/\rho(O)] = [\rho(H) - \rho(O)]/\rho(O) \qquad (1)$$

where $\rho(H)$ and $\rho(O)$ are the resistances or resistivities at a given temperature in the presence and absence of a magnetic field, H, respectively. Magnetoresistance (MR) can be negative or positive. Most metals show a small MR (only a few percent). In non-magnetic pure metals and alloys MR is generally positive and MR shows a quadratic dependence on H. MR can be negative in magnetic materials because of the suppression of spin disorder by the magnetic field. Suppression of quantum mechanical interference can give rise to negative MR in highly resistive alloys. Large magnetoresistance, referred to as giant magnetoresistance (GMR), was first observed on the application of magnetic fields to atomically engineered magnetic superlattices (e.g. Fe/Cr)[1] and in magnetic semiconductors[2]. Besides some of the bimetallic or multimetallic layers, comprising ferromagnetic and antiferromagnetic or non-magnetic metals, GMR is also found in ferromagnetic granules dispersed in paramagnetic metal films (e.g. Co/Cu).[1] The phenomenon of GMR is of considerable interest because of its potential technological applications in magnetic recording, actuators and sensors. GMR in layered and granular magnetic materials arises from the ability of magnetic fields to change and control the scattering of conduction electrons through the modification of the electron-orbit and spin-orbit interactions. GMR is a result of the reduction of the extra resistance due to the scattering of electrons by the non-aligned ferromagnetic components in zero magnetic field.

The discovery of negative GMR in rare-earth manganates,[3,4] $Ln_{1-x}A_xMnO_3$ (Ln = rare-earth, A = a divalent cation such as an alkaline earth) with the perovskite structure, has attracted wide attention. The magnitude of GMR in these materials can be very large, close to 100% as per eq. (1). For this reason, many workers prefer to call this colossal magnetoresistance (CMR), as distinct from GMR in layered or granular metallic materials. It is customary to give %MR by multiplying MR from eq. (1) by 100. Many workers give MR by using $\rho(H)$ in the denominator of eq. (1) instead of $\rho(O)$, giving rise to very large numbers. In metallic multilayers or granular alloys, the mechanism involves spin-polarized transport. In the manganates also, spin-polarized transport is responsible for the large negative MR, but it is distinctly different from what happens in metallic multilayers.

A variety of manganese oxides, in the form of polycrystalline powders, thin films and single crystals, have been investigated for CMR. These studies have provided valuable insight into the CMR phenomenon in the manganates and have also revealed several other fascinating features of these oxides. In this article, we present some of the highlights of the results pertaining to CMR and related properties of the manganates obtained hitherto and point out the significant general features. We examine certain novel aspects of the manganates, in particular, electron-lattice or Jahn-Teller interactions, charge-ordering and the associated properties arising from them. We also refer to other systems exhibiting CMR and indicate some future directions.

2. Introduction to Rare Earth Manganates

2.1. *Insulating parent manganates*

Without hole doping, parent compounds such as $LaMnO_3$, $PrMnO_3$ and $NdMnO_3$, are insulators at all temperatures. For example, in $LaMnO_3$ the optical gap is ~ 1.1eV. These oxides undergo an antiferromagnetic (AFM) transition with a Neél temperature, $T_N \approx 150K$. The AFM ordering is of A-type where ferromagnetically aligned layers are coupled antiferromagnetically. The insulating nature of the parent compounds as well as the anisotropic magnetic interaction are related to their structure, in particular the Jahn-Teller (JT) distortion around Mn^{3+} ions. When these insulators are hole-doped, (i.e., Ln is partly substituted by a divalent ion), the Mn^{4+} ions decrease the cooperative JT distortion. The structure plays a crucial role in determining the electron transport and magnetic properties of these oxides.

There are two characteristic distortions which influence the perovskite structure of manganates. One consists of the cooperative tilting of MnO_6 octahedra which is essentially established below ~ 1000K. This distortion is a consequence of the mismatch of the ionic radii. For perovskites containing small central cations, the Goldschmidt tolerance factor, t, is typically < 1. The tilting of the octahedra is common in perovskites with t < 1. The other distortion, arising from the Jahn-Teller effect due to Mn^{3+}, distorts the MnO_6 octahedra in such a way that there are long and short Mn-O bonds. This occurs below a characteristic temperature (T_{JT}). The most effective distortion is the basal-plane distortion (called Q_2 mode) with one diagonally opposite O-pair displaced outwards and the other pair displaced inward. It has recently been shown that a JT distortion involving a displacement of oxygen ions $\gtrsim$ 0.1 Å can split the e_g band of the manganate (which forms the conduction band) and opens a gap at the Fermi level.

In Fig. 1. We show a schematic of the band diagram of $LaMnO_3$ to elucidate how JT distortion splits the conduction band and makes the material insulating. We will see below that hole doping reduces JT distortion and makes the material metallic for hole filling ≥ 0.18. In Table 1 we show the lattice constants, distortion and the characteristic T_{JT} values. The perovskite structure has an orthorhombic distortion ($b > a > c/\sqrt{2}$) and the unit cell consisting of 4-formula units can be mapped into *Pbnm* (or *Pnma*) symmetry.

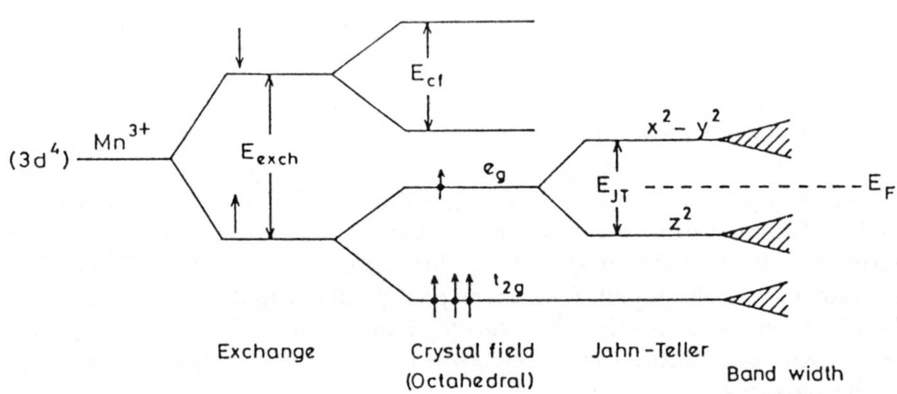

Fig. 1. Energy level scheme for $LaMnO_3$. The values of E_{ex}, E_{cf} and E_{JT} from density functional calculations are around 3eV, 2V and 1.5eV respectively. Typical bandwidth is 1.5 - 2.0eV. Fermi level lies between JT split orbitals. (S. Satpathy et al, J. Appl. Phys. 79, 4555 (1996); I. Solovyev et al, Phys. Rev. Lett. 76, 4825 (1996).

Electronic structure calculations show that in the absence of JT distortion, if the crystal would have been cubic, the nearest neighbor exchange $J^n_{ab} = J^n_c \cong 30$meV. With lattice distortion in $LaMnO_3$, the exchange is reduced and becomes anisotropic with $J^n_{ab} \approx 9$meV and $J^n_c \approx 3$meV. This is comparable to the next nearest neighbor exchange (J^{nn}_c)

which is AFM type giving $J_{total} = 2J_c^n + 8J_c^{nn} \cong 19$meV, which is close to K_BT_N of the material. The ferromagnetic J_{ab}^n exchange and AFM J_{total} give rise to the A-type AFM ordering in these materials. With doping, the structure changes along with the magnetic ordering.

Table 1: Properties of parent manganates

Material	a Å	b Å	$c/\sqrt{2}$ Å	$\tilde{D}^{(a)}$	$V(Å)^3$	T_{JT} (K)
$LaMnO_3$ (b)	5.532	5.742	5.442	1.8%	244.5	875
$PrMnO_3$ (c)	5.454	5.809	5.365	3.2%	240.4	820
$NdMnO_3$ (d)	5.410	5.762	5.334	3.1%	235.2	1100

(a) $\tilde{D} = 1/3 \sum_{i=1}^{3} |(a_i - \tilde{a})| / \tilde{a}_i, \tilde{a} = (abc/\sqrt{2})^{1/3}; \; a_1 = a, a_2 = b, a_3 = c/\sqrt{2}$

(b) J.B.A. Elemao et al, *J. Solid State Chem.*, 3, 288 (1971).

(c) K. Knizek et al, *J. Solid State Chem.*, 100, 292 (1992).

(d) H. Taguchi et al, *J. Solid State Chem.*, 76, 284 (1988).

2.2. *Hole-doped manganates*

$LaMnO_3$ with a small proportion of Mn^{4+} ($\lesssim$ 5%) becomes antiferromagnetically ordered at low temperatures ($T_N \sim 150$K). When the La^{3+} in $LaMnO_3$ is progressively substituted by a divalent cation as in $La_{1-x}A_xMnO_3$ (A = Ca, Sr or Ba), the proportion of Mn^{4+} increases and the orthorhombic distortion decreases. The material becomes ferromagnetic with a well-defined Curie temperature at a finite x, and metallic below T_c.[5] The saturation moment is typically $\approx 3.8\ \mu_B$ which is close to the theoretical estimate based on localized spin-only moments. This suggests that the conduction electrons are fully spin-polarized. An important quantity characterizing a spin wave is the stiffness constant D. From neutron scattering measurements on $La_{0.67}Ca_{0.33}MnO_3$, D has been found to be $\approx$ 150-170 meV.Å^2.[6] The simple Heisenberg model predicts a D value of $\approx$ 18-20 meV.Å^2 and a high T_c in the 900-1000K region. It has been argued that the lattice distortion brings down T_c to below 400K. Below T_c, the manganate exhibits metal-like conductivity. Fig. 2 illustrates how the ferromagnetism and the insulator-metal (I-M) transition occur around T_c.

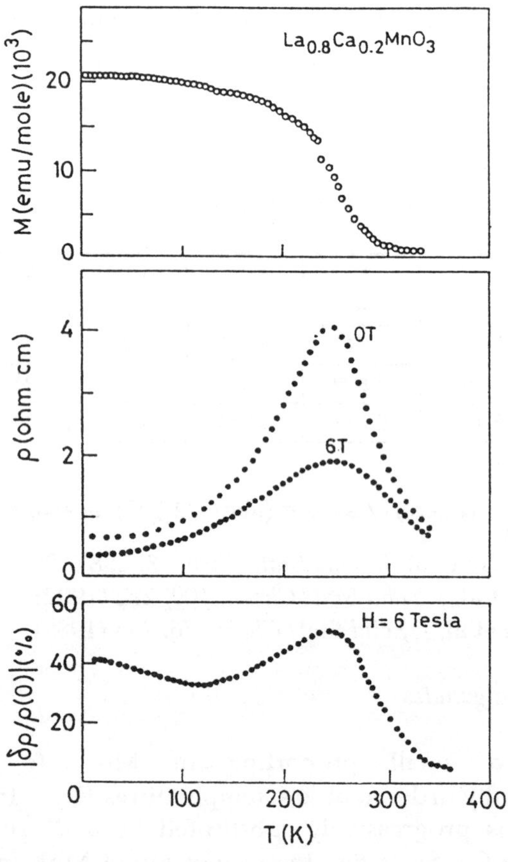

Fig. 2. Temperature variation of magnetization, resistivity (at H = 0T and 6T) and magnetoresistance of polycrystalline $La_{0.8}Ca_{0.2}MnO_3$ (adapted from Mahendiran et al[11]).

The simultaneous observation of itinerant electron behavior and ferromagnetism in the manganates is explained by Zener's double-exchange mechanism.[7] The basic process in this mechanism is the hopping of a d-hole from Mn^{4+} (d^3, t_{2g}^3, $S = 3/2$) to Mn^{3+} (d^4, $t_{2g}^3e_g^1$, $S = 2$) via the oxygen, so that the Mn^{4+} and Mn^{3+} ions change places:

$$Mn^{3+}O^{2-}Mn^{4+} \rightarrow Mn^{4+}O^{2-}Mn^{3+} \quad (2)$$

This involves the transfer of an electron from the Mn^{3+} site to the central oxide ion and simultaneously the transfer of an electron from the oxide ion to the Mn^{4+} site. Such a transfer is referred to as *double-exchange (DE)*. The integral defining the exchange energy in such a system is nonvanishing only if the spins of the two d-shells are parallel. That is, the lowest energy of the system is one with a parallel alignment of the spins on the Mn^{3+} and Mn^{4+} ions. This is indeed a novel situation where the lining up of the spins of the incomplete d orbitals of the adjacent Mn ions is accompanied by an increase in the rate of hopping of electrons and therefore by an increase in electrical conductivity. Thus, the mechanism which leads to enhanced electrical conductivity requires a ferromagnetic coupling. It is assumed here that the intra-atomic exchange, J, is large compared to the transfer integral, t_{ij}, between the two Mn sites.

The relation between the electrical conductivity and ferromagnetism by the DE mechanism as originally proposed by Zener is obtained as follows. The magnitude of the exchange energy, ϵ, is given by

$$\epsilon = h\nu/2 \tag{3}$$

where ν is the frequency of oscillation of the electron between two Mn sites. The diffusion coefficient for an Mn^{4+} ion is defined by,

$$D = \underline{a}^2\epsilon/h \tag{4}$$

where a is the lattice parameter. Making use of the Einstein equation relating conductivity, σ, and D, $\sigma = ne^2D/kT$, where n is the number of Mn^{4+} ions per unit volume, we obtain

$$\sigma = xe^2\epsilon/ahkT \tag{5}$$

Here x is the fraction of Mn^{4+} ions in $La_{1-x}A_xMnO_3$. Since the ferromagnetic Curie temperature T_c is related to the exchange energy by the approximate relation, $\epsilon \approx kT_c$,

$$\sigma \approx (xe^2/ah)(T_c/T) \tag{6}$$

Eq. (6) relates the electrical conductivity to ferromagnetic T_c and the fraction of Mn^{4+} ions. We would therefore expect the I-M transition in the manganates to occur at T_c. Double-exchange is strongly affected by structural parameters such as the Mn-O-Mn angle or the Mn-Mn transfer integral.

The $t_{2g}{}^3$ electrons of the Mn^{3+} ion are localized on the Mn site giving rise to a local spin of 3/2, but the e_g state, which is hybridized with the oxygen 2p state, can be localized or itinerant and only those electrons which have their spins parallely aligned give rise to conductivity in the hopping process. There is strong Hund's rule interaction between the e_g and the $t_{2g}{}^3$ electrons. Goodenough[8] pointed out that ferromagnetism is governed not only by double exchange, but also by the nature of the superexchange interactions. While delineating the nature of the double exchange and superexchange interactions, de Gennes[9] suggested that a non-collinear magnetic structure forms at intermediate concentrations between the antiferromagnetic and the ferromagnetic states. It may be recalled that the Mn^{3+}-O-Mn^{4+} superexchange interaction is ferromagnetic while the Mn^{3+}-O-Mn^{3+} and Mn^{4+}-O-Mn^{4+} interactions are both antiferromagnetic. The resulting magnetic exchange is strongly dependent on the structural distortion, as indeed shown by electronic structure calculations. While recent investigations bring out the essential role of DE, they also suggest the crucial role of the lattice and the electron-lattice interactions. In understanding the novel transport properties, we need to consider the overall lattice symmetry as well as local lattice distortions.

The $x = 0.0$ and the $x = 1.0$ members of the $La_{1-x}A_xMnO_3$ system (viz. $LaMnO_3$ and $AMnO_3$) are antiferromagnetic (AFM) insulators at low temperatures, with A- and G-type ordering, respectively.[10] In A type ordering, there are ferromagnetically aligned spins in the planes, but the planes are aligned antiferromagnetically. Addition of A ions reduces the JT distortion, in particular the Q_2 mode distortion which gives rise to long and short Mn-O distances, leading to anisotropic exchange. Removal of the distortion reduces the AFM interaction and this in turn destroys the AFM order. Ferromagnetic behavior starts to manifest itself when x is ~ 0.1, and the compositions up to $x \sim 0.3$ have both AFM and FM characteristics . The $x = 0.3$ composition is clearly ferromagnetic, while the $x > 0.5$ compositions are antiferromagnetic. The FM transition in the $La_{1-x}A_xMnO_3$ compositions is quite sharp[11] as can be seen from Fig. 2. Although manganates with $x > 0.5$ are essentially antiferromagnetic, ferromagnetic Mn^{3+}-O-Mn^{4+} clusters would be present in the AFM medium. Fig. 3 illustrates how the electrical and magnetic properties of $La_{1-x}Ca_xMnO_3$[12] and $La_{1-x}Sr_xMnO_3$[4] change with the composition. The phase diagrams are even richer than indicated in Fig. 3, giving rise to the fascinating properties and phenomena of these manganates. The material is generally a ferromagnetic metal (FMM) below T_c when $\sim 0.2 < x < 0.5$, becoming a paramagnetic insulator (PMI) when $T > T_c$. In the small x regime ($x < \sim 0.2$), a canted spin antiferromagnetic insulating state (CSI) or a ferromagnetic insulating state (FMI) is often encountered. Charge ordering can also occur in this composition range, but more commonly in the $0.3 < x < 0.7$ range. When $x > 0.5$, the materials are generally antiferromagnetic insulators (AFMI) becoming paramagnetic metals (PMM) or insulators above T_N. It should be noted that $CaMnO_3$ and $SrMnO_3$ (x = 1.0) are antiferromagnetic insulators.

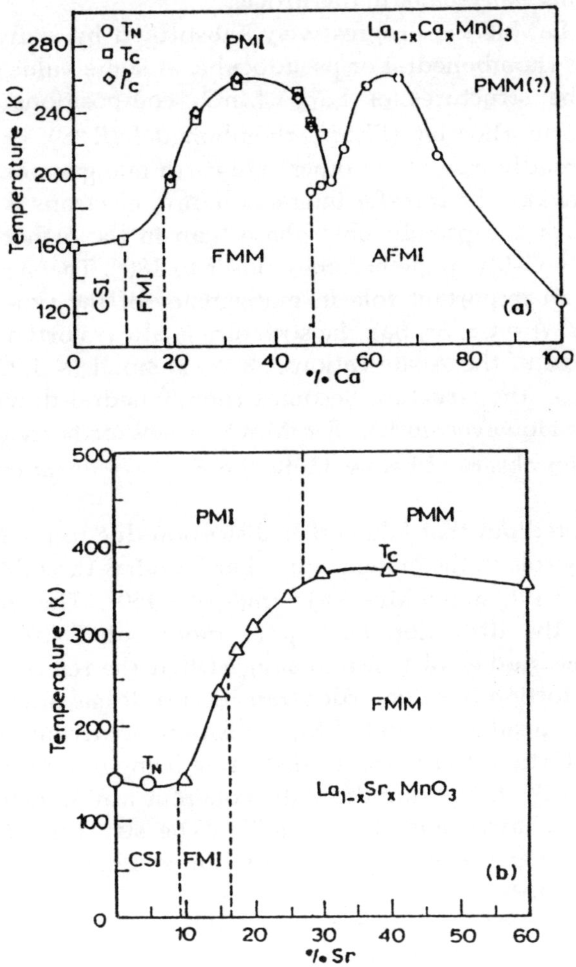

Fig. 3. Temperature-composition diagram of $La_{1-x}A_xMnO_3$ (a) = Ca (b) = Sr; CSI, canted-spin insulator; FMI, ferromagnetic insulator; FMM, ferromagnet metal; AFMI, antiferromagnetic insulator; PMI, paramagnetic insulator, PMM, paramagnetic metal (adapted from Schiffer et al[12] and Urushibara et al[4]).

The extent of the FMM regime depends on the A site cation, being greater when A = Sr than when A = Ca. Thus, many of the manganates exhibit charge ordering, its nature depending on the size of the A-site cations. We shall discuss this aspect later in this article.

As La in $LaMnO_3$ is progressively substituted by a divalent ion, the oxide becomes rhombohedral or pseudocubic at some value of x (typically $x \gtrsim 0.2$). The structures of $La_{1-x}A_xMnO_3$ compositions are therefore described as orthorhombic (*Pbnm*), rhombohedal ($R\bar{3}c$), or pseudocubic. The same is broadly true of the other rare-earth manganates, $Ln_{1-x}A_xMnO_3$ (Ln = Pr, Nd etc.). The transfer interaction of e_g electrons is greater in the rhombohedral or the pseudocubic phase than in the orthorhombic phase because the Mn-O-Mn angle becomes closer to 180°. The size of the A-site cations plays an important role in manganates of the type $Ln_{1-x}A_xMnO_3$ (Ln = La, Pr, Y; A = Ca, Sr, Ba); the structure is always orthorhombic when the average size of the A-site cations, $\langle r_A \rangle$, is small (< 1.22Å). With the increase in $\langle r_A \rangle$, the structure becomes rhombohedral down to very low temperatures. However, in $La_{0.7}Ba_{0.3}MnO_3$, a new orthorhombic structure (*Imma*) has been observed below 150K, the $R\bar{3}c \rightarrow Imma$ transition being first order.[13]

It was pointed out that Jahn-Teller distortion due to the Mn^{3+} ions ($t_{2g}^3 e_g^1$) plays a key role in the manganates. $LaMnO_3$ has three Mn-O distances (1.91, 1.96 and 2.19Å) and a Mn-O-Mn angle of ~ 150°. The creation of Mn^{4+} ions removes the distortion leading to more cubic structures. X-ray crystallographic studies of $LaMnO_3$ suggest that the removal of the static Jahn-Teller distortion is a first order transition.[14] It has indeed been found that across the insulator-metal (I-M) transition occurring around T_c, the Jahn-Teller distortion decreases, the distortion being more prominent in the insulating phase.[15] Increasing the static coherent MnO_6 distortion favours the insulating behavior and decreases T_c. The structural parameters, in particular the oxygen thermal parameters, show significant changes across the I-M transition.[16]

As mentioned earlier, as-prepared $LaMnO_3$ generally contains some Mn^{4+}. The origin of Mn^{4+} has been examined in some detail. Since perovskites cannot accommodate excess oxygen, the defect chemistry of $LaMnO_3$ must involve the presence of cation vacancies in the La and Mn sites.[17] Accordingly, $LaMnO_3$ can be represented as $La_{1-\delta}Mn_{1-\delta}O_3$. If the Mn^{4+} content in $LaMnO_3$ is 33%, the formula will approximately be $La_{0.945}Mn_{0.945}O_3$. $LaMnO_3$ becomes rhombohedral and then pseudocubic as the Mn^{4+} content is increased by chemical or electrochemical means.

Creation of Mn^{4+} in $LaMnO_3$ leads to an increase of the tolerance factor, t, because of the smaller radius of the Mn^{4+} ions. For $\delta \approx 0.02$ (24% Mn^{4+}), $t \gtrsim 0.90$ and the transition occurs from the orthorhombic to the rhombohedral phase.[18] The $La_{1-\delta}MnO_3$ compositions exhibit ferromagnetism and the I-M transition up to $\delta = 0.3$, but $LaMn_{1-\delta'}O_3$ compositions do so only up to $\delta' = 0.05$.[19]

Since the presence of ~ 30% Mn^{4+} is essential for the ferromagnetism and the insulator-metal (I-M) transition, it is desirable to independently determine the Mn^{4+} content by means of redox titrations (Fe(II) sulphate + $KMnO_4$). It can be somewhat misleading to assume that the Mn^{4+} content is simply dictated by the formula $La_{1-x}A_xMnO_3$, since it can vary with the method and conditions of preparation. Clearly, Mn^{4+} plays a crucial role in this material. The role is essentially two-fold. It provides the double-exchange needed for ferromagnetism and metallic behavior and it also helps to remove the Jahn-Teller distortion of Mn^{3+} ions and makes the structure closer to cubic.

3. Transport Properties and Colossal Magnetoresistance (CMR) of Rare Earth Manganates

Most of the $La_{1-x}A_xMnO_3$ compositions are paramagnetic insulators at ordinary temperatures and exhibit an increase in electrical resistivity with a decrease in temperature. Compositions that are ferromagnetic show insulating behavior above T_c, but the resistivity decreases with decreasing temperature, as in metals, when they are cooled below T_c. This insulator-metal (I-M) transition is therefore associated with a peak in resistivity at a temperature T_{im}. A typical I-M transition in a polycrystalline $La_{1-x}Ca_xMnO_3$ sample was shown in Fig. 2. Such a transition is also found in $LaMnO_3$ with a sufficient proportion of Mn^{4+} (Fig. 4). Generally, T_{im} is somewhat lower than T_c.[11] The sharpness of the transition in polycrystalline samples and films is often dependent on the sample quality. For samples with grain size in excess of 5 μm, the transition is relatively sharp.

3.1. *Transport in the paramagnetic phase (T>Tc)*

The temperature dependence of the resistivity, ρ, of the manganates with $0.2 \leq x \leq 0.4$ in the paramagnetic phase generally shows two types of behavior. The manganates doped with larger cations such as Sr, Ba and Pb (i.e. large A-site cation radius, $\langle r_A \rangle$) have a larger bandwidth. As a result the paramagnetic phase can be metallic (PMM) or insulating (PMI) with a small gap. This leads to a weak dependence of ρ on T. For manganates

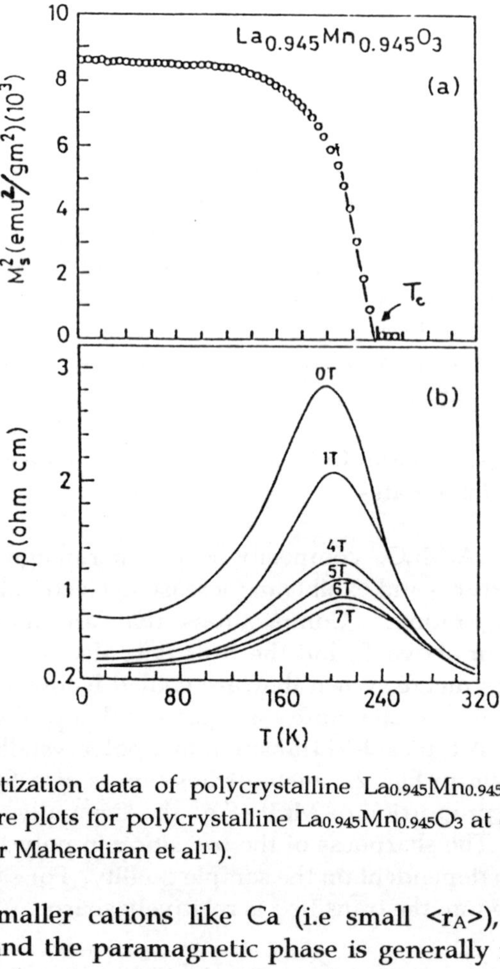

Fig. 4. (a) Magnetization data of polycrystalline $La_{0.945}Mn_{0.945}O_3$ (b) resistivity-temperature plots for polycrystalline $La_{0.945}Mn_{0.945}O_3$ at different magnetic fields (after Mahendiran et al[11]).

with relatively smaller cations like Ca (i.e small $\langle r_A \rangle$), the bandwidth would be small and the paramagnetic phase is generally insulating (PMI) with a well-defined activation energy E_a in the range ≈ 0.1 - 0.25eV. There is an interesting dependence of E_a on the Mn-O distance[11] wherein it increases with the Mn-O distance. Thus the behavior above T_c (whether the high temperature phase is PMM or PMI) entirely depends on the bandwidth.

The important issue that remains is what makes the $T > T_c$ phase insulating. Is it lattice polaron formation from strong electron-lattice coupling due to JT effect or is it some other type of localization? This question has been investigated mainly by fitting ρ above T_c to different expressions. Some investigators find that a variable range hopping (VRH) type of expression $\rho \approx \rho_{oh} \exp (T_o/T)^{1/4}$ holds good above T_c. These investigations have been carried out in a limited temperature range above T_c. The problem, however, is that one obtains physically unrealistic values of T_o. Some authors argue in favor of lattice polaron formation and have fitted the resistivity to expressions of the type, $(\rho/T) \approx \rho_{op} \exp (E_a/k_BT)$. Here, E_a is the small polaron hopping energy. From the resistivity data taken over a limited temperature range, it is often difficult to distinguish between VRH, small polaron hopping or the simple semiconductor form $\rho \sim \exp (E/k_BT)$. Analysis of the high temperature ρ data along with thermopower data tend to support the small polaron hopping picture. However, there is need for further investigations to settle the nature of electrical transport for $T > T_c$.

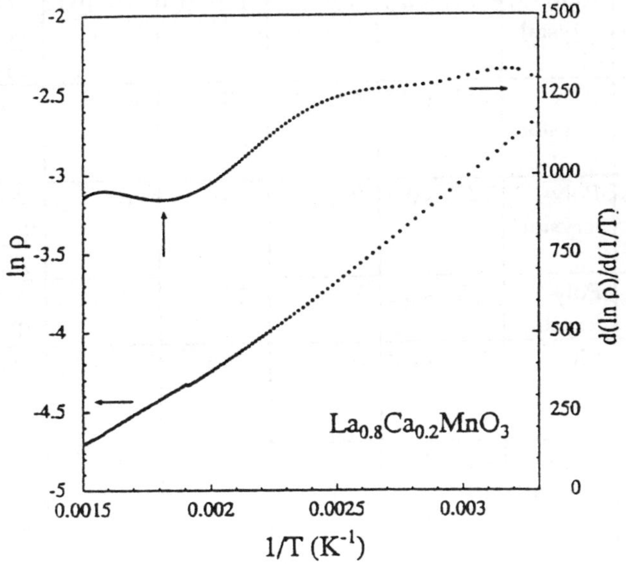

Fig. 5. Temperature dependence of the resistivity, ρ, ($T > 300K$) and $d\ln\rho/d(1/T)$ for $La_{0.8}Ca_{0.2}MnO_3$.

There are certain interesting observations when ρ is investigated at higher temperatures. In Fig. 5 we plot $d\ln\rho/d(1/T)$ vs $1/T$ in $La_{0.8}Ca_{0.2}MnO_3$. There are two regions with different activation energies. The two regions are demarkated by an enhancement of the activation energy ($d\ln\rho/d(1/T)$ α activation energy) at $T \simeq 500K - 600K$. A similar observation has been made in a single crystal of $La_{0.67}Ca_{0.33}MnO_3$ at $T \simeq 750K$. This change in activation energy can be due to the formation of polarons due to lattice distortion. This issue needs further investigation.

3.2. *The ferromagnetic metallic state ($T < T_c$)*

The electrical resistivity in the ferromagnetic region has been investigated by a number of authors. In the region just below T_c, ρ varies rapidly with T. For $T < 0.5\ T_c$, typically the variation is much less rapid, but it is

Table 2. Properties of $La_{1-x}A_xMnO_3$

Material	Physical State	Temp. Range	$\rho(T) = \rho_0 + \rho_2 T^n$	ρ_0 (mΩcm)	ρ_2/ρ_0 (10^{-6} K^{-n})	Reference
$\underline{Ln_{1-x}A_x}$ $La_{0.67}Ca_{0.33}$	film/ crystal	$T/T_c < 0.4$	$n \cong 2$	0.1 - 0.3	60 - 100	J.G.Snyder et al Phys. Rev. B 54, 14434 (1996)
$La_{0.67}Ca_{0.33}$	Poly-crystal	$T/T_c < 0.5$	$n \simeq 2.5$	15	10	J.M. De Teresa et al, Phys. Rev. B54, 1187 (1996)
$La_{0.75}Ca_{0.25}$	Poly-crystal	$T/T_c < 0.5$	$n \simeq 2.5$	3	6	P. Schiffer et al Phys. Rev. Letts. 75, 3336 (1995)
$La_{0.8}Ca_{0.2}$	Poly-crystal	$T/T_c < 0.2$	3	24	12	A. Biswas (Private Comm.)
$La_{0.67}Sr_{0.33}$	film	$T/T_c < 0.4$	$n \simeq 2$	0.15	55	J.G. Snyder et al Phys. Rev. B 53, 14434 (1996)
$La_{0.7}Sr_{0.3}$	Single crystal	$T/T_c < 0.5$	$n \simeq 2$	0.1	150	A. Urushibara et al, Phys. Rev. B 51, 14103 (1995)
$La_{0.67}Ba_{0.33}$	film	$T/T_c < 0.4$	$n \simeq 2$	0.34	52	S.E. Lofland et al Phys. Rev. B 52, 15058 (1995)

different from what is generally seen in a metallic ferromagnet. In Table 2 we give a brief summary of the temperature dependence of ρ seen by different authors. The table is not exhaustive and gives typical data. In general, for $T < 0.2\ T_c$, a T^2 dependence seems to make the major contribution. At somewhat high temperatures, $T \simeq 0.5\ T_c$, contribution from an extra term faster than T^2 is needed. However, ρ_0 and ρ_2 of different authors show wide variations even for the same composition. Clearly, this is an indication that we are not dealing with the intrinsic temperature variation of ρ. For $0.5\ T_c < T < T_c$, the variation of ρ with T depends on the physical state of the samples. An interesting issue that needs to be settled in this region is how a temperature-dependent density of states at the Fermi level affects ρ (T).

3.3 *Effect of magnetic field, CMR*

Application of a magnetic field (up to, say, 6 Tesla) causes a significant decrease in the resistivity of the $La_{1-x}A_xMnO_3$ samples, particularly in the compositions with $0.1 < x < 0.5$; these materials are generally ferromagnetic with well-defined T_cs. The magnitude of the decrease in resistivity (i.e. the magnetoresistance, MR) is highest in the region of T_c or T_{im}. In Fig. 2 we compare the temperature variation of the resistivity of $La_{1-x}Ca_xMnO_3$ at 6T with that in zero field. Magnetoresistance in the manganates is generally negative and is highest around T_c or T_{im}. CMR close to 100% has been observed in many polycrystalline and single crystal $La_{1-x}A_xMnO_3$ compositions, but the applied field is quite high (5-6 Tesla). In $La_{1-x}Pb_xMnO_3$, high CMR is found around room temperature or above as shown in Fig. 6.[20] Self-doped $LaMnO_3$ samples with sufficient Mn^{4+} content (i.e. $La_{1-\delta}Mn_{1-\delta}O_3$) exhibit CMR similar to that observed in the $La_{1-x}A_xMnO_3$ compositions. In Fig. 4 we show the behavior of $La_{0.945}Mn_{0.945}O_3$, with 33% Mn^{4+}.[11] La-deficient, $La_{1-\delta}MnO_3$ compositions also exhibit CMR, but Mn-deficient $LaMn_{1-\delta'}O_3$ samples do not when $\delta' > 0.05$.[19] In Fig. 7 we show the resistivity behavior of La- and Mn- deficient compositions to illustrate how deficiency in La is tolerated readily without affecting the magnetotransport properties and CMR, unlike deficiency in Mn. This is understandable since the active electrical and magnetic network is made up of Mn-O-Mn bonds. As a result, any defect in the Mn lattice will affect the transport properties greatly. Substitution of other trivalent ions such as Fe^{3+} and Al^{3+} for Mn^{3+} adversely affects the system, lowering the T_c and increasing the resistivity.

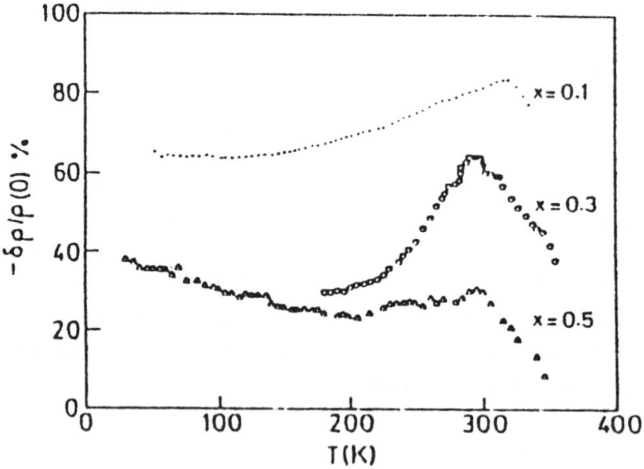

Fig. 6. Magnetoresistance of polycrystalline $La_{1-x}Pb_xMnO_3$ at 6T (from Mahendiran et al[20]).

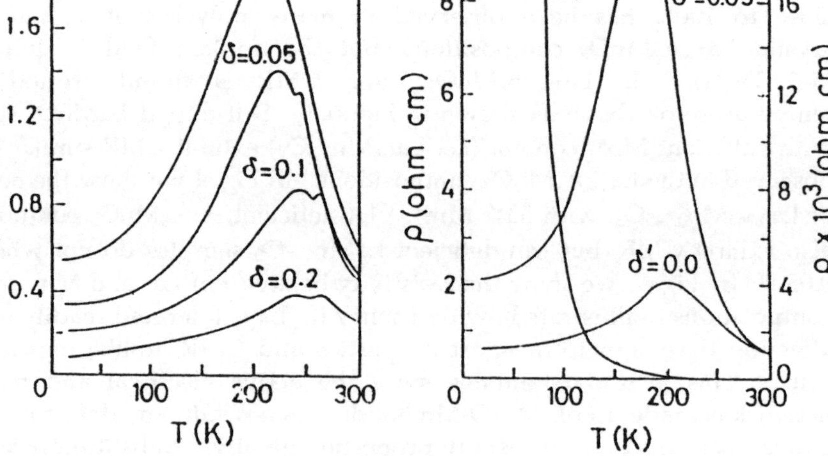

Fig. 7. Resistivity behavior of $La_{1-\delta}MnO_3$ and $LaMn_{1-\delta'}O_3$ (from Arulraj et al[19]).

The variation of MR with the magnetic field at different temperatures is shown in Fig. 8.[11] The dependence of MR on H is a function of temperature of the measurement. The change is gradual close to T_c. For $(T/T_c) < 0.5$, the MR in polycrystalline manganates exhibits two distinct regimes as a function of magnetic field. For H < 1T, the MR changes rapidly with H followed by a gradual change as H is increased further. There are two aspects to the field dependence. One aspect has to do with ferromagnetic domain wall movement as in all ferromagnets and the other has to do with grain boundaries which contribute substantially to the MR of manganates at $T \ll T_c$. The magnitude of MR has a close relation to the magnetization (M) even in polycrystalline samples, showing a quadratic dependence.[21] It is interesting that the two distinct regimes seen in the MR - H curve for polycrystalline samples at $T \ll T_c$ is generally absent in single crystals. In single crystals, the magnitude of MR increases almost linearly with H.

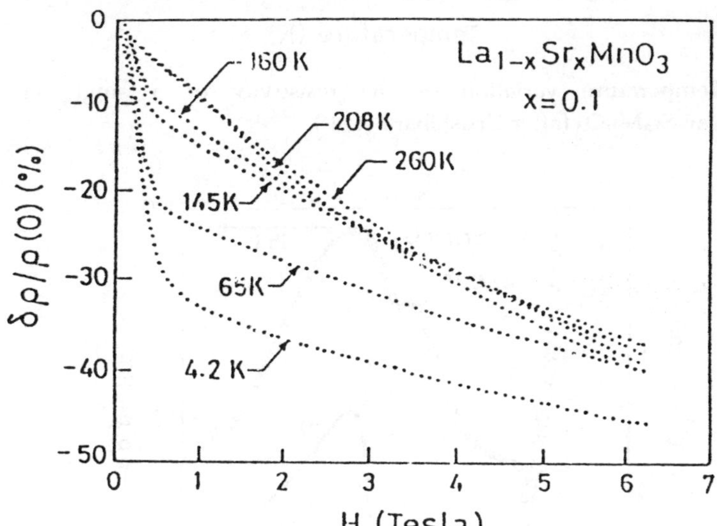

Fig. 8. Variation of magnetoresistance of polycrystalline $La_{1-x}Sr_xMnO_3$ with magnetic field, H (from Mahendiran et al[11]).

CMR studies of single crystals of rare-earth manganates exhibit features in the transport properties which are not very different from those of the polycrystalline samples, although there are some subtle differences. For example, the I-M transition does not always show a distinct resistivity peak

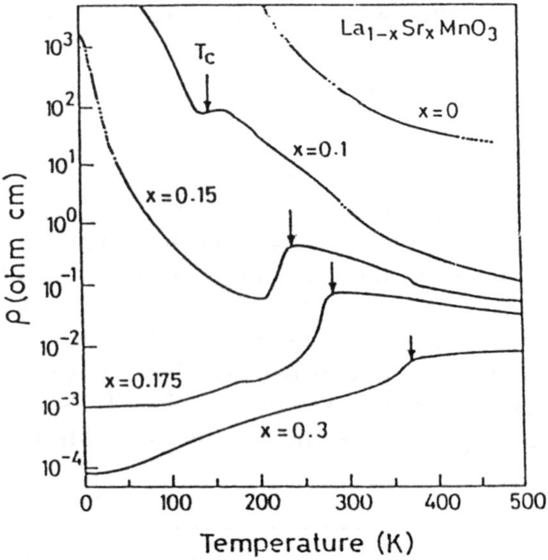

Fig. 9. Temperature variation of the resistivity of a single crystal of $La_{1-x}Sr_xMnO_3$ (after Urushibara et al[4]).

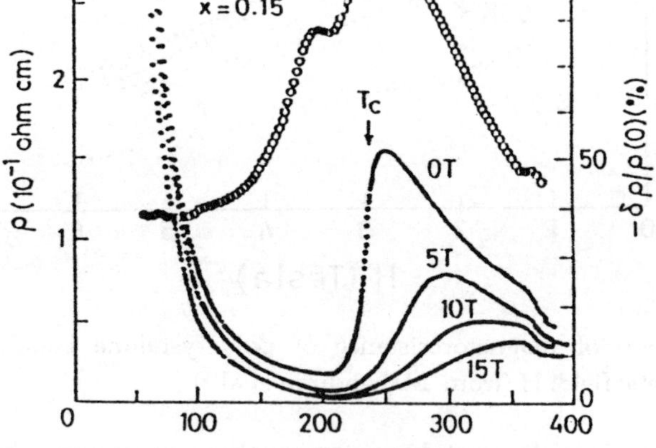

Fig. 10. Resistivity-temperature plots of a single crystal of $La_{0.85}Sr_{0.15}MnO_3$ at different magnetic fields (after Urushibara et al[4]).

as in the polycrystalline samples. In Figs. 9 and 10, we show the resistivity behavior of $La_{1-x}Sr_xMnO_3$ crystals at zero field and the effect of magnetic field on the resistivity.[5] In single crystals and good epitaxial films, CMR is substantial only near T_c and near zero as $T \rightarrow 0$. The reduction in the resistivity of $La_{1-x}Sr_xMnO_3$ crystal scales with the field-induced magnetization, M, as:

$$-\Delta\rho/\rho = C(M/M_s)^2 \qquad (7)$$

for $M/M_s < 0.3$, where M_s is the saturation magnetization. This dependence on magnetization is the same as that found in polycrystalline materials.

3.4. *Role of electron-lattice coupling*

Lattice polaron formation and local structural distortion have important implications on electron transport in these materials. In particular, they play a role in carrier localization and can be a source of the high resistivity seen in these materials. CMR is observed over a wide range of compositions of $La_{1-x}Ca_xMnO_3$ ($x < 0.5$), but the material shows lattice effects, particularly for low x. When $x = 0.2$, a local structural distortion occurs due to the formation of lattice polarons which arises from the strong electron-lattice coupling. EXAFS studies reveal significant changes in the local structure in the $x = 0.3$ composition in the 80-300K range which are attributed to the formation of small polarons due to the Jahn-Teller distortion when $T > T_c$.[22,23] These studies have shown that small polarons delocalize as the magnetization increases in the manganates which show the I-M transition. Channeling experiments close to T_c in $Nd_{0.7}Sr_{0.3}MnO_3$, $Pr_{0.7}Ba_{0.3}MnO_3$ and $La_{0.7}Ba_{0.3}MnO_3$ films have provided evidence for Jahn-Teller distortion.[24]

The spin and lattice dynamics of $La_{0.7}Sr_{0.3}MnO_3$ have been examined by neutron scattering.[25] The spin dynamics and the magnetic critical scattering seem to be comparable to those of typical metallic ferromagnets. In $La_{1-x}Sr_xMnO_3$ ($x = 0.17$), the local spin moments and charge carriers couple strongly to changes in the structure and the structure can be switched by the application of a magnetic field depending on the temperature. It is noteworthy that the magnetic field which causes this crossover from the orthorhombic to the rhombohedral structure is actually very low (~ 2T). This observation is significant because it clearly shows that an extremely small energy difference exists between these two structures.[26a] Single crystal neutron diffraction experiments have confirmed that the structural transition in $La_{1-x}Sr_xMnO_3$ can be induced at constant temperature by the application of a magnetic field.[26b] For $0.1 \le x \le 0.17$, ferromagnetic ordering suppresses the lattice distortion at and below T_c.[27a] Magnetostriction and linear thermal expansion experiments establish the strong correlation between magnetotransport and magnetovolume properties.[27b] When $x = 0.1$ and 0.15, a polaron ordered phase is formed and the polaron lattice has a tendency to lock into a commensurate structure when $x = 0.125$.[28] Spontaneous formation of localized 12Å magnetic clusters at $T > T_c$ are found to grow in size but decrease in number on application of the magnetic field in $(La_{1-x}Y_x)_{0.67}Ca_{0.33}MnO_3$, showing thereby that magnetic polarons are present in the insulating phase.[29] While the role of lattice polarons seems clear, that of magnetic polarons requires further study.

3.5. *Effect of external and internal pressure*

Hydrostatic pressure stabilizes the ferromagnetic metallic state of $La_{1-x}A_xMnO_3$ (i.e. increases the T_c).[30] The effect of hydrostatic pressure is to enhance the Mn-O-Mn transfer integral through a change in the Mn-O-Mn angle. A similar effect can be brought about by increasing the radius of the A-site cation. There is a direct relationship between T_c and the average radius of the A-site cation, $\langle r_A \rangle$, the T_c increasing with $\langle r_A \rangle$.[31-33] The equivalence of increasing $\langle r_A \rangle$ and increasing hydrostatic pressure can be seen in Fig. 11. It should be noted that small $\langle r_A \rangle$ gives rise to a distortion of the MnO_6 octahedra by bending the Mn-O-Mn bond and in turn causing the narrowing of the e_g bandwidth to a greater extent. Accordingly, yttrium substitution in $La_{1-x}A_xMnO_3$ decreases T_c, and enhances the magnitudes of resistivity as well as magnetoresistance.[34,35] In $La_{1-x}A_xMnO_3$, magnetic frustration has been shown to increase with a decrease in $\langle r_A \rangle$ or the bending of the Mn-O-Mn angle; the electronegativity of the A ion could also be a contributing factor.[36a] In Fig. 12 we show the resistivity behavior of $La_{0.5}(Ca_{0.5-z}Y_z)MnO_3$. In $La_{(2-x)/3}Nd_{x/3}La_{1/3}MnO_3$, the Mn-O-Mn angle bending increases with Nd content and the T_c decreases along with it.[36b] By plotting the T_c against $\langle r_A \rangle$, one obtains a phase diagram separating the ferromagnetic metal and paramagnetic insulator regimes.[31,37] In Fig. 13 we show a typical phase diagram obtained in this manner. In the left hand bottom corner, we see the ferromagnetic insulator regime. CMR generally decreases with increase in $\langle r_A \rangle$ just as the peak resistivity at T_c or at the insulator-metal transition as shown in Fig. 14. CMR is therefore favoured by low T_c and high peak resistivity . Enhanced CMR can be considered to arise from a reduced mobility of the doping holes and an increase in the

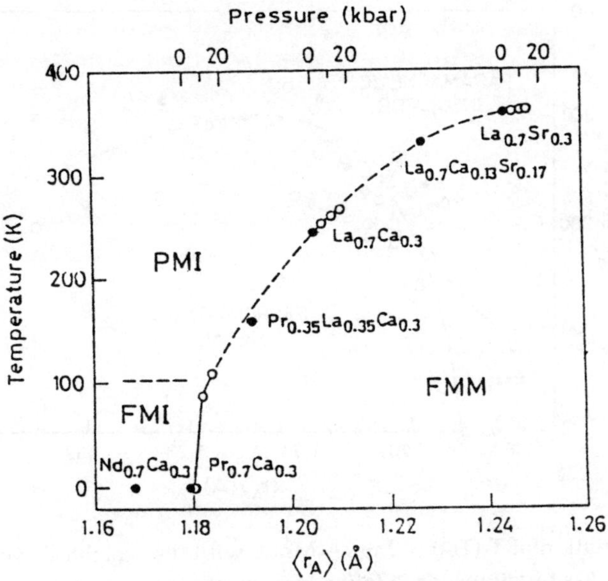

Fig. 11. Variation of $T_c(T_{im})$ of $Ln_{0.7}A_{0.3}MnO_3$ with hydrostatic pressure and the weighted average radius of the A-site cations (from Hwang et al[31]).

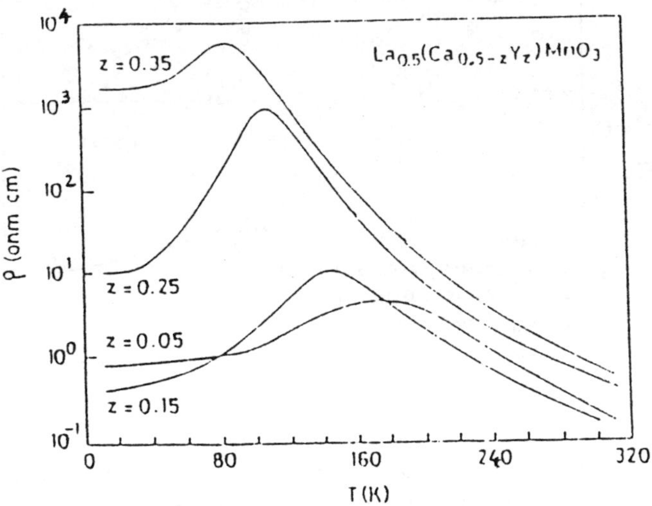

Fig. 12. Temperature variation of resistivity of $La_{0.5}(Ca_{0.5-z}Y_z)MnO_3$ (from Mahendiran et al[35]).

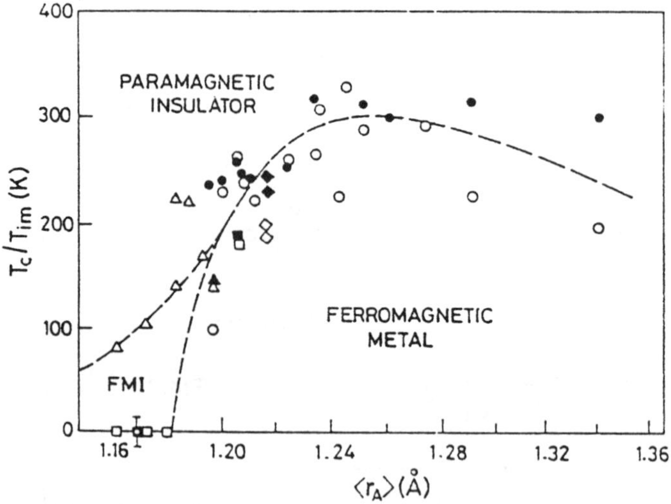

Fig. 13. Variation of $T_c(T_{im})$ of $Ln_{1-x}A_xMnO_3$ with the weighted average radius of the A-site cations, <r_A> (after Mahesh et al[37]).

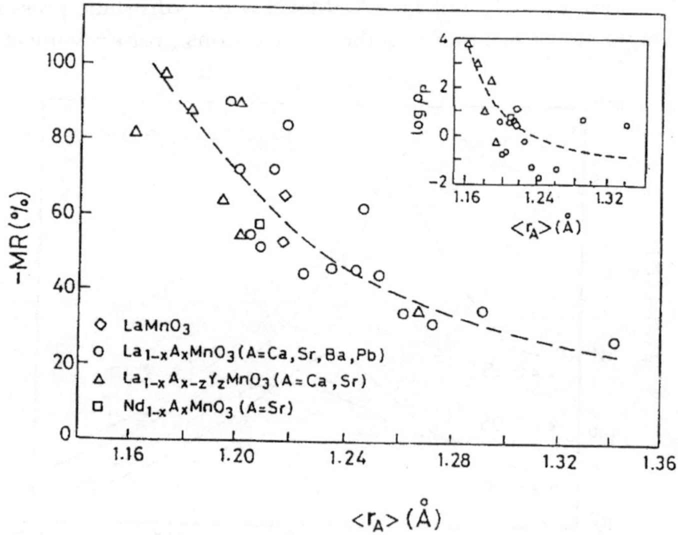

Fig. 14. Variation of the magnetoresistance of $Ln_{1-x}MnO_3$ with the weighted average radius of the A site cations, <r_A>; the inset shows the variation of the logarithm of peak resistivity with <r_A> (after Mahesh et al[37]).

coupling between the localized and itinerant electrons.[33] The enhanced CMR in materials with low T_c (and hence small bandwidth) can also arise from a stronger effect of the magnetic field on the electron-lattice interaction. The magnetic field in these materials can lead to significant carrier delocalization by lowering the lattice distortion or increasing the transfer integral.

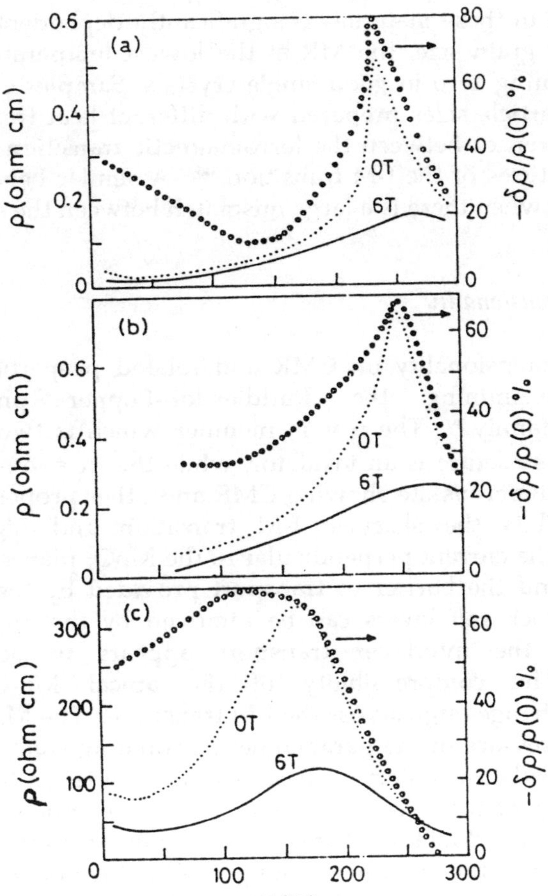

Fig. 15. Temperature variation of resistivity (at 0T and 6T) and magnetoresistance of $La_{0.7}Ca_{0.3}MnO_{0.3}$ with particle diameters of (a) 3.5μm (b) 1.5μm and (c) 0.025μm (from Mahesh et al[38a]).

3.6. *Effect of grain boundaries*

The effect of particle size on the electron transport and magnetic properties of polycrystalline $La_{0.7}Ca_{0.3}MnO_3$ has been investigated.[38a] Although T_c decreases with decreasing particle size, the magnetoresistance is insensitive to particle size. We show this feature in Fig. 15 where the I-M transition becomes broader when the particle size is small, but the %MR is nearly constant over a wide temperature range. Fig. 15 also shows that the low temperature MR in these materials is significantly dependent on the grain size. For larger grain size, the MR at the lowest temperature decreases eventually becoming zero in good single crystals. Samples of manganates with different particle sizes prepared with different heat treatments often show large differences between the ferromagnetic transition temperatures and the temperatures of the I-M transition.[38b] A similar behavior appears to manifest itself when there is a large mismatch between the sizes of the A-site cations.[18b,37]

3.7. *Effect of Dimensionality*

The effect of dimensionality on CMR and related properties have been studied by examining the Ruddlesdon-Popper phases, (SrO) $(La_{1-x}Sr_xMnO_3)_n$ family.[39] The n = 1 member which is two dimensional with the K_2NiF_4 structure is an insulator, while the $n = \infty$ member is the three-dimensional perovskite showing CMR and other properties. The n = 2 member exhibits the sharpest I-M transition and high MR. In $La_{1.4}Sr_{1.6}Mn_2O_7$, the current perpendicular to the MnO_2 planes is large close to T_c (Fig. 16) and the barrier to transport provided by the intervening, insulating SrO rock salt layers can be removed by the application of a magnetic field; the interlayer transport appears to occur through tunneling.[40] The compressibility of the apical Mn-O bonds in $La_{1.2}Sr_{1.8}Mn_2O_7$ changes sign across the I-M transition. The Mn-O-Mn angle contracts with pressure in the paramagnetic insulating state, unlike in the ferromagnetic insulating state.[41a] These features are taken to support the presence of exchange striction due to a competition between superexchange and double-exchange. Neutron scattering studies of $La_{1.2}Sr_{1.8}Mn_2O_7$ show the existence of long-lived antiferromagnetic clusters along with ferromagnetic critical fluctuations.[41b] It appears that at least in two dimensions, Anderson localization has to be taken into account. Investigations of $Nd_{1+x}Sr_{2-x}Mn_2O_7$ (x = 0.0, 0.1) have revealed them to be biphasic, consisting of two closely similar Ruddlesdon-Popper phases, but they show CMR.[42]

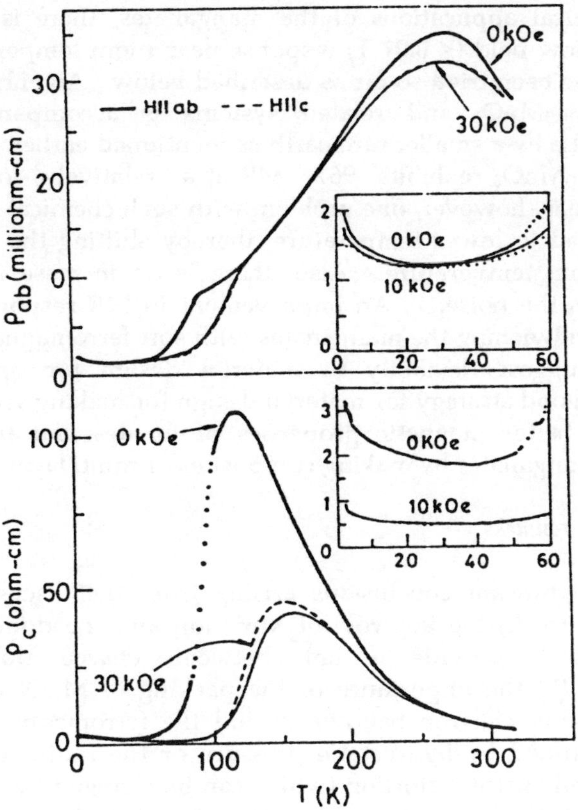

Fig. 16. Temperature variation of the resistivity of $La_{1.4}Sr_{1.6}Mn_2O_7$ in the ab plane and along the c axis under magnetic fields. Inset shows expanded low-temperature data (After Kimura et al[40]).

3.8. *Other observations*

Zero-field muon spin relaxation and resistivity experiments on $La_{0.7}Ca_{0.3}MnO_3$ show that the sublattice magnetization ν_μ (T) is well described for $T \leq T_c$ by $(1-T/T_c)^\beta$, where $\beta = 0.345 \pm 0.015$.[43] Below T_c, ν_μ and the zero-field resistivity ρ_o are correlated, with $\nu_\mu \propto -\ln \rho_o$. The rather unusual relaxational dynamics suggest spatially inhomogeneous Mn-ion correlation times.

For practical applications of the manganates, there is a need for enhancing the low field (< 0.01 T) response near room temperature. Two approaches have been tried so far as described below. An enhancement of MR in $La_{0.7}Ca_{0.3}MnO_3$ and related systems is accompanied by the substitution of La by a smaller rare earth as mentioned earlier. Thus, bulk $La_{0.33}Nd_{0.33}Ca_{0.33}MnO_3$ exhibits 96% MR at a relatively low field of ~ 0.7T.[44a] There is, however, one problem with such chemical substitution. The T_c is pushed to lower temperature, thereby shifting the peak in MR well below room temperature. Also, there is an increase in resistivity which increases the noise. An improvement in MR response has been obtained by sandwiching the manganates with soft ferromagnets.[44b] These observations suggest flexibility in material design for optimal CMR properties. A good strategy for material design for making sensors will be to couple the better magnetic properties of ferrites and the transport properties of manganates by making composites or multi layer films.

3.9. *Concluding remarks*

Some of the significant conclusions arising from the discussion in this section pertain to: (i) the key role of Mn^{4+} ion and the double-exchange interaction which provide a link between charge transport and magnetization; (ii) the importance of the packing of MnO_6 octahedra in determining the e_g electron bandwidth and the ferromagnetic exchange, which can be tuned by hydrostatic pressure or the radius of the A-site cations; (iii) local lattice distortion (which can be changed by temperature, chemical substitution as well as magnetization) which can give rise to polaron formation and charge localization; (iv) existence of structures with small energy differences so that a small pressure or a magnetic field can lead to a structural transition; (v) a substantial contribution of grain boundaries in MR of polycrystalline materials (in particular, at $T \ll T_c$).

4. Novel Features of Rare Earth Manganates

Electron transport, magnetic and other measurements on $La_{1-x}A_xMnO_3$ systems have revealed rather unusual features with respect to the charge carriers in these oxides. The manganates exhibit high resistivities even in the metallic state, particularly at low temperatures.[45] The values of resistivities are considerably higher than Mott's value of maximum metallic resistivity,[11] the resistivity reaching constant values as high as 10^3-10^4 ohm cm at low temperatures, in the so-called metallic state. Coey et al[45] argue that T < T_c , the e_g electrons are delocalized on an atomic scale but the spatial fluctuations in the Coulomb and spin dependent potentials tend to localize the e_g electrons in wave packets larger than the Mn-Mn distance. A clear picture of the nature of the electrons responsible for such high resistivity is yet to emerge. It is also surprising that large MR occurs in manganates in a small magnetic field at a relatively high temperature. The scale of Zeeman energy associated with a field of 1T in these spin systems is ≈ 5K. This is much too small compared to the scale of thermal energy (200-300K) where one sees such a large MR.

Optical conductivity measurements on $La_{1-x}Sr_xMnO_3$ in the range 0-10eV show a band of 1.5eV at moderate doping and the spectral weight is transferred from this band with decreasing temperature. The 1.5eV band is

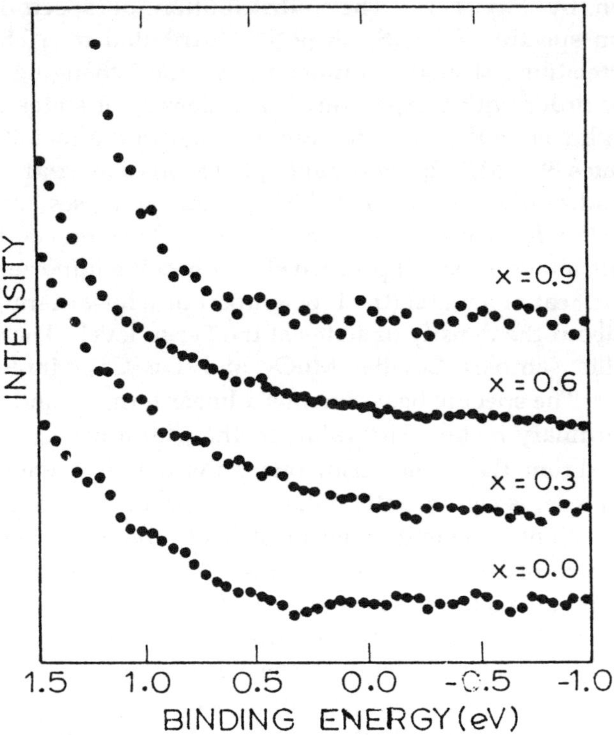

Fig. 17. Photoelectron spectra of $La_{1-x}Sr_xMnO_3$ in the valence region at room temperature (from Saitoh et al[49b]).

due to interband transitions between the exchange split spin polarized e_g bands. At $T < T_c$, intraband transitions in the e_g band dominate the spectrum.[46] Double-exchange alone cannot explain the optical conductivity data. A transfer of spectral weight from high energy to low energy occurs at $T < T_c$ in $Nd_{0.7}Sr_{0.3}MnO_3$.[47] A recent study of tunneling density of states by low temperature STM in $La_{0.8}Ca_{0.2}MnO_3$ has shown that at $T < T_c$, a large number of states become available near E_t.[48] Photoemission studies of $La_{1-x}Sr_xMnO_3$[49] have provided valuable information on the electronic structure of the manganates. An aspect that needs attention is concerned with the negligible density of states near the Fermi level in the metallic state of these systems. Very close to E_F, the density of states rises rapidly with energy. In Fig. 17 we show typical photoemission spectra to illustrate this feature. There is a transfer of spectral weight from the unoccupied states to occupied states with decrease in hole doping. (Note: This aspect is not shown by Fig. 17). The redistribution of spectral weight in photoemission spectra of $La_{0.6}Sr_{0.4}MnO_3$ is attributed to a change in the electron correlation strength induced by the changing degree of ferromagnetic order with temperature. The density of states at the Fermi level is much lower in the insulator regime than in the metallic regime of the manganates.[50] Mn 2p resonant photoemission and O 1s x-ray absorption studies have shown that the band gap collapses around T_c with an increase in the density of states near the Fermi level on cooling.[50] In the insulating state above T_c, small polaron effects become important.

The specific heat of a metal (the low temperature linear term, γ) provides information about the density of states at the Fermi level. The specific heat of the metallic samples $La_{0.7}Ba_{0.3}MnO_3$ and $La_{0.8}Ca_{0.2}MnO_3$ have been measured.[45,51a] The specific heat contains a linear term (γ) and a cubic term (β) as in an ordinary metal. The value of the γ term lies in the region 4-6 mJ/mole K^2. From the cubic term, one finds a Debye temperature $\theta_D \approx 400$K. The γ term is significantly enhanced from the free electron and the band values, the enhancement being most likely due to correlation effects and strong electron-lattice interaction. This γ value is comparable to that in a number of similar metallic ABO_3 perovskites. In short, the specific heat does not appear to be unusual and does not show any indication of depletion of the density of states at the Fermi level unlike that seen in photoemission studies. Low-temperature specific heat measurements on $La_{0.7}Sr_{0.3}MnO_3$ have shown that it is low-carrier density metal with conventional ferromagnetic spin waves.[51b] Raman scattering experiments on $La_{0.7}Sr_{0.3}MnO_3$ show a band at 2100 cm^{-1} whose intensity decreases with increasing temperature and this is taken to imply a reduction in the density of states at the Fermi energy with increasing temperature.[51c]

The Hall coefficient has been measured in epitaxial films of $La_{0.67}Ca_{0.33}MnO_3$ and $La_{0.67}Sr_{0.33}MnO_3$ grown by MOCVD and the carrier concentration estimated from the Hall effect is more than that calculated with the assumption that each Mn^{4+} contributes one hole and that all the holes are mobile.[52a] For the Sr-doped sample, it is 2.1 holes/Mn and for Ca-doped samples it is 0.9 holes/Mn. This observation is not compatible with the low density of states found from photoemission experiments.[49] It is to be noted that while specific heat and Hall measurements are bulk properties, photoemission measurements are sensitive to surface effects. The density of states at the Fermi level and the carrier concentration in the manganates are yet to be settled quantitatively. In particular, how these change as the samples are cooled through T_c need further investigation.

The Seebeck coefficient, S, in several rare earth manganates show a positive peak in the composition range and at a temperature where MR also peaks (around T_c), and becomes more negative with the increase in Mn^{4+} content.[11,52b] Seebeck coefficient measurements on $La_{1-x}Ca_xMnO_3$ along with self-doped $LaMnO_3$ with comparable Mn^{4+} contents carried out by us have shown the nature of hole doped states to be different in the two cases, being more localized in the latter.[52b] Measurements of S in $La_{1-x}Sr_xMnO_3$ show a change in sign from positive to negative at the I-M transition boundary or at $T < T_c$ depending on the composition.[53a] An anomalous change in electronic structure with spin-polarization may occur when $x = 0.2$-0.3. Based on the pressure dependence of S, it has been suggested that there is a crossover from a polaronic to extended state electronic conduction at T_c.[53b] While a positive S in the ferromagnetic state indicates transport involving polarized states, it is not clear why the values of S are so small in a regime where the resistivity changes by orders of magnitude.[54]

Furukawa[55] considers the origin of the large magnetoresistance observed near T_c in the manganates to essentially lie in the forced alignment of the local t_{2g} spins by the application of a magnetic field causing a reduction in the spin disorder scattering of the e_g electrons. This arises from the strong spin-charge coupling (Hund coupling) between the t_{2g} and e_g electrons on the same Mn atom . This intra-atomic ferromagnetic interaction is considerably large compared to the bandwidth and causes a splitting of the e_g band. The double-exchange model essentially explains the behavior of magnetoresistance with respect to magnetization (- ρ/ρ =

$C(M/M_s)^2)$. Based on a study of the $LnMn_{1-x}Cr_xO_3$ system where isoelectronic Cr^{3+} replaces Mn^{4+}, it has indeed been shown recently that double-exchange is essential for the occurrence of large GMR in the manganates.[56] However, there is the basic question why these manganates are insulating when $T > T_c$. The anomalous resistance has been explained by Millis et al[57] by considering electron-phonon interaction arising from the Jahn-Teller splitting of the Mn d-levels. Jahn-Teller distortion localizes conduction electrons as polarons. As the temperature decreases, the transfer integral t_{ij} increases and the ratio of the Jahn-Teller energy to t_{ij} decreases. The coupling varies with temperature and composition and increases across the metal-insulator transition.[58] The large effect of local lattice distortion and crystal structure symmetry on electron transport and magnetism show that one cannot explain the unusual behavior without taking the lattice into account. The observed T_c ($\approx$ 250-400K) in most manganates is considerably lower than that estimated from the measured spin-wave stiffness constant[6] which implies that the ferromagnetic exchange arising from double exchange mechanism is not the only factor that affects T_c. This has led to the proposal of Jahn-Teller distortion as another factor.[57] The importance of electron-phonon interaction is also underscored by the observation of the large oxygen isotope shift of T_c in $La_{0.8}Ca_{0.2}MnO_3$.[59] Besides lattice effects, correlation effects also yield a large distribution of spectral weights.[60] Clearly, it is difficult to understand the electrical resistivity of the rare earth manganates across the I-M transition, especially at low temperatures based on double-exchange or electron-phonon interaction alone. We do not yet have a theory which explains most of the important features of the electronic transport in these materials. In particular, why CMR actually occurs has not yet been settled although we know some of the ingredients of the basic physics (like double-exchange, lattice distortion, magnetic localization etc.) which can lead to CMR. In terms of the chemistry of $Ln_{1-x}A_xMnO_3$ type manganates, there is always the problem of homogeneity. Thus, there would be ferromagnetic clusters at large or small x, eventhough the material may be considered to be antiferromagnetic. In $La_{(2-x)/3}Nd_{x/3}Ca_{1/3}MnO_3$, the presence of La-rich ferromagnetic domains along with semiconducting Nd-rich domains has been considered.[36b] The ordering of Ln and A ions *vis a vis* the ordering of Mn^{3+} and Mn^{4+} ions could also be a factor in determining some of the finer aspects of structure and properties.

5. Charge-ordering in Rare Earth Manganates

The study of CMR in $Ln_{1-x}A_xMnO_3$ has brought forth novel features related to the charge ordering in these oxides. Charge-ordering in the manganates is interesting because double-exchange gives rise to metallicity along with ferromagnetism while the charge-ordered state is associated with insulating and antiferromagnetic (or paramagnetic) behavior. What we have is a competition between ferromagnetism with metallic behavior and cooperative Jahn-Teller effect with charge-ordering. Lattice distortion plays a crucial role in charge-ordering, orthorhombic distortion stabilizing the charge-ordered state. The charge-ordered state can be melted into a metallic ferromagnetic state, by the application of a magnetic field. An examination of the observation on the rare earth manganates[61] reveals two types of scenarios, as exemplified by $Nd_{0.5}Sr_{0.5}MnO_3$ and $Pr_{1-x}Ca_xMnO_3$ $(0.3 \leq x \leq 0.5)$. In $Nd_{0.5}Sr_{0.5}MnO_3$, a ferromagnetic metallic state ($T_c \sim 250K$) gives over to an antiferromagnetic charge-ordered state around 150K (T_{CO}) accompanied by changes in lattice parameters[62] as shown in Fig. 18. The competition between the ferromagnetic double-exchange interaction and the antiferromagnetic charge-ordering instability has been examined in $(Nd,Sm)_{0.5}Sr_{0.5}MnO_3$.[63] This system exhibits large magnetoresistance which is strongly coupled to the lattice striction. The CMR phenomenon here is nothing but a first order I-M transition induced by a magnetic field, accompanied by a metamagnetic transition and a structural change. In $Pr_{1-x}Ca_xMnO_3$ ($x = 0.35$), an insulating charge-ordered state ($T_{CO} \sim 200K$) becomes antiferromagnetic around 140K (T_N) and then undergoes a transformation to a canted-spin antiferromagnetic state at a still lower temperature (~ 110K).[64] The application of a magnetic field melts the charge-ordered state in most of the charge-ordered manganates giving rise to metallic behavior (Fig. 19). The first-order I-M transition induced by magnetic fields is generally accompanied by considerable hysteresis as shown in Fig. 20.[62] The I-M transitions in the manganates where the one-electron bandwidth is controlled by the composition, show interesting electronic phase diagrams[61] in the temperature-magnetic field (T-H) plane as depicted in Fig. 21 for $Pr_{1-x}Ca_xMnO_3$ and $Nd_{0.5}Sr_{0.5}MnO_3$. The charge-ordering gap (0.3 eV) in $Nd_{0.5}Sr_{0.5}MnO_3$ has been measured by tunneling spectroscopy[65b] as shown in Fig. 22. Application of pressure increases the ferromagnetic T_c of $Nd_{0.5}Sr_{0.5}MnO_3$ and suppresses charge ordering ; in the case of $Pr_{0.7}Ca_{0.3}MnO_3$, pressure increases the metallicity and suppresses the charge ordering transition.[65a]

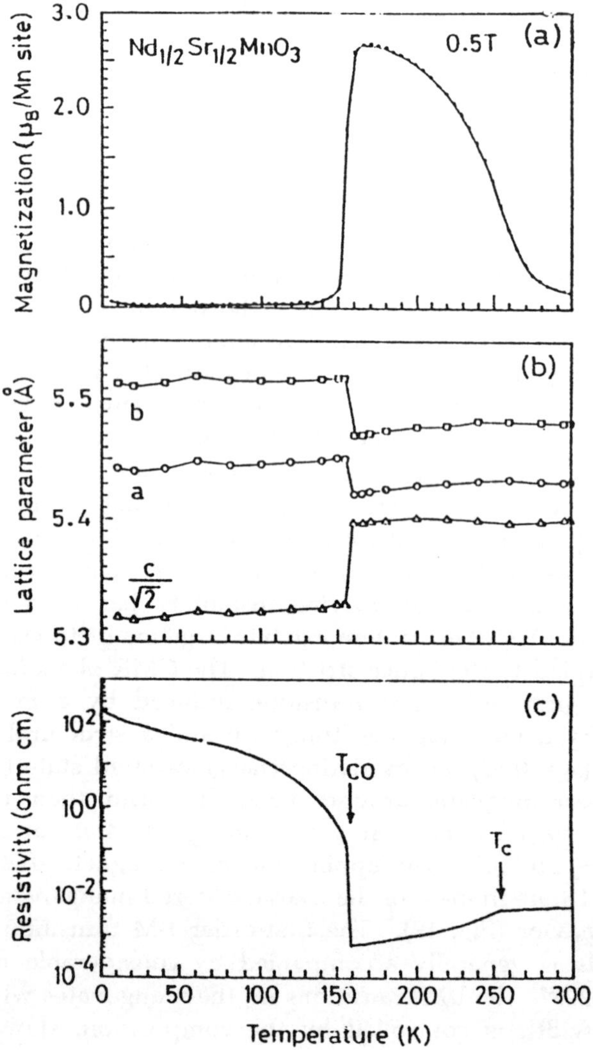

Fig. 18. The temperature variation of (a) magnetization (b) lattice parameters and (c) resistivity of $Nd_{0.5}Sr_{0.5}MnO_3$ single crystal (from Kuwahara et al[62]).

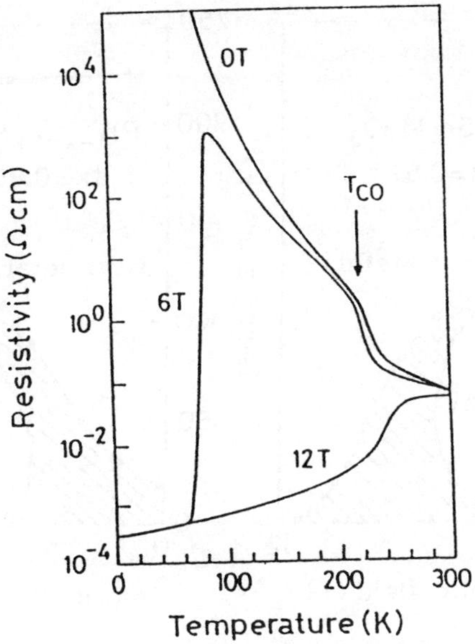

Fig. 19. The temperature variation of resistivity under H = 0T, 6T and 12T in $Pr_{1-x}Ca_xMnO_3$ (x = 0.35) single crystal (from Tomioka et al[64]).

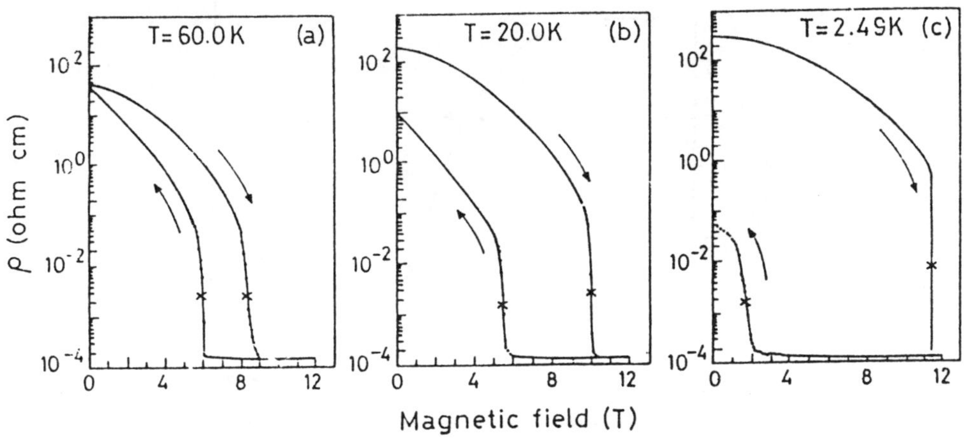

Fig. 20. Changes in the resistivity of $Nd_{0.5}Sr_{0.5}MnO_3$ with increasing and decreasing magnetic fields. Notice the sharp jump in resistivity at the lower and upper critical fields. (from Kuwahara et al[62]).

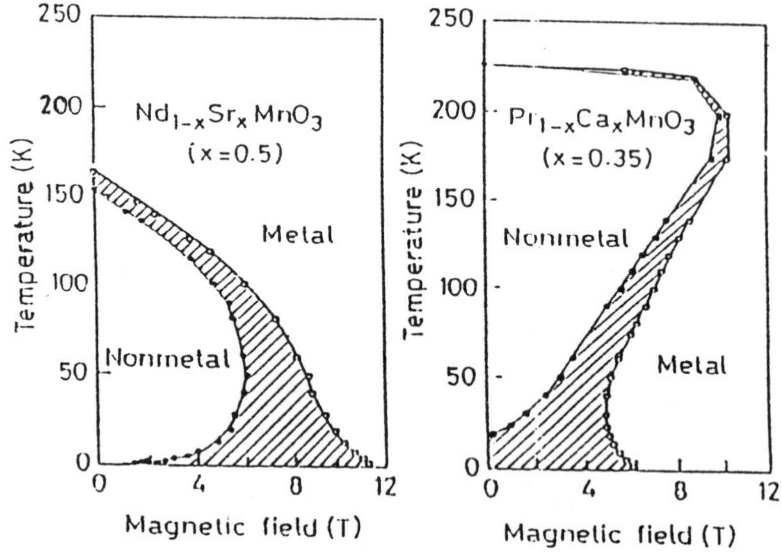

Fig. 21. The electronic phase diagrams in the T-H plane of (a) $Nd_{1-x}Sr_xMnO_3$ (x = 0.5) and (b) $Pr_{1-x}Ca_xMnO_3$ (x = 0.35) (from Tokura et al[61]).

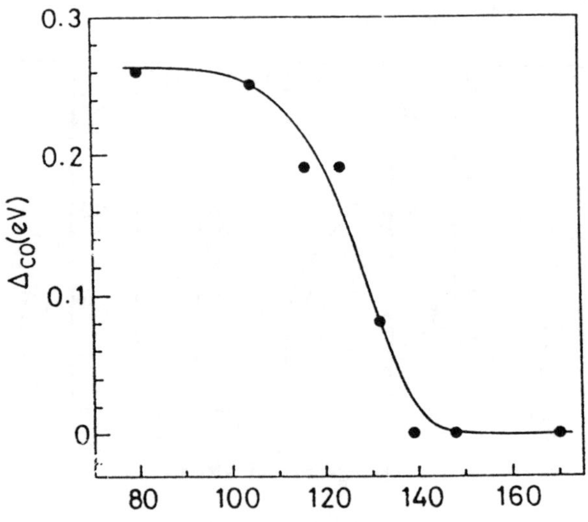

Fig. 22. Temperature dependence of the charge-ordering gap, Δ_{CO}, in $Nd_{0.5}Sr_{0.5}MnO_3$ (from Biswas et al[65b]).

Charge-ordering in the manganates is governed by the width of the e_g band which is directly determined by the weighted average radius of the A-site cations $<r_A>$, or the tolerance factor. This is because a distortion of the Mn-O-Mn bond angle affects the transfer interaction of the e_g conduction electrons (holes). We can describe the spin and charge ordering phenomena in the rare-earth manganates in terms of the generalized phase diagram shown in Fig. 23a. The diagram shows that when $<r_A>$ is large as in region A (e.g. $La_{1-x}Sr_xMnO_3$), only ferromagnetism and the associated I-M transition occur with no charge-ordering. With a slight decrease in $<r_A>$ (region B), the ferromagnetic (metallic) state transforms to the

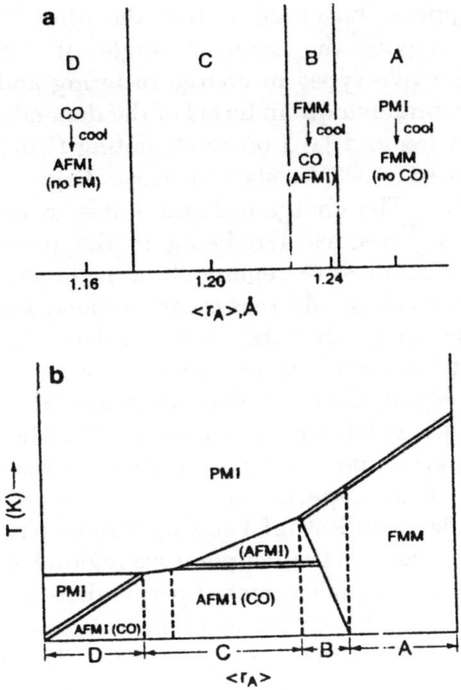

Fig. 23. (a) Schematic diagram showing the prevalence of charge-ordered (CO) and ferromagnetic states in $Ln_{1-x}A_xMnO_3$ depending on the weighted average radius of the A-site cation, $<r_A>$ or the e_g bandwidth (b) Schematic temperature $<r_A>$ diagram for $Ln_{0.5}A_{0.5}MnO_3$ where double line represents a second-order transition and a single line represents a first-order transition (from Kumar and Rao[66]).

antiferromagnetic charge-ordered state on cooling as in $Nd_{0.5}Sr_{0.5}MnO_3$ ($<r_A>$ = 1.236Å). When $<r_A>$ is very small as in region D as exemplified by $Pr_{0.7}Ca_{0.3}MnO_3$ and $Nd_{0.5}Ca_{0.5}MnO_3$ ($<r_A>$~1.17Å), no ferromagnetism is encountered and charge-ordering occurs in the paramagnetic state. The ferromagnetic metallic state can only be created by the application of the magnetic field in the charge-ordered state, depending on $<r_A>$. In $Y_{0.5}Ca_{0.5}MnO_3$ with $<r_A>$ = 1.13Å, the charge ordered state is not perturbed by magnetic fields.

The dependence of charge ordering in $Ln_{1-x}A_xMnO_3$ on the radius of the A-site cations, $<r_A>$, shown in Fig. 23a is instructive and suggests at least two types of charge ordered states in regions B and D. It is remarkable that the range in $<r_A>$ giving such a vast variety of properties is only ~ 0.1Å. What actually happens, however, is that the Mn-O-Mn angle varies by ~ 20° over this region, the smallest angle of 148° being found in $Y_{0.5}Ca_{0.5}MnO_3$. The two types of charge ordering and the other features of Fig. 23a can be understood in terms of the dependence of the exchange coupling constants J_{FM} and J_{AFM} on $<r_A>$, noting that the single ion Jahn-Teller energy is invariant.[66] We show a phase diagram worked out in this manner in Fig. 23b. The charge-ordered states in region D show some interesting features. Besides T_{CO} being in the paramagnetic state, the manganates in this small $<r_A>$ regime show negligible volume change at T_{CO}, and the unit cell volume decreases continuously from T_{CO} down to low temperatures. Furthermore, the bond valence sum shows charge localization only at low temperatures as shown in Fig. 24.[67]

Manganates in region C in the phase diagram (Fig. 23) are expected to show rather complex behavior. Thus, in $La_{1-x}Ca_xMnO_3$ ($0.63 \leq x \leq 0.67$) charge-ordering is accompanied by an increase in the sound velocity and anomalies in the heat capacity and derivative of the logarithm of resistivity.[68] Detailed studies[69] of $La_{0.5}Ca_{0.5}MnO_3$ have shown that Jahn-Teller distortion develops in the temperature region between 225K (T_c) and 155K (T_N), with the d_{z^2} orbitals oriented perpendicular to the orthorhombic b-axis. Magnetic domain boundaries seem to break the coherence of the spin ordering on the Mn^{3+} sites while preserving the coherence of the spin ordering on the Mn^{4+} sublattice as well as the identity of the two sublattices. The similarity between these structures and the structural charge ordering and discommensuration domain boundaries suggests that in $La_{0.5}Ca_{0.5}MnO_3$, commensurate long-range charge ordering coexists with quasi-commensurate orbital ordering.

By subtle variations of $<r_A>$, one can bring manganates in the regions B and D of the phase diagram in Fig. 23 to region C. Thus, in

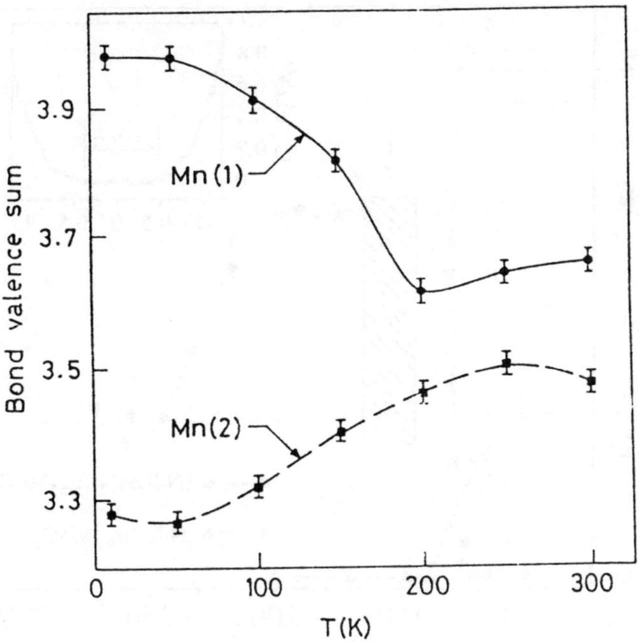

Fig. 24. Variation of bond valence sums in $Nd_{0.5}Ca_{0.5}MnO_3$ with temperature (from Vogt et al[67]).

$Nd_{0.5-x}Sm_xSr_{0.5}MnO_3$, a charge ordered state transforms to a ferromagnetic metallic state on cooling.[63] In $Nd_{0.25}La_{0.25}Ca_{0.5}MnO_3$ ($\langle r_A \rangle$ = 1.19Å), however, a novel reentrant ferromagnetic transition occurs from an incipient charge-ordered state on cooling ; the transition is accompanied by a collapse of the charge order gap (Fig. 25), driven by a first order transition [70]. The incipient nature of the charge-ordered state is verified by the absence of superstructure peaks in the diffraction pattern.

Charge ordering has been found in quasi two-dimensional $L_{0.5}Sr_{1.5}MnO_4$ as well. Here, T_{CO} is 217K and T_N is 110K.[71] Below T_N, the magnetic moments get ordered with a $2\sqrt{2}\,a \times 2\sqrt{2}\,a \times 2c$ cell. Superlattice reflections corresponding to alternating Mn^{3+} and Mn^{4+} ions are seen below T_c. The size of the A-site cations would have negligible effect on the charge-ordering in these insulating two-dimensional manganates. It may be noted that in the perovskites, $Ln_{1-x}A_xMnO_3$ ($x \approx 0.5$), T_{CO} initially increases as $\langle r_A \rangle$ decreases and then become nearly constant. Thus,

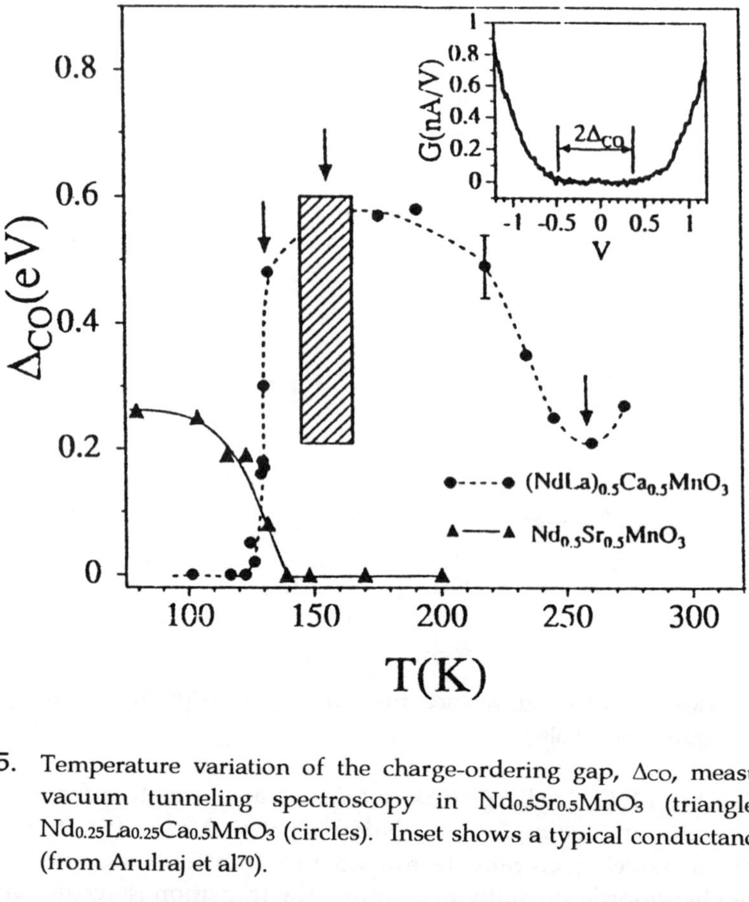

Fig. 25. Temperature variation of the charge-ordering gap, Δ_{CO}, measured by vacuum tunneling spectroscopy in $Nd_{0.5}Sr_{0.5}MnO_3$ (triangles) and $Nd_{0.25}La_{0.25}Ca_{0.5}MnO_3$ (circles). Inset shows a typical conductance curve (from Arulraj et al[70]).

T_{CO} = 0K in $La_{0.5}Sr_{0.5}MnO_3$ ($\langle r_A \rangle$ = 1.23Å), 220K in $La_{0.5}Ca_{0.5}MnO_3$ ($\langle r_A \rangle$ = 120Å) and 240K in $Nd_{0.5}Ca_{0.5}MnO_3$ ($\langle r_A \rangle$ = 1.17Å). The near constancy of T_{CO} is in the insulating regime.

Investigations on charge-ordering have established the intimate connection of the lattice distortion to CO. It is the lattice distortion associated with orbital ordering that appears to localize the charge and initiate charge-ordering. Eventually, the Coulomb interaction wins over the kinetic energy of the electrons to form long range CO state. The scale of the energy involved with CO as measured by the charge-ordering gap is around 0.5 - 1 eV. This is similar to the unscreened bare nearest neighbour

Coulomb repulsion. This is also close to the approximate energy required to create ~ 1% orthorhombic distortion. It is likely that both the energy scales along with magnetic exchange decide the energy scale of the charge-ordering gap.

6. Other CMR Materials

CMR is found in $Tl_2Mn_2O_7$ which has only Mn^{4+} ions and hence without double-exchange. The material is an itinerant electron ferromagnet showing a FMM-PMM transition.[72] Recently, CMR has been found in a thiospinel, $Fe_{1-x}Cu_xCr_2S_4$, where there can be no double exchange.[73] The occurrence of CMR in nonstoichiometric EuO has been known for sometime[74] ; in this oxide also, there is an insulator-metal transition around the ferromagnetic Curie temperature. Clearly, there is need to explore a variety of oxides, chalcogenides and other materials with or without double-exchange. Composites of manganates with other materials (see section 3.8) deserve further attention.

7. Future Prospects

Besides search for new and better CMR materials, there is scope of careful investigations of the various properties and phenomena exhibited by the manganates. These include a study of charge and orbital ordering[75] and the effect of electric fields and radiation on the charge ordered states. A transition from the insulating (charge ordered) state to the ferromagnetic metallic state has been induced in $Pr_{0.7}Ca_{0.3}MnO_3$ by x-ray illumination.[76] It has been possible to trigger the collapse of the charge-ordered state in $Pr_{0.7}Ca_{0.3}MnO_3$ by the application of an electric field as well.[77a] Laser irradiation is also found to induce an insulator-metal transition in this manganate.[77b] Structural studies of charge ordered manganates under magnetic fields would be interesting. Theoretical studies directed towards understanding the transport properties of manganates as well as charge-ordering[75] and related aspects are necessary.

Acknowledgement:

The authors thank the Science Office of the European Community, the Department of Science and Technology, Government of India and the CSIR (India) for support.

References

1. (a) P.M. Levy, *Solid State Phys.* **47**, 367 (1994).
   (b) P.M. Levy and S. Zhang, *J. Mag. Mag. Mater.* **151**, 315 (1995).
2. N.B. Brandt and V.V. Moschalkov, *Adv. Phys.* **33**, 193 (1984).
3. (a) R. von Helmolt et al, *Phys. Rev. Lett.* **71**, 2331 (1993).
   (b) K. Chahara et al, *Appl. Phys. Lett.* **63**, 1990 (1993).
   (c) M. McCormack et al, *Appl. Phys. Lett.* **64**, 3045 (1994).
   (d) R. Mahesh et al, *J. Solid State Chem.* **114**, 297 (1995).
   (e) C.N.R. Rao et al, *Chem. Mater.*, **8**, 2421 (1996) and the references therein.
4. A. Urushibara et al, *Phys. Rev.* **B 51**, 14103 (1995).
5. (a) G.H. Jonker and J.H. van Santen, *Physica* **16**, 377 (1950).
   (b) J.H. van Santen and G.H. Jonker, *Physica* **16**, 599 (1950).
6. J.W. Lynn et al, *Phys. Rev. Lett.* **76**, 4046 (1996).
7. C. Zener, *Phys. Rev.* **82**, 403 (1951).
8. J.B. Goodenough, *Prog. Solid State Chem.* **5**, 149 (1971).
9. P.G. de Gennes, *Phys. Rev.* **118**, 141 (1960).
10. E.O. Wollan and W.C. Koehler, *Phys. Rev.* **100**, 545 (1955).
11. R. Mahendiran et al, *Phys. Rev.* **B 53**, 3348 (1996).
12. P. Schiffer et al, *Phys. Rev. Lett.* **75**, 3336 (1995).
13. P.G. Radaelli et al, *J. Solid State Chem.* **122**, 444 (1996).
14. I.O. Troyanchuk, *Sov. Phys. JETP* **75**, 132 (1992).
15. (a) J.L. Garcia-Munoz et al, *J. Phys. Rev.* **B 55**, 34 (1997).
    (b) L.M. Rodriguez-Martinez and J.P. Attfield, *Phys. Rev.* **B 54**, 15622 (1996).
16. P.G. Radaelli et al, *Phys. Rev.* **B 54**, 8992 (1996).
17. (a) J.A.M. van Roosmalen and E.H.P. Cordfunke, *J. Solid State Chem.* **110**, 109 (1994).
    (b) M. Hervieu et al, *Euro J. Solid State Chem.* **32**, 79 (1995).
18. R. Mahesh et al, *J. Solid State Chem.* **114**, 294 (1995).
19. A. Arulraj et al, *J. Solid State Chem.* **127**, 87 (1996).
20. R. Mahendiran et al, *J. Phys. D. Appl. Phys.* **28**, 1743 (1995).
21. R. Mahendiran et al, *Appl. Phys. Lett.* **66**, 233 (1995).
22. S.J.L. Brillinge et al, *Phys. Rev. Lett.* **77**, 715 (1996).
23. T.A. Tyson et al, *Phys. Rev.* **B 53**, 13985 (1996).
24. R.P. Sharma et al, *Phys. Rev.* **B 54**, 10014 (1996).
25. M.C. Martin et al, *Phys. Rev.* **B 53**, 14285 (1996)

26. (a) A. Asamitsu et al, *Phys. Rev.* **B 54**, 1716 (1996).
(b) A.J. Campbell et al, *Phys. Rev.* **B 55**, 8622 (1997).
27. (a) H. Kaivano et al, *Phys. Rev.* **B 53**, 14709 (1996).
(b) J.M. De Teresa et al, *J. Appl. Phys.* **B 79**, 5175 (1996).
28. Y. Yamada et al, *Phys. Rev. Lett.* **77**, 904 (1996).
29. J.M. De Teresa et al, *Nature* **386**, 256 (1997).
30. (a) Y. Moritomo et al, *Phys. Rev.* **B 51**, 16491 (1995).
(b) J.J. Neumeier et al, *Phys. Rev.* **B 52**, 7006 (1995).
31. H.Y. Hwang et al, *Phys. Rev. Lett.* **75**, 914 (1995).
32. B. Raveau et al, *J. Solid State Chem.* **117**, 424 (1995).
33. J. Fontcuberta et al, ***Phys. Rev. Lett***. **76**, 1122 (1996).
34. (a) M.R. Ibarra et al, *Phys. Rev. Lett.* **75**, 3541 (1995).
(b) A. Maignan et al, *J. Appl. Phys.* **79**, 7891 (1996).
35. R. Mahendiran et al, *Phys. Rev.* **B 53**, 12160 (1996).
36. (a) J.L. Garcia-Munoz et al, *Phys. Rev.* **B 55**, 668 (1997).
(b) G.H. Rao et al, *Phys. Rev.* **B 55**, 3742 (1997).
37. R. Mahesh et al, *J. Solid State Chem.* **120**, 204 (1995).
38. (a) R. Mahesh et al, *Appl. Phys. Lett.* **68**, 2291 (1996).
(b) R. Mahendiran et al, *Solid State Commun.* **99**, 149 (1996).
39. (a) Y. Moritomo et al, *Nature* **380**, 141 (1996).
(b) R. Mahesh et al, *J. Solid State Chem.* **122**, 448 (1996).
40. T. Kimura et al, *Science* **274**, 1698 (1996).
41. (a) D.N. Argyriou et al, *Phys. Rev. Lett.* **78**, 1568 (1997).
(b) T.G. Perring et al, *Phys. Rev. Lett.* **78**, 3197 (1997).
42. P.D. Battle et al, *Phys. Rev.* **B 54**, 15967 (1997).
43. R.H. Heffner et al, *Phys. Rev. Lett.* **77**, 1869 (1996).
44. (a) G.H. Rao et al, *Appl. Phys. Lett.* **69**, 424 (1996).
(b) H.Y. Hwang et al, *Appl. Phys. Lett.* **68**, 3494 (1996).
45. J.M.D. Coey et al, *Phys. Rev. Lett.* **75**, 3910 (1995).
46. (a) Y. Okimoto et al, *Phys. Rev. Lett.* **75**, 109 (1995).
(b) Y. Okimoto et al, *Phys. Rev.* **B 55**, 4206 (1997).
47. S.G. Kaplan et al, *Phys. Rev. Lett.* **77**, 2081 (1996).
48. A. Biswas and A.K. Raychaudhuri, *J. Phys. Condens. Matter*. **8**, L739 (1996).
49. (a) D.D. Sarma et al, *Phys. Rev.* **B 53**, 6873 (1996).
(b) T. Saitoh et al, *Phys. Rev.* **B 51**, 13942 (1995).
50. (a) J.H. Park et al, *J. Appl. Phys.* **79**, 4558 (1996).
(b) J.H. Park et al, *Phys. Rev. Lett.* **76**, 4215 (1996).

51. (a) J.J. Hamilton et al, *Phys. Rev.* **B 54**, 14926 (1996).
(b) B.F. Woodfield et al, *Phys. Rev. Lett.* **78**, 3201 (1997).
(c) R. Gupta et al, *Phys. Rev.* **B 54**, 14899 (1996).
52. (a) J.G. Snyder et al; *Phys. Rev.* **B53**, 14434 (1996).
(b) R. Mahendiran et al, *Solid State Commun.* **98**, 701 (1996); also see R. Mahendiran et al, *Phys. Rev.* **B 54**, 9604 (1996).
53. (a) A. Asamitsu et al, *Phys. Rev.* **B 53**, 2952 (1996).
(b) W. Archibald et al, *Phys. Rev.* **B 53**, 14445 (1996).
54. B. Fisher et al, *Phys. Rev.* **B 54**, 9359 (1996).
55. N. Furukawa, *J. Phys. Soc. Jpn.* **63**, 3214 (1994).
56. R. Gundakaram et al, *J. Solid State Chem.* **127**, 354 (1996).
57. A.J. Millis et al, *Phys. Rev. Lett.* **74**, 5144 (1995).
58. H. Roder et al, *Phys. Rev. Lett.* **76**, 1356 (1996).
59. G. Zhao et al, *Nature* **381**, 676 (1996).
60. S. Satpathy et al, *Phys. Rev. Lett.* **76**, 960 (1996).
61. Y. Tokura et al, *J. Appl. Phys.* **79**, 5288 (1996).
62. H. Kuwahara et al, *Science* **270**, 961 (1995).
63. Y. Tokura et al, *Phys. Rev. Lett.* **76**, 3184 (1996).
64. Y. Tomioka et al, *Phys. Rev.* **B 53**, 1689 (1996).
65. (a) Y. Moritomo et al, *Phys. Rev.* **B 55**, 7549 (1997).
(b) A. Biswas et al, *J. Phys. Condens. Matter* **9**, L355 (1997).
66. N. Kumar and C.N.R. Rao, *J. Solid State Chem.* **129**, 363 (1997).
67. T. Vogt et al, *Phys. Rev.* **B 54**, 15303 (1996).
68. A.P. Ramirez et al, *Phys. Rev. Lett.* **76**, 3188 (1996).
69. P.G. Radaelli et al, *Phys. Rev.* **B 55**, 3015 (1997).
70. A. Arulraj et al, To be published.
71. B.J. Sternlieb et al, *Phys. Rev. Lett.* **76**, 2169 (1996).
72. (a) M.A. Subramanian et al, *Science* **273**, 81 (1996).
(b) D.K. Seo et al, *Solid State Commun.* **101**, 417 (1997).
73. A.P. Ramirez et al, *Nature* **386**, 186 (1997).
74. Y. Shapira et al, *Phys. Rev.* **B 8**, 2299, 2316 (1973).
75. C.N.R. Rao and A.K. Cheetham, *Science* **276**, 911 (1997).
76. V. Kiryukhin et al, *Nature* **386**, 813 (1997).
77. (a) A. Asamitsu et al, *Nature* **388**, 50 (1997).
(b) K. Miyano et al, *Phys. Rev. Lett.* **78**, 4257 (1997).

THE IMPORTANT ROLE OF CRYSTAL CHEMISTRY UPON THE CMR PROPERTIES OF MANGANITES

B. RAVEAU, A. MAIGNAN, C. MARTIN and M. HERVIEU
Laboratoire CRISMAT, UMR 6508 associée au CNRS, ISMRA et Université de Caen
6, Boulevard du Maréchal Juin, 14050 CAEN Cedex - France

The recent exploration of the manganites with the perovskite structure has shown the important role of crystal chemistry upon the colossal magnetoresistance (CMR) of these materials. Two types of manganites are distinguished, type I manganites characterized by a ferromagnetic metallic (FMM) - paramagnetic insulating (PMI) transition, and type II manganites involving two transitions antiferromagnetic insulating (AFMI) - ferromagnetic metallic (FMM) - paramagnetic insulating (PMI). The different parameters which govern the CMR properties of these materials are reviewed : mixed valence of manganese (hole carrier density), size of the A interpolated cation, mismatch effect (between the A-site cations), structural transitions, doping of the manganese sites with various elements.

The discovery of negative giant magnetoresistance (GMR) properties in manganese perovskite thin films some years ago [1-5] has demonstrated for the second time after the through break about high T_c superconducting cuprates, that oxides offer a very promising field for the investigation of new materials with specific properties susceptible to be involved in device applications. Soon after the publication of these results, in 1994-1995, many papers were published almost simultaneously about the magnetoresistance properties of manganites with the generic formula $Ln_{1-x}A_xMnO_3$ with A = Ca, Sr, Ba, Pb [6-16].

The ability of these materials to be used in devices can be defined by their magnetoresistance ratio, that is generally defined as $\Delta R/R_0 = [R_0-R_H]/R_0$ where R_0 is the zero field resistance and R_H is the resistance in the applied field H. Magnetoresistance ratios ranging from 10 to 99% are typically observed for such oxides. In fact the resistance ratio (RR) $RR = R_0/R_H$ is more realistic since it shows clearly how many orders of magnitude the resistance can be decreased by applying a magnetic field. The first series of investigations made in 94-95 on thin films [1-5] and on sintered ceramics [6-16], showed rapidly that extremely high RR ratios, up to 10^{11}, can be reached, so that these oxides are generally named as colossal magnetoresistance (CMR) manganites.

From a fundamental point of view the CMR effect seems to be closely related to the double exchange mechanism between the Mn^{3+} and Mn^{4+} species developed by de Gennes [17] and Goodenough [18]. Nonetheless recent studies made by Millis [19] suggest that Jahn-Teller type electron phonon interactions are greatly involved in the appearance of such properties. Clearly, the knowledge of the crystal structure and of the chemical bonding of these compounds is of capital importance to progress in the understanding of the CMR properties of these phases. The optimisation of the CMR properties of these phases and the discovery of new properties are intimately linked to their complex crystal chemistry. We review herein the main features which govern the relationships between the crystal chemistry and magnetotransport properties of these materials.

1 The two types of CMR manganites

Although the phenomenon is fundamentally the same from the viewpoint of magnetoresistance properties for all manganites $Ln_{1-x}A_xMnO_3$, it is convenient in a first approach to distinguish two types of behavior, based on the evolution of the resistance of these oxides versus temperature in a zero magnetic field.

The type I manganites generally synthesized for $x < 0.50$ exhibit a $R_0(T)$ curve characterized by a peak as shown for instance for $Pr_{0.7}Sr_{0.05}Ca_{0.25}MnO_3$ (Fig. 1a) that shows a jump of resistance of about five orders of magnitude at $T_{max} \cong 85$ K [20]. Such an evolution, characteristic of an insulator to metal (I - M) transition with decreasing temperature coincides with an abrupt change of the magnetic state. One indeed observes for the same compound a magnetization curve M(T) (Fig. 1b), registered in a magnetic field of 1.4 T, characteristic of a paramagnetic (PM) to a ferromagnetic (FM) transition with decreasing temperature, with $T_C = T_{max}$. Thus the behavior of this first class of manganites can be described by a PMI to FMM transition with decreasing temperature.

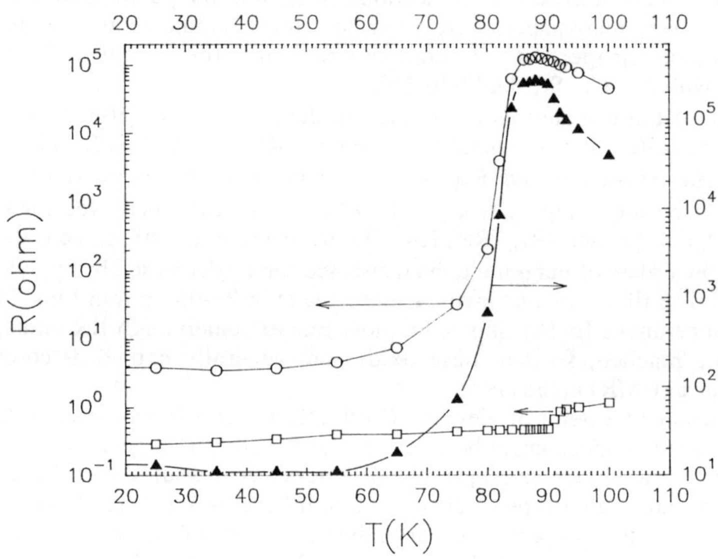

Figure 1a : T dependence of the resistance (R) for the ceramic $Pr_{0.7}Ca_{0.25}Sr_{0.05}MnO_3$.
(○) : B = 0, (□) B = 5 T and (▲) : R/R_{5T}.

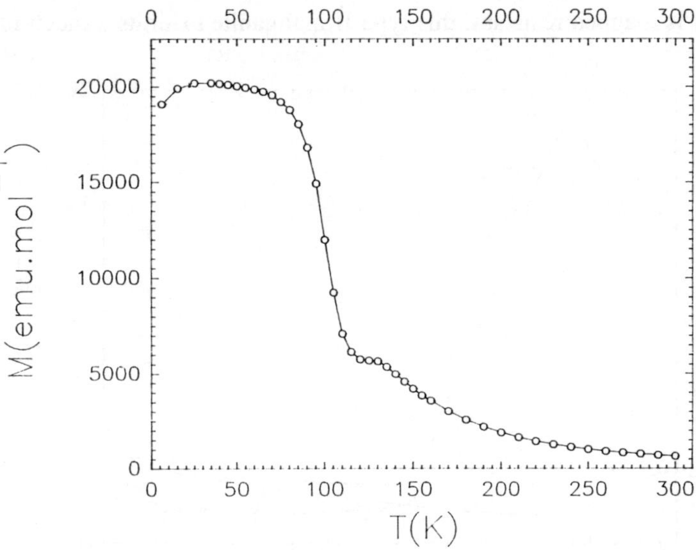

Figure 1b : Corresponding magnetization (M) versus T curve.

The type II manganites, synthesized for x = 0.50 exhibit more complex $R_0(T)$ curves. This is the case of the manganite $Pr_{0.5}Sr_{0.5}MnO_3$ whose CMR properties have been described for the first time by Tomioka et al. [12]. This second kind of manganite exhibits two transitions at 250 K and 140 K from the $R_0(T)$ curve (Fig. 2a), indicating the existence of a metallic state between 140 K and 250 K, sandwiched between two insulating states for T < 140 K and T > 250 K respectively. The corresponding magnetization curve M(T) registered in a low magnetic field of 100 G (Fig. 2b) shows that the first transition at 250 K corresponds to the appearance of the ferromagnetic state with decreasing temperature (right branch of the curve), whereas the second transition at 140 K results from the appearance of an antiferromagnetic (AFM) state (left branch of the curve). In other words this type II manganite exhibits like the type I manganites a PMI to FMM transition with decreasing temperature, with $T_C \cong 250$ K, but differs from the type I, by the existence of a FMM to AFMI transition when the temperature is further decreased, with $T_N = 140$ K.

Both types of manganites exhibit negative CMR properties when submitted to a magnetic field. One indeed observes a resistance ratio $R_0/R_H = 2.10^5$ near T_C (85 K) in magnetic field of 5 T for the type I phase $Pr_{0.7}Ca_{0.25}Sr_{0.05}MnO_3$ (Fig. 1a). The type II phase shows a different behavior : the FMM to AFMI transition temperature is decreased down to 80 K by applying a magnetic field of 7 T (Fig.

2a). Although it is magnetoresistant, this type II manganite exhibits a much lower

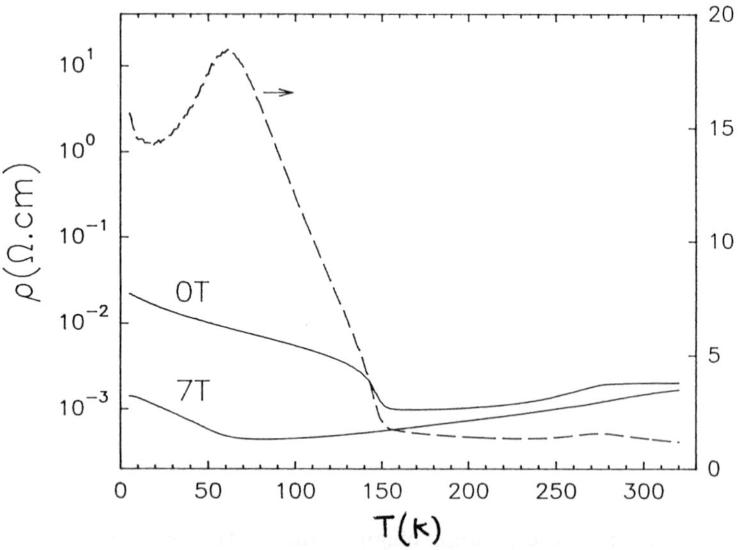

Figure 2a : ρ(T) curves of $Pr_{0.5}Sr_{0.5}MnO_3$.

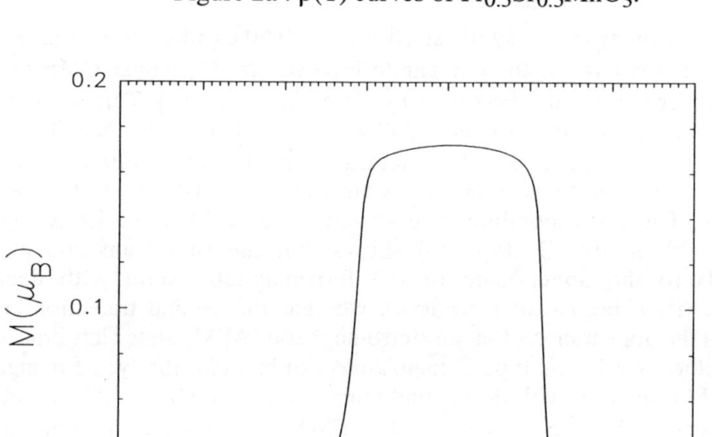

Figure 2b : Corresponding M(T) curve registered under 100 Gauss.

resistance ratio ; the maximum value R_0/R_H reaches scarcely 20 at 60 K under 7 T (Fig. 2a).

Thus the two types of manganites differ one from the other by the nature of the transition but also by the value of the resistance ratio that can be reached. Finally, it must be emphasized that the existence of a metal-insulator transition is not absolutely necessary for the appearance of CMR effect. This is for instance the case of the manganite $Pr_{0.7}Ca_{0.26}Sr_{0.04}MnO_3$ [15], that exhibits a semi-conducting behavior in a zero magnetic field, and is transformed into a ferromagnetic metal when submitted to a magnetic field of 5 T (Fig. 3), so that a resistance ratio of 10^{11} can be reached at 30 K for this compound.

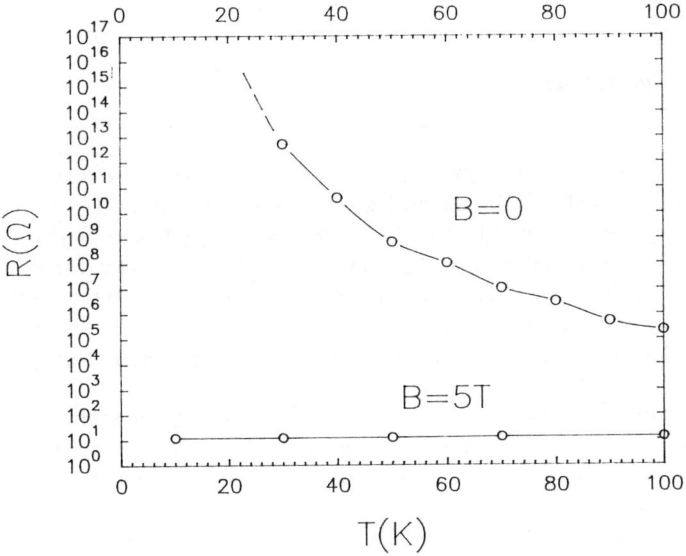

Figure 3 : R(T) curves of $Pr_{0.7}Ca_{0.26}Sr_{0.04}MnO_3$.

2 Mn mixed valence and size of the A site cation : two key factors for CMR properties

In order to explain the particular transport and magnetic properties of the manganese perovskites, two parameters have to be taken into consideration, the hole carrier density and the overlapping of the manganese and oxygen orbitals. Such an approach is common to all oxides and has previously been applied in the high T_c superconducting cuprates.

The hole carrier density corresponds to the amount of tetravalent manganese that can be introduced in the trivalent manganese perovskite $LnMnO_3$. In other words, the hole carrier density corresponds to the mixed valence Mn(III) - Mn(IV) and can be controlled by the ratio Mn(III)/Mn(IV). From the chemical viewpoint, the hole carrier density can then be increased by replacing partly the lanthanide cation Ln^{3+} by a divalent cation A^{2+} (Ca, Sr, Ba), leading to the generic formula $Ln_{1-x}A_xMnO_3$. But other possibilities should also be considered, based on the ability of the perovskite structure to exhibit non stoichiometry. Holes may indeed be introduced by a lanthanide deficiency according to the formula $Ln_{1-x}MnO_3$.

The overlapping of the Mn-d orbitals and O-p orbitals can be modified by varying the Mn-Mn distance between two adjacent octahedra, but also by changing the Mn-O-Mn angle. Clearly the overlapping of Mn and O orbitals can be modified by varying the size of the interpolated cation (A, Ln).

2.1 The type I manganites

The effect of the size of the interpolated cation on the magnetoresistance of type I CMR manganites was shown by several groups simultaneously [7, 13-16, 21-25, 30]. Such lattice effects are illustrated for the series $Pr_{0.7}Ca_{0.3-x}Sr_xMnO_3$ [21], where R(T) curves (Fig. 4a) and M(T) curves (Fig. 4b) show that for x > 0.04 a transition from the FMM to PMI state takes place. Note that the temperature T_{max} which characterizes the resistivity peak coincides with the Curie temperature T_C deduced from the M(T) curves. From these curves two important features can be emphasized :

(i) The resistance ratio $R_{(Tmax)}/R_{(10K)}$ which characterizes the resistivity peak is very high for a small doping level with strontium. But most important, this ratio decreases significantly as x increases, from 3.10^5 for x = 0.05 to 170 for x = 0.10.

(ii) The Curie temperature T_C, and by the way T_{max}, increases regularly as x increases.

Figure 5 shows the evolution of T_C versus the average ionic radius $\langle r_A \rangle$ of the interpolated cation (Pr^{3+}, Sr^{2+}, Ca^{2+}) using the ionic radii of the interpolated cations in sixfold coordination according to Shannon [26]. One observes that for the series $Pr_{0.7}Ca_{0.3-x}Sr_xMnO_3$, characterized by a constant hole carrier density (0.3 hole per Mn mole), T_C increases dramatically almost linearly versus $\langle r_A \rangle$. A linear behavior with an identical slope is also observed for the series $Pr_{0.66}Ca_{0.34-x}Sr_xMnO_3$, that exhibits a different hole carrier density (0.34 hole per Mn). The comparison of these two straight lines shows that the influence of the hole carrier density upon T_C is spectacular: for a same $\langle r_A \rangle$ value, each oxide of the series $Pr_{0.66}Ca_{0.34-x}Sr_xMnO_3$

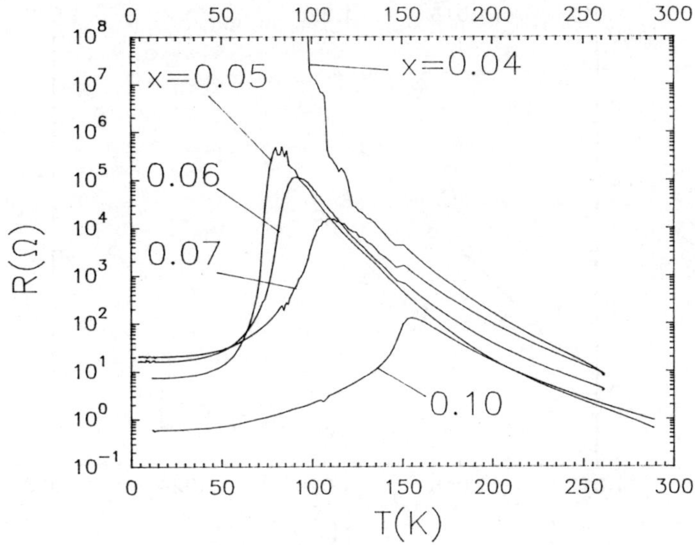

Figure 4a : R(T) curves of the $Pr_{0.7}Ca_{0.30-x}Sr_xMnO_3$ series.

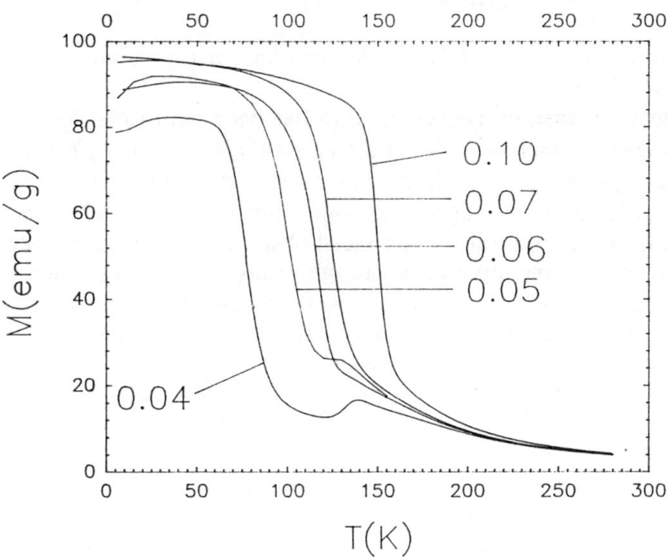

Figure 4b : Corresponding M(T) curves.

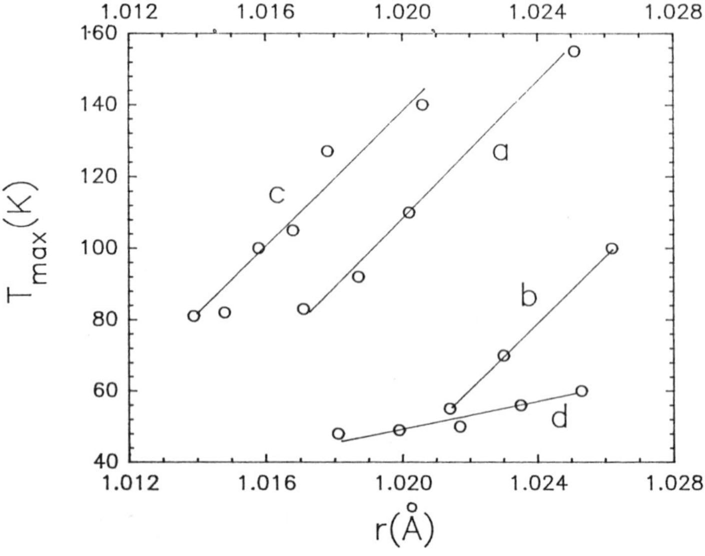

Figure 5 : T_{max} (R(T) maximum value) versus $<r_A>$ for the series : (a) $Pr_{0.7}Ca_{0.30-x}Sr_xMnO_3$, (b) $Pr_{0.66}Ca_{0.34-x}Sr_xMnO_3$, (c) $Pr_{0.7-x}La_xCa_{0.3}MnO_3$ and (d) $Pr_{0.7}Ca_{0.30-x}Ba_xMnO_3$.

exhibits a Curie temperature 50 K smaller than the corresponding oxide of the series $Pr_{0.7}Ca_{0.3-x}Sr_xMnO_3$.

This linear increase of T_C versus $<r_A>$ has been confirmed for different other series of manganites such as $Pr_{0.70-x}La_xCa_{0.3}MnO_3$ [21], $La_{0.7-x}Y_xCa_{0.3}MnO_3$ [27] and $Nd_{0.7}Ca_{0.3-x}Sr_xMnO_3$ [23]. In the latter series, the influence of the average size of the interpolated cation upon the amplitude of the CMR effect has been demonstrated for the first time. One indeed observes that the resistance ratio RR = R_0/R_{5T} for a given temperature goes through a maximum as $<r_A>$ increases (Fig. 6). For instance, at 60 K, the RR ratio increases dramatically as x decreases from 0.14 to 0.08 (Fig. 6a) reaching 10^7 for x = 0.08, and then decreases abruptly as x decreases from 0.065 to 0 (Fig. 6b) down to 10 for x = 0.

In conclusion, these results show that the type I manganites CMR properties are dramatically influenced by the size of the interpolated cation and by the hole carrier density. For a fixed size of the A site cation, T_C and correlatively the MI transition temperature decreases significantly as the hole carrier density increases, i.e. as the Mn(IV)/Mn(III) ratio increases. For a fixed valence of manganese, T_C increases linearly as $<r_A>$ increases and correlatively the magnetoresistance varies significantly.

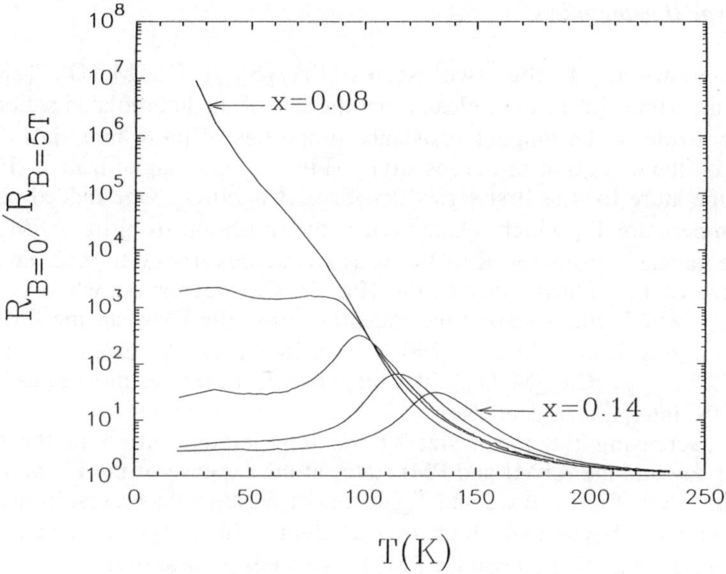

Figure 6a : T dependence of the resistance ratios R_0/R_{5T} for the series $Nd_{0.7}Ca_{0.30-x}Sr_xMnO_3$. From top to bottom : x = 0.08, 0.09, 0.10, 0.12 and 0.14.

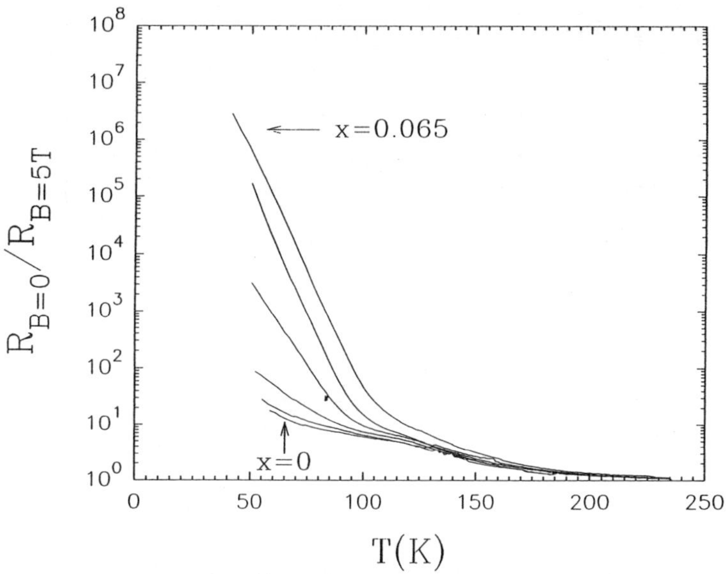

Figure 6b : Following x values from top to bottom : x = 0.065, 0.05, 0.04, 0.03, 0.02 and 0.

2.2 *The type II manganites*

The investigation of the two series $Pr_{1/2}Sr_{1/2-x}Ca_xMnO_3$ [28] and $Pr_{1/2-x}Y_xCa_{1/2}MnO_3$ [29] shows clearly that the size of the interpolated cation plays an important role in the magnetoresistance properties of these materials. This is illustrated by the evolution of the resistivity (Fig. 7a) and magnetization (Fig. 7b) versus temperature for the first series $Pr_{1/2}Sr_{1/2-x}Ca_xMnO_3$. One indeed observes that the temperature T_N which characterizes the transition from the AFMI to the FMM state increases from 136 K to 190 K as x increases from 0 to 0.12, i.e., as the average size of the interpolated cation (Pr, Sr, Ca) decreases, whereas on the opposite T_C, which characterizes the transition from the FMM to the PMI state, decreases slightly from 270 K to 240 K as x increases. A similar evolution is observed for $Pr_{1/2-x}Y_xCa_{1/2}MnO_3$ [29], except that T_C decreases more drastically as the size of the interpolated decreases.

Thus, decreasing the mean size of the interpolated cation in the type II manganites favours the AFMI and PMI states at the expense of the FMM state, so that in a first step, T_N increases and T_C decreases as $\langle r_A \rangle$ decreases. In fact, there exists a $\langle r_A \rangle$ value beyond which the two antagonist effects, type I (T_C) and type II (T_N) come near each other so that the FMM state tends to disappear.

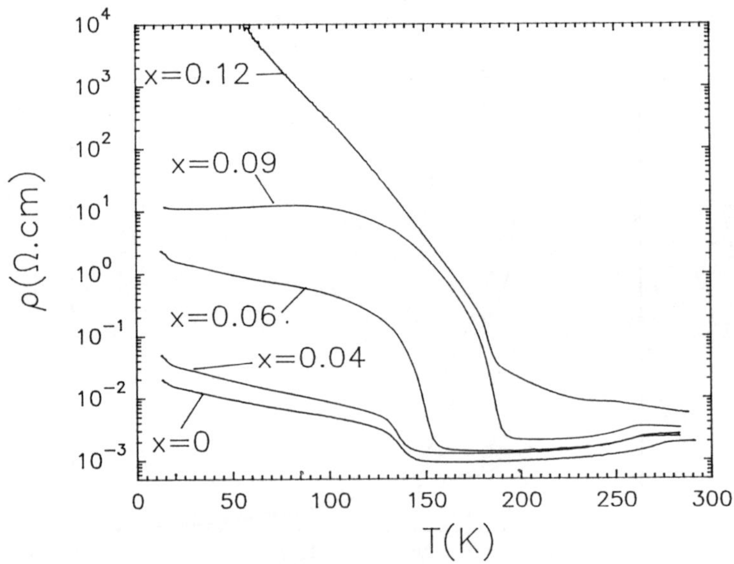

Figure 7a : $\rho(T)$ curves for the $Pr_{0.5}Sr_{0.5-x}Ca_xMnO_3$ series.

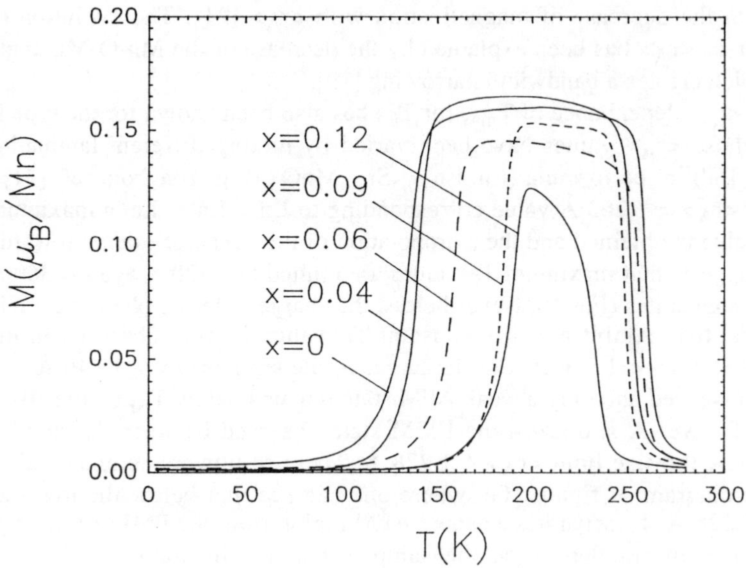

Figure 7b : Corresponding M(T) curves (10^{-2}T).

The application of a magnetic field modifies significantly the shape of the M(T) and ρ(T) curves as illustrated for the manganite $Pr_{0.5}Sr_{0.41}Ca_{0.09}MnO_3$ (Fig. 8a and 8b). Thus, a negative magnetoresistance effect is observed for all these oxides. Moreover the value of the maximum resistance ratio close to 3000, i.e., much higher than that of $Pr_{0.5}Sr_{0.5}MnO_3$ (≅ 17), shows that the size of the interpolated cation also affects the amplitude of the CMR effect.

2.3 Magnetic phase diagrams

Taking into consideration the above size effects, magnetic phase diagrams can be established for different series of manganites [30-31, 46].

For the type I manganites $Ln_{0.7}A_{0.3}MnO_3$, a T_{max} ($<r_A>$) diagram has been established by H.Y. Hwang et al. [30] (Fig. 9). One can see on this graph that the PMI-FMM transition temperature T_{max} goes through a maximum value of 360 K at $<r_A>$ = 1.24 Å ; T_{max} slightly decreases on the right side corresponding to larger $<r_A>$ values whereas it is more severely depressed for $<r_A>$ < 1.24 Å reaching T_{max} = 100 K for $<r_A>$ = 1.19 Å. Moreover below that size, the FMM state disappears at the benefit of a FMI one. Note that for $<r_A> \leq$ 1.18 Å, which corresponds to the $<r_A>$ value of $Pr_{0.7}Ca_{0.3}MnO_3$, the CMR effect tends to disappear, this is in good agreement with the decrease of magnetization with $<r_A>$ [21]. The evolution of magnetism versus $<r_A>$ has been explained by the decrease of the Mn-O-Mn angle with $<r_A>$, which creates a bandwidth narrowing [19].

A similar $<r_A>$ dependence of T_{max} (or T_C) has also been shown for the type II manganites whose $<r_A>$ values have been varied by mixing different lanthanide elements (Ln, Ln') in the formula $(Ln, Ln')_{0.5}Sr_{0.5}MnO_3$ (Fig. 10a from ref. [31]). For the larger $<r_A> = 1.263$ Å value corresponding to Ln = Ln' = La, a maximum $T_C = 310$ K value is obtained and the extrapolation of the T_C ($<r_A>$) curve towards larger $<r_A>$ suggests that maximum T_C values are limited to ~ 320 K against 360 K for the type I manganites (Fig. 9). Nevertheless, the charge ordering phenomenon in these materials, that exhibit a nearly constant T_N value, is only observed in the range 1.228 Å $< <r_A> <$ 1.246 Å. For the larger A site sizes ($<r_A> >$ 1.246 Å) the FMM state is favoured and only a weak AFM state remains below T_{AFM} (Fig. 10a). On the side of lower A site sizes, the FMM state observed between T_C and T_N disappears below the size limit $<r_A> <$ 1.228 Å. These results are summarized on the schematic diagram of figure 10b, where one can see that below the low size limit $<r_A> =$ 1.228 Å, T_C separates a canted AFM region from the PMI one as $<r_A>$ decreases. For even smaller $<r_A>$, the samples become insulators with weak ferromagnetism (WFMI) below T_C ($\approx$ 90 K).

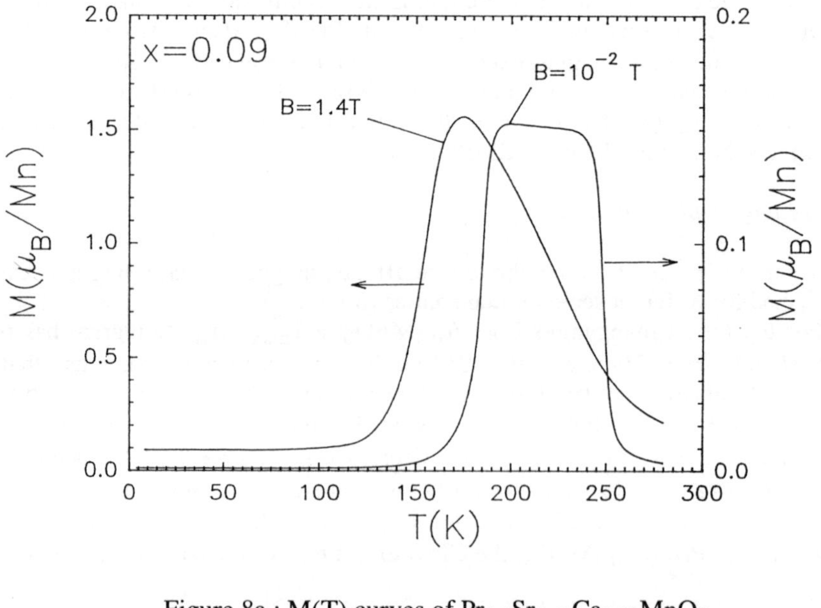

Figure 8a : M(T) curves of $Pr_{0.5}Sr_{0.41}Ca_{0.09}MnO_3$.

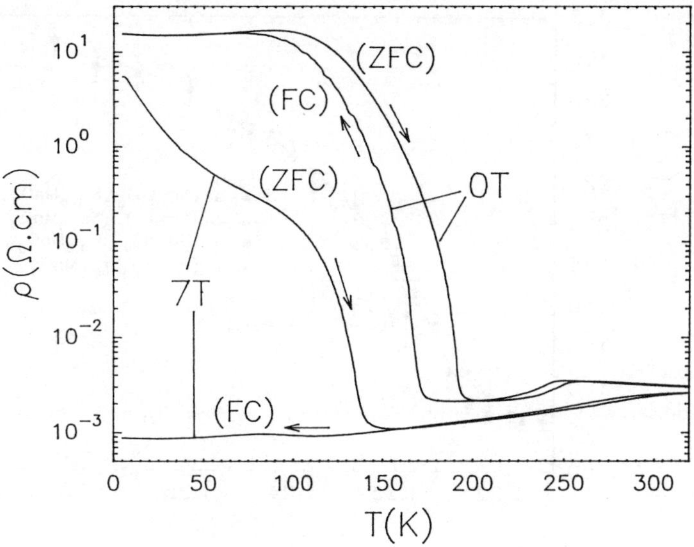

Figure 8b : Corresponding ρ(T) curves.

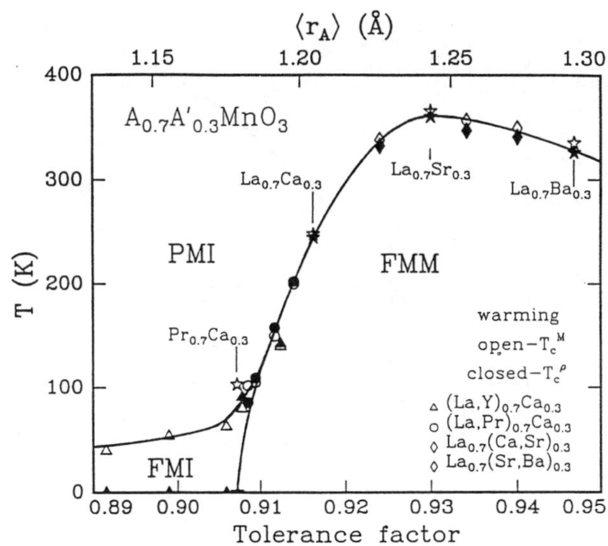

Figure 9 : $<r_A>$ (or tolerance factor t) dependence of T_C (or T_{max}) for the $A_{0.7}A'_{0.3}MnO_3$ manganites (H.Y. Hwang et al.[30]).

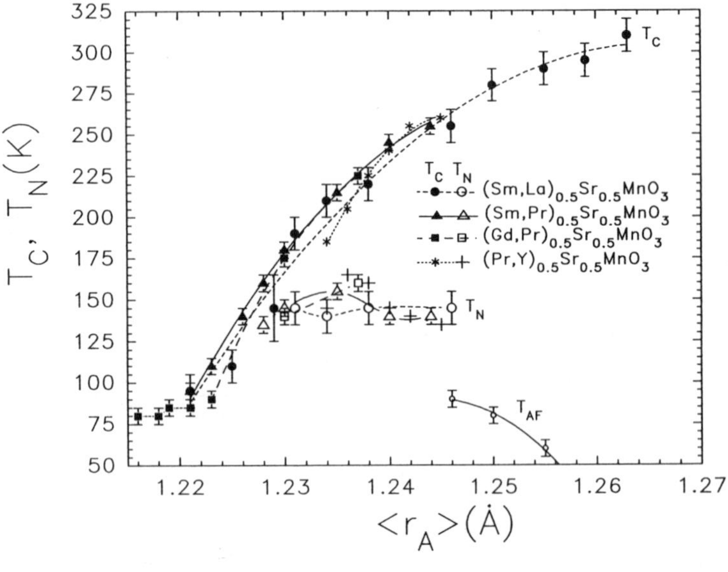

Figure10a : $<r_A>$ dependence of both T_C and T_N for the $(Ln,Ln')_{0.5}Sr_{0.5}MnO_3$ samples.

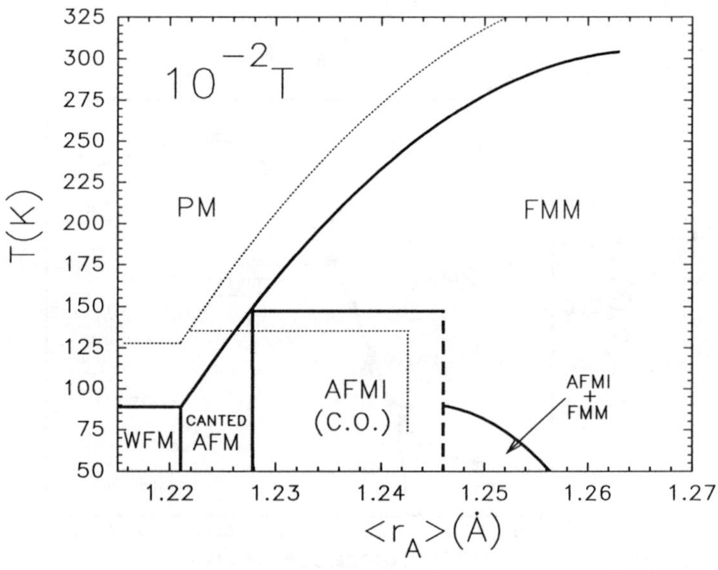

Figure 10b : Schematic magnetic phase diagram function of cationic size and temperature of the manganites $Ln_{0.5}Sr_{0.5}MnO_3$ established for 100 Gauss. Dotted lines would correspond to higher magnetic field values.

Finally, the equivalent diagram deduced from the R(T) and M(T) curves of the $Pr_{0.5}(Sr,Ca)_{0.5}MnO_3$ samples deduced in 2.2 section is not superimposable to that of the $(Ln,Ln')_{0.5}Sr_{0.5}MnO_3$ compounds (Fig. 11). The $<r_A>$ dependence of T_C for "$Pr_{0.5}$" phases is less severe than for "$Sr_{0.5}$" phases, yielding a smoother curve, so that T_N can reach up to 190 K on the lower $<r_A>$ side. However in contrast to the "$Sr_{0.5}$" phases, T_N can decrease down to 135 K as $<r_A>$ increases for the "$Pr_{0.5}$" phases. Thus T_C and T_N evolve in a different manner versus $<r_A>$ in the $Pr_{0.5}(Sr,Ca)_{0.5}MnO_3$ manganites than in the $(Ln,Ln')_{0.5}Sr_{0.5}MnO_3$ ones. This unexpected result suggests that another factor should be taken into consideration : the size difference between the different cations that are located in the A sites, called mismatch effect.

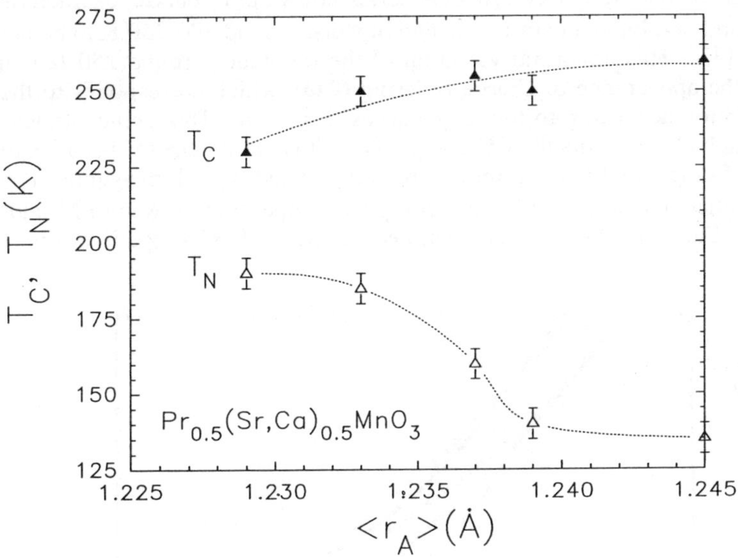

Figure 11 : $<r_A>$ dependence of both T_C and T_N for $Pr_{0.5}(Sr,Ca)_{0.5}MnO_3$ perovskites.

3 Role of A-site size mismatch

The effect of random disorder of Ln^{3+} and A^{2+} on the magnetoresistive properties of manganese oxide perovskites has been first investigated for the $Ln_{0.7}A_{0.3}MnO_3$ samples by L.M. Rodriguez-Martinez et al. [32]. They have shown that for the same A-site average size $<r_A>$ = 1.23 Å, T_{max} decreases as the cationic disorder increases. To quantify this effect they have introduced the σ^2 variance of the A-cations radii distribution, $\sigma^2 = \Sigma y_i r_i^2 - \langle r_A \rangle^2$, where y_i are the fractional occupancies of the different i cations of r_i radii. For instance from $La_{0.70}Ca_{0.11}Sr_{0.19}$ (σ^2 = 0.0016 Å^2) to $Sm_{0.70}Ba_{0.30}$ (σ^2 = 0.0240 Å^2) T_C decreases from 360 K to 60 K, demonstrating the crucial effect of the mismatch effect on the properties.

This new parameter, whose intensity is given by σ^2, must be thus minimized in order to study the T_m or T_C ($\langle r_A \rangle$) diagrams for manganites. At this stage, it is really interesting to study the magnetotransport properties of the $Pr_{1-x}Ca_xMnO_3$ series, since negligible disorder is realized due to the similar size of Pr^{3+} and Ca^{2+} ($r_{Pr^{3+}} \sim r_{Ca^{2+}} \sim$ 1.18 Å). Moreover, for x = 0.3, we have previously seen that the size limit of 1.18 Å below which a MI is no more observed corresponds to $Pr_{0.7}Ca_{0.3}MnO_3$.

Thus, for this small size of the interpolated cation of 1.18 Å, the very low value of σ^2 obtained for the $Pr_{1-x}Ca_xMnO_3$ series allows ρ(T) curves characteristic of insulators and involving an abrupt change of slope around 200-250 K to be obtained (Fig. 12a) [33]. This abnormal variation of the resistance around 250 K is in fact related to the appearance of charge ordering (CO), which corresponds to the PM-AFM transition according to the ac-χ curves (Fig. 12b). This demonstrates that a low mismatch effect favors the CO states so that it is reasonable to speculate that the size limit of $\langle r_A \rangle$ = 1.18 Å common to both type I and type II manganites is related to the CO phenomena. It results in a complex competition between FM and AFM for the $Pr_{1-x}Ca_xMnO_3$ samples as illustrated for the x = 0.35 ac-χ curve (Fig. 13).

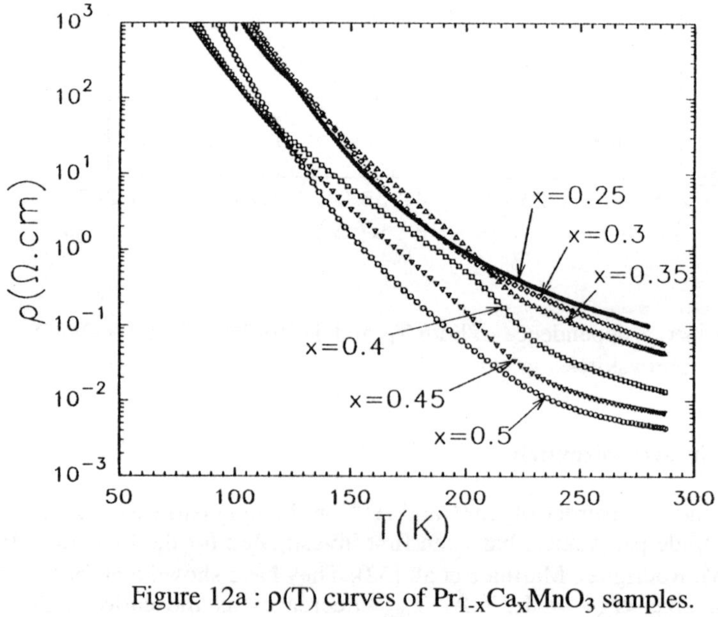

Figure 12a : ρ(T) curves of $Pr_{1-x}Ca_xMnO_3$ samples.

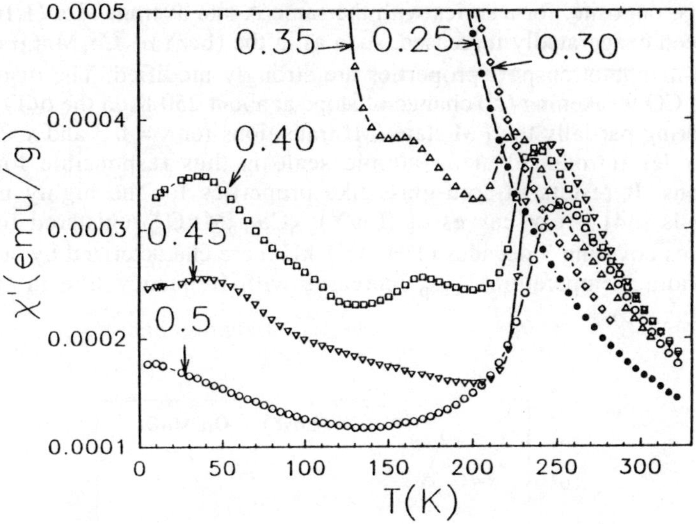

Figure 12b : Corresponding ac-χ(T) curves.

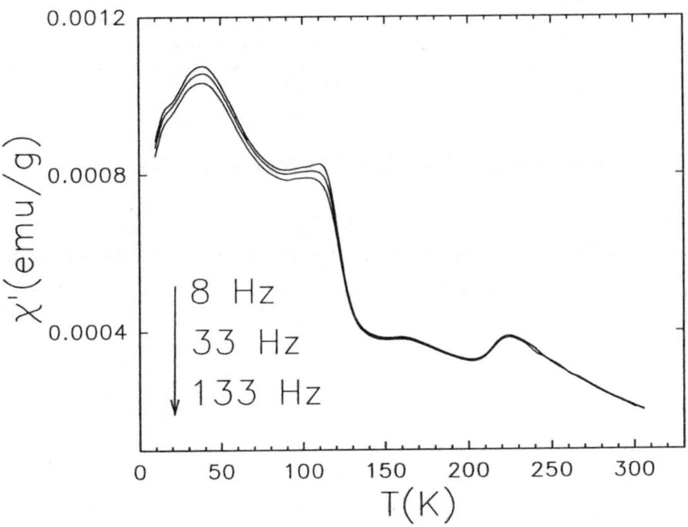

Figure 13 : χ'(T) curve of $Pr_{0.65}Ca_{0.35}MnO_3$.

On the opposite, for a series with the same A-site average size (1.18 Å) where σ^2 has been exageratedly increased, such as in the $(La/Y)_{1-x}Ca_xMnO_3$ compounds [34], the magnetotransport properties are strongly modified. The disorder effect induces a CO weakening (no change of slope at about 250 K on the ρ(T) curves Fig. 14), restoring partially the FM state (MI transitions for x = 0.3 and x = 0.25). The size disorder introduced at an atomic scale is thus responsible for magnetic frustrations. It results in spin-glass like properties for the highly mismatched compounds [34]. Ac-χ curves of $(Ln/Y)_{0.5}Ca_{0.5}MnO_3$ registered for different frequencies covering 3 decades (1 Hz to 1 kHz) are characterized by cusps, whose corresponding temperature T_{cusp} increases with frequency like in typical spin glasses (Fig. 15).

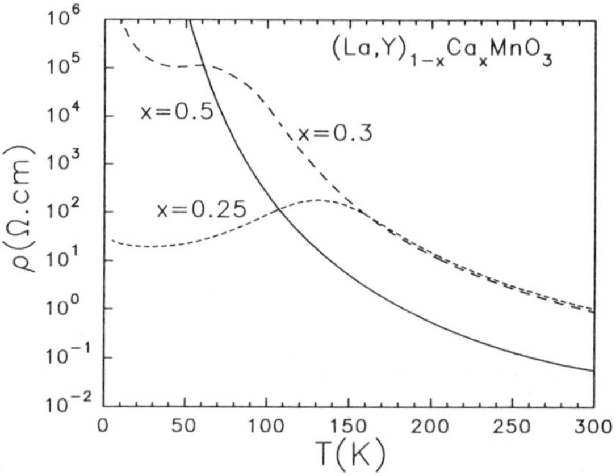

Figure 14 : ρ(T) curves of $(La,Y)_{1-x}Ca_xMnO_3$ samples. ($\langle r_A \rangle = 1.18$ Å).

The strong influence of the A-site cationic size mismatch on the magnetotransport properties of those manganites must also be invoked to explain the differences between $T_m(\langle r_A \rangle)$ phase diagrams previously described for $(Ln,Ln')_{0.5}Sr_{0.5}MnO_3$ and $Ln_{0.5}(Sr,Ca)_{0.5}MnO_3$ perovskites. This apparent discrepancy is further illustrated in figure 16 where T_C and T_N versus $\langle r_A \rangle$ are reported for different $(Ln,Ln')_{0.5}(Ca,Sr)_{0.5}MnO_3$ series of very different σ^2 values. In order to understand these results one must first consider the σ^2 ($\langle r_A \rangle$) diagram of the corresponding compounds (Fig. 17). It clearly appears that σ^2 and $\langle r_A \rangle$ are not univocally related. For instance for the $Pr_{0.5}(Sr,Ca)_{0.5}MnO_3$ series, σ^2 increases as $\langle r_A \rangle$ increases, whereas for $(Gd,Pr)_{0.5}Sr_{0.5}MnO_3$ σ^2 decreases as $\langle r_A \rangle$ increases. Consequently, measuring T_N and T_C for each point of the latter figure a new magnetic diagram in the (σ^2, $\langle r_A \rangle$) plane can be established as shown on figure 18. Four regions can be distinguished : weak ferromagnetic (WFM), ferromagnetic (FM) and two charge ordered (CO) types. The low mismatch region of this diagram is limited to the $La_{0.5}(Sr,Ca)_{0.5}MnO_3$ series (broadest line) and the highest σ^2 = 0.014 values correspond to the limit of purity for these phases.

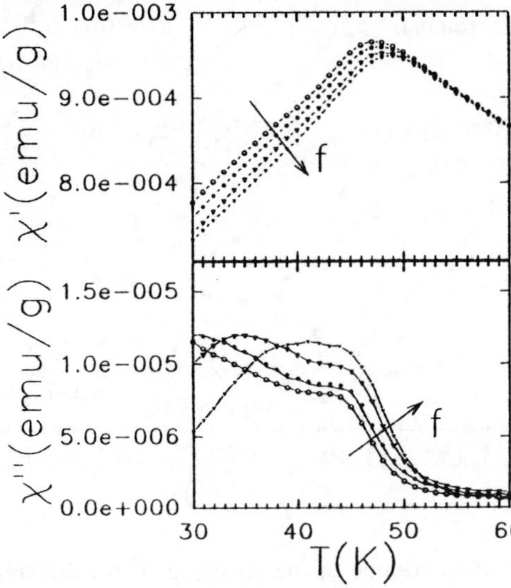

Figure 15 : $\chi'(T)$ and $\chi''(T)$ curves registered at 1, 10, 100 Hz and 1 kHz for $(La,Y)_{0.5}Ca_{0.5}MnO_3$.

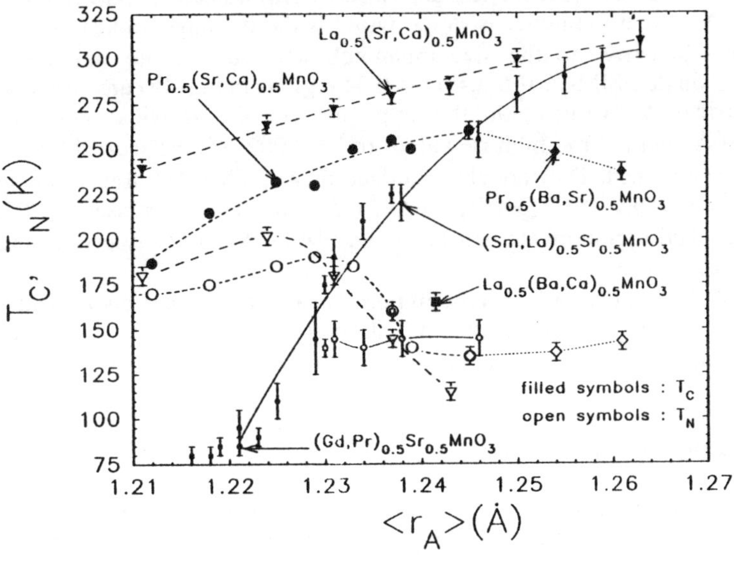

Figure 16 : $\langle r_A \rangle$ dependence of T_C and T_N for six series of composition $(Ln,Ln')_{0.5}(Sr,Ca)_{0.5}MnO_3$.

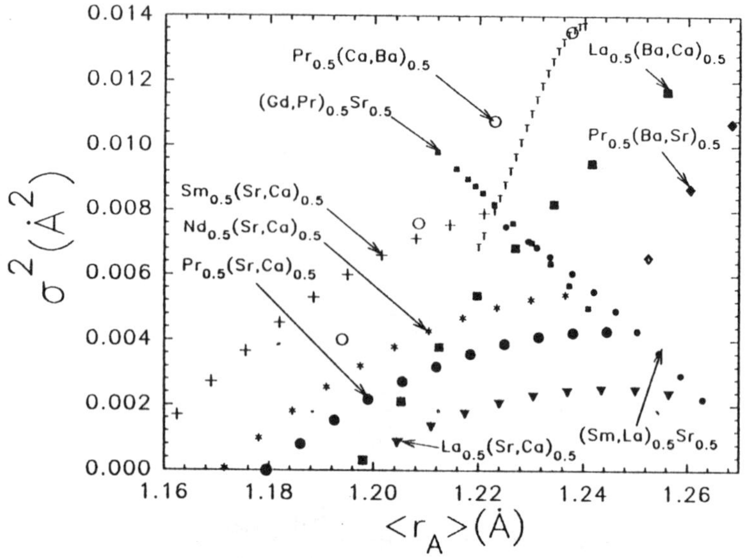

Figure 17 : $<r_A>$ dependence of the variance σ^2 for different CO series.

Two kinds of CO compounds are encountered. The first ones corresponding to the large sizes (1.21-1.26 Å) are characterized by two transitions AFMI-FMM and FMM-PMI. Note that the larger is $<r_A>$, wider is the σ^2 range, indicating that CO becomes less sensitive to the size mismatch when the A-site size increases. Nevertheless single FMM-PMI transitions (FM regions in Fig. 18) are only obtained beyond critical σ^2 values. On the opposite, for the smallest $<r_A>$ values, $<r_A> < 1.19$ Å, and not too high σ^2 values ($\sigma^2 < 0.002$ Å^2) the second kind of CO compounds is obtained. They are characterized by an AFMI-PMI transition (see the curves for x = 0.5 in $Pr_{1-x}Ca_xMnO_3$, Fig. 12). As soon as σ^2 increases CO tends to disappear (dashed region on the left part) and FM or WFM behaviors are induced like for $(La/Y)_{0.5}Ca_{0.5}MnO_3$.

Thus this diagram emphasizes the important role of both A site size, and A site mismatch effect [35]. In particular, all attempts to study $<r_A>$ effects on the properties of manganites without controlling σ^2 are doomed to failure.

4 Role of structural transitions

An important issue concerns the relationships between the structure and the magnetotransport properties of the manganites. Are the magnetic transitions related to structural transitions ? Does the Jahn-Teller effect of manganese play a role in these properties ? Is the charge ordering phenomenon, i.e. the ordering of Mn^{3+} and Mn^{4+} species, closely related to the magnetic transitions ? These are questions that have to be answered in order to understand the mechanisms which govern the CMR properties of these materials.

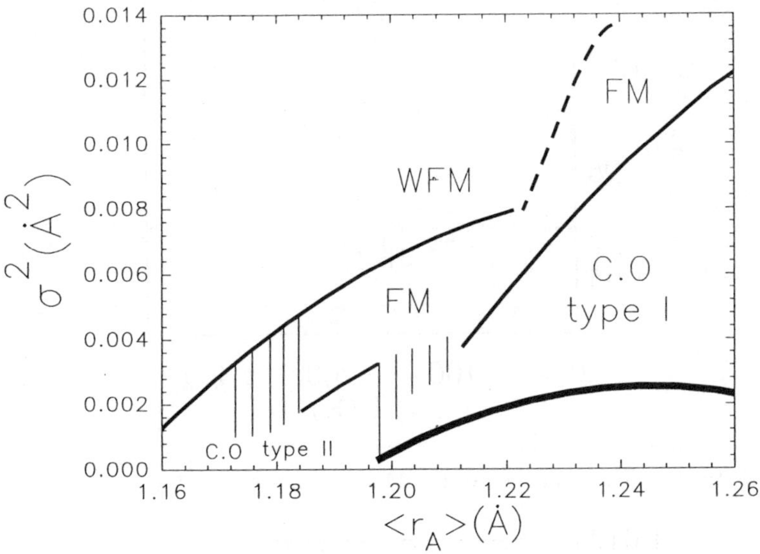

Figure 18 : Magnetic phase diagram in the σ^2 - <r_A> plane of $Ln_{0.5}A_{0.5}MnO_3$ phases.

The neutron powder diffraction (NPD) study of the type I CMR manganite $Pr_{0.7}Ca_{0.2}Sr_{0.1}MnO_3$ which exhibits a PMI-FMM transition at 170 K [36] clearly shows that there exists a significant contraction of the lattice below the FMM transition. An abrupt variation of the Mn-O distances is observed at the transition (Fig. 19a). In the PMI state, the Jahn-Teller distortion increases as T decreases. One observes elongated octahedra with two larger apical bonds (1.971-1.976 Å) and four shorter equatorial bonds (1.958-1952 Å). In the ferromagnetic state, the distortion still exists but does not vary significantly at low temperature, with two apical distances of 1.96 Å, and four equatorial distances ranging from 1.955 to 1.966 Å. Clearly, the PMI to FMM transition coincides exactly with the decrease of the Jahn-Teller distortion (Fig. 19b). A rather similar decrease of the Jahn-Teller distortion below T_C has also been observed in the canted ferromagnetic insulator $La_{0.875}Sr_{0.123}MnO_3$ by Argyriou et al. [37]. Although they correspond to a static effect, these correlations between Jahn-Teller and magnetic ordering of manganese are in agreement with the theory developed by Millis [19] which emphasizes the important role of the dynamic Jahn-Teller effect in the properties of these materials.

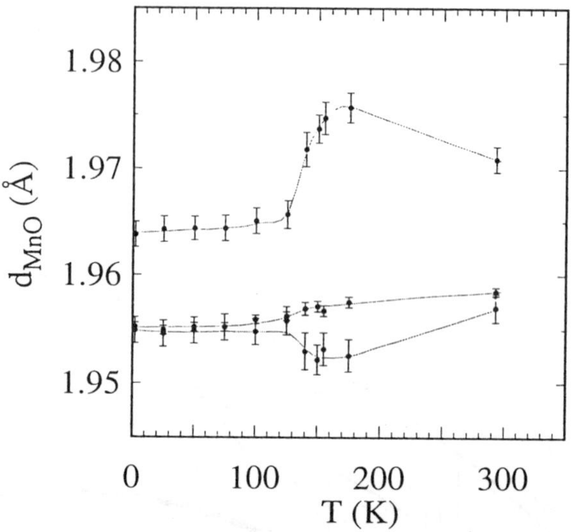

Figure 19a : T dependence of the Mn-O distances for $Pr_{0.7}Ca_{0.2}Sr_{0.1}MnO_3$.

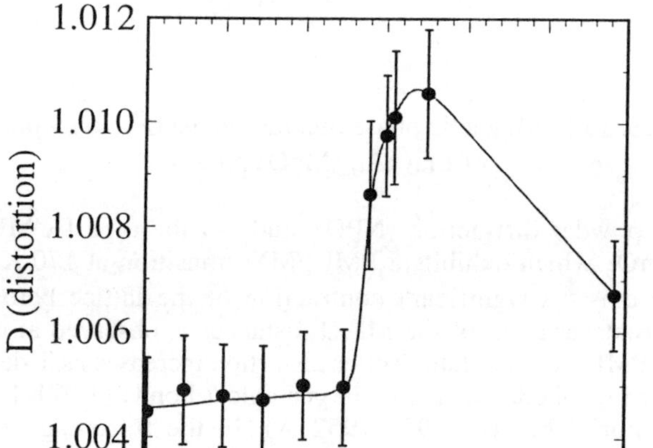

Figure 19b : Jahn-Teller distortion (d_{Mn}-O_{apical}/d_{Mn}-$O_{equatorial}$).

But the HREM study of this phase at room temperature [38], shows that the relationship between structure and properties is certainly more complex than expected from NPD data. The NPD study of $Pr_{0.7}Ca_{0.25}Sr_{0.05}MnO_3$ [39] has indeed evidenced for this phase an orthorhombic symmetry "$a_p\sqrt{2}$ x $2a_p$ x $a_p\sqrt{2}$", space group Pnma, whereas HREM observations show the existence in this matrix of monoclinic domains ($a_m \sim 2a_p$, $b_m \sim a_p\sqrt{2}$, $\beta \sim 90.3°$), whose structure can be simulated in the space group $P2_1/c$. Moreover the $[101]_0$ HREM images of these phases show the existence of "point defects" which can be correlated to clustering phenomena, i.e., to a local ordering of the Mn(III) and Mn(IV) species, so that a variation of the Mn-O distances around the manganese atoms would be involved. This is illustrated on Fig. 20a where the bright dots are correlated to the positions of Pr, Ca, Sr and the smaller ones to Mn atom. Then we notice that contrary to the (Ln,A) sites, the intensity of the spots corresponding to the Mn sites varies considerably within a limited area (see darker dots indicated by white arrows). This local variation of the contrast (Fig. 20b) can be easily simulated (Fig. 20c, d, e) by considering $CaMn^{IV}O_3$ and $PrMn^{III}O_3$ "clusters" in which the Mn-O distance varies significantly, depending on the manganese charge. Thus, local charge ordering is demonstrated in these materials from HREM observations.

The manganite $Pr_{1/2}Sr_{1/2}MnO_3$ which has been shown to exhibit the AFMI-FMM-PMI transitions (Fig. 2) by Tomioka et al. [12] is itself from the structural viewpoint, still subject to controversy. For instance Argyriou et al. [40] found out for the room temperature of this phase a F4/mmc symmetry whereas Kawano et al. [41] and Knizek et al. [42] published a Pnma space group. Moreover no data were available about the structural transitions versus temperature for this compound. A recent study [43], considering NPD and EM (electron microscopy) sheds light on the structural behaviour of this important compound versus temperature. The NPD diffraction study confirms the structure of the room temperature evidenced by Argyriou et al. [40], although the structure is described in a different cell "$a_p\sqrt{2}$ x $2a_p$ x $a_p\sqrt{2}$", space group I4/mcm (Fig. 21a). One observes almost regular MnO_6 octahedra, and a double tilting of the octahedra along the [110] and $[1\bar{1}0]$ directions, whereas along $\vec{c}$ the octahedra are not tilted. But the most important result deduced from the NPD data concerns the existence of a structural transition at low temperature which coincides exactly with the magnetic FMM-AFMI transition at 140 K as shown by comparing the $\chi'(T)$ and $\rho(T)$ curves with structural evolution (Fig. 22). The low temperature form exhibits also an "$a_p\sqrt{2}$ x $2a_p$ x $a_p\sqrt{2}$" cell, but with a different space group, Fmmm. The MnO_6 octahedra of this low temperature phase are significantly distorted, but exhibit a similar double tilting (Fig. 21b) compared to the room temperature one. One important feature deals with the fact that in both structures, I4/mcm and Fmmm, there is only one crystallographic Mn site, so that it tends to demonstrate that no ordering between the Mn^{3+} and Mn^{4+} species takes place whatever the temperature. Moreover no structural transition accompanies the FMM-PMI transition at 254 K according to NPD data.

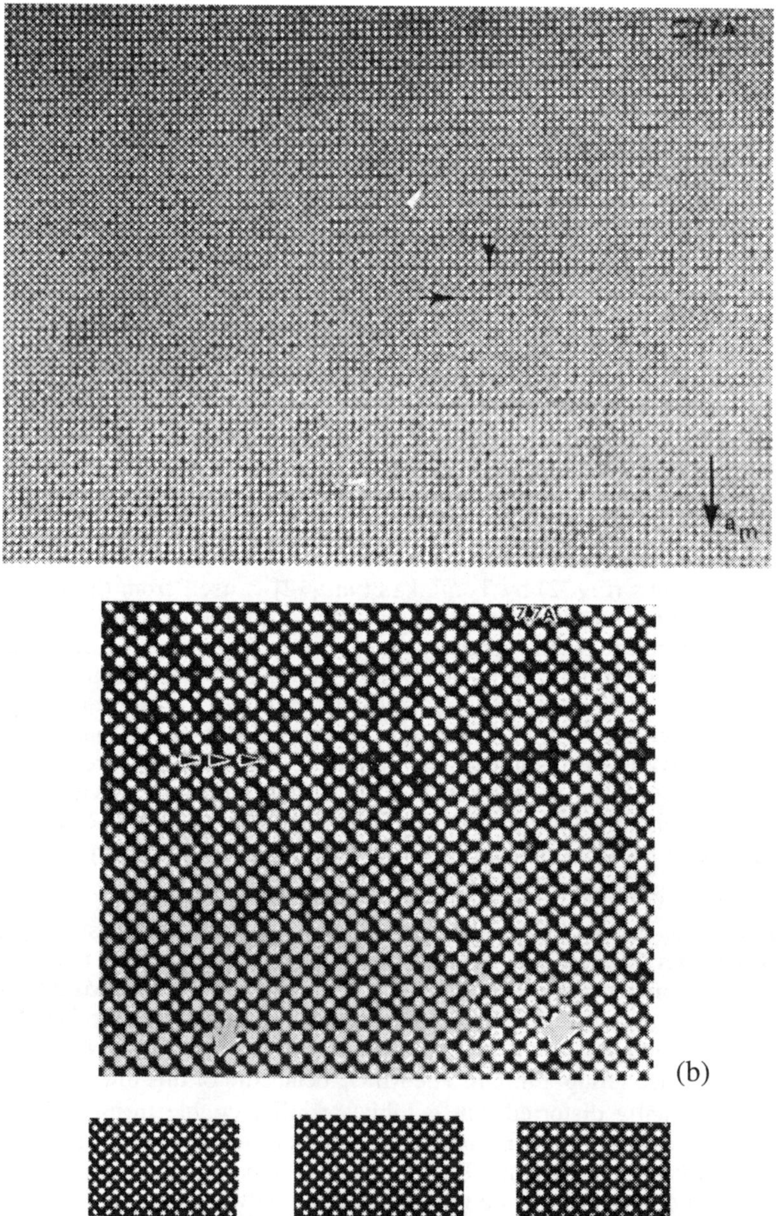

(a)

(b)

Pr (c) **Pr,Ca** (d) **Ca** (e)

Figure 20 : Typical $[101]_0$ a) overall, b) enlarged images of point like defects in $Pr_{0.7}Ca_{0.25}Sr_{0.05}MnO_3$. Comparison between the calculated images for c) $PrMnO_3$, d) $Pr_{0.7}Ca_{0.25}Sr_{0.05}MnO_3$ and e) $CaMnO_3$.

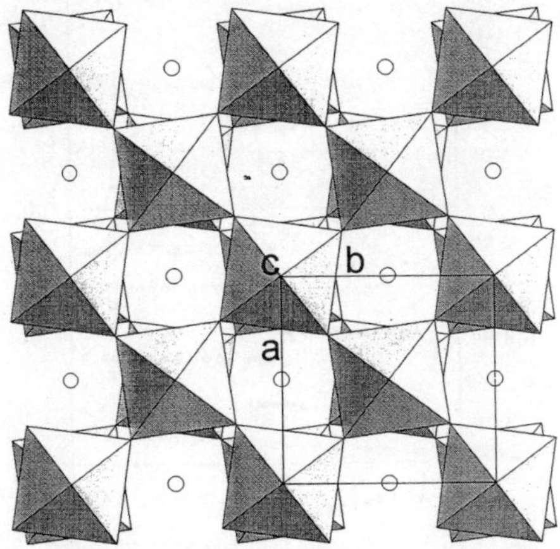

Figure 21a : [001] projection of the room temperature I4/mcm schematic structure for $Pr_{0.5}Sr_{0.5}MnO_3$.

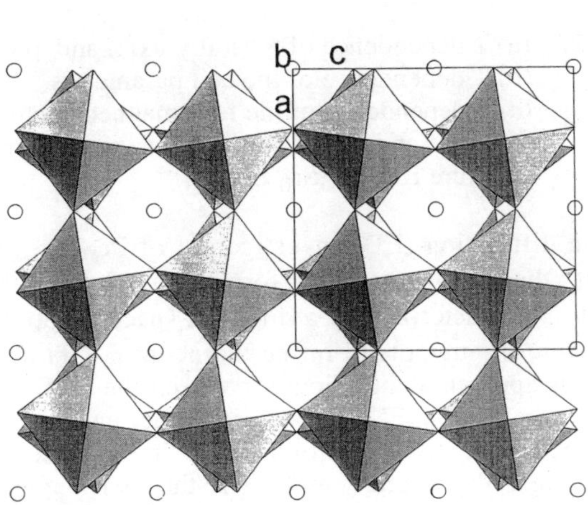

Figure 21b : [010] projection of the low temperature Fmmm schematic structure of $Pr_{0.5}Sr_{0.5}MnO_3$.

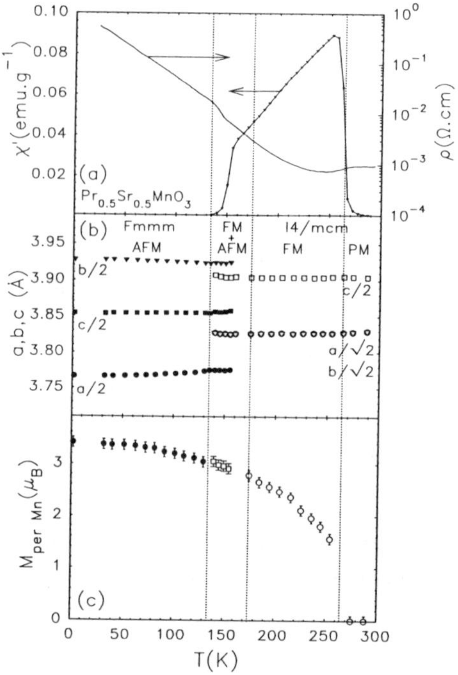

Figure 22 : (a) T dependence of χ' (left y-axis) and ρ (right y-axis).
(b) T dependence of the cell parameters.
(c) T dependence of the ferromagnetic component in the I4/mcm structure and of the antiferromagnetic component in the Fmmm structure for $Pr_{0.5}Sr_{0.5}MnO_3$.

The electron diffraction (ED) investigation of $Pr_{1/2}Sr_{1/2}MnO_3$ brings a very surprising result : the room temperature form, although it exhibits a similar "$a_p\sqrt{2}$ x $2a_p$ x $a_p\sqrt{2}$" cell is characterized by a different space group, i.e., Imam instead of I4/mcm. Moreover, all attempts to refine the neutron data in this space group are unsuccessful. This apparent contradiction between the NPD and ED results can be explained by the fact that the FMM phase could be sensitive to the magnetic field. There exists indeed a magnetic field of 1 or 2 T in the electronmicroscope, so that the M(T) curve (Fig. 23) is strongly modified ; thus at room temperature the sample is in the FMM state and submitted to a field of 1 or 2 T in the electron microscope. Such a field might modify the structure of the phase. The reversible transition at around 430 K, from the Imam to the I4/mcm space group deduced from the ED patterns (Fig. 24) supports strongly this hypothesis, since at 430 K, in a magnetic field of 1 or 2 T, the sample exhibits a paramagnetic behaviour. In contrast, the antiferromagnetic state is not sensitive to the magnetic field, since by decreasing the temperature the E.D. patterns show a reversible transition at 140 K from the Imam (Fig. 25a) to the Fmmm (Fig. 25b) which coincides exactly with the FMM-AFMI transition. Moreover it is remarkable that the "AFMI, Fmmm" low temperature state exhibits twins (Fig. 25c at 90 K), which disappear (Fig. 25d) above 140 K, and reappear by cooling (Fig. 25e).

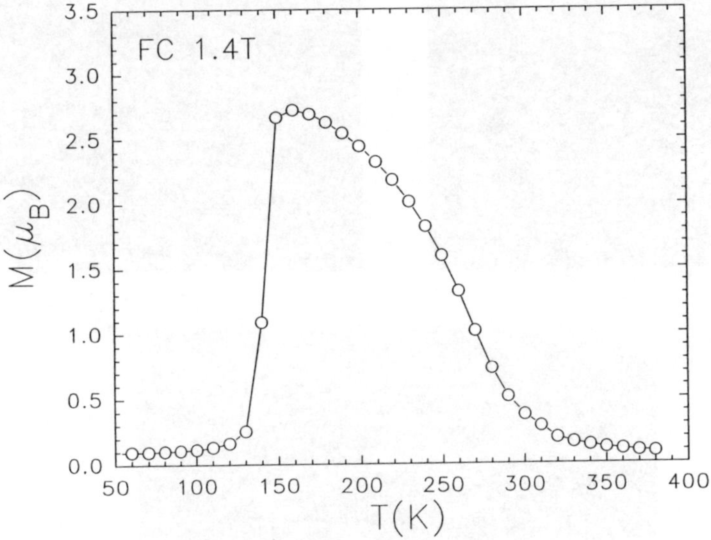

Figure 23 : $Pr_{0.5}Sr_{0.5}MnO_3$ M(T) curve ; B =1.4 T

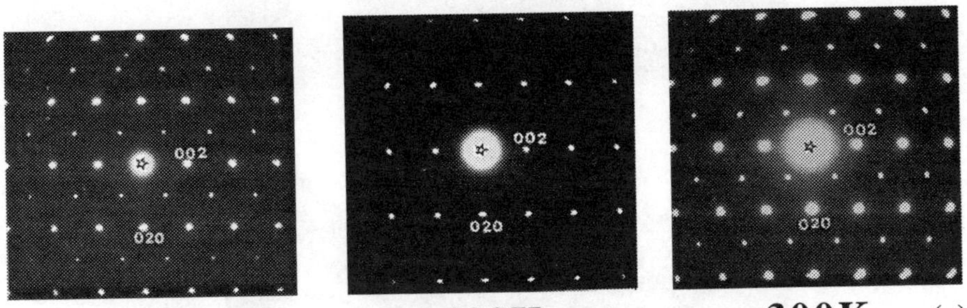

Figure 24 : [100] ED pattern recorded at RT (a), 430 K (b) and RT (c).

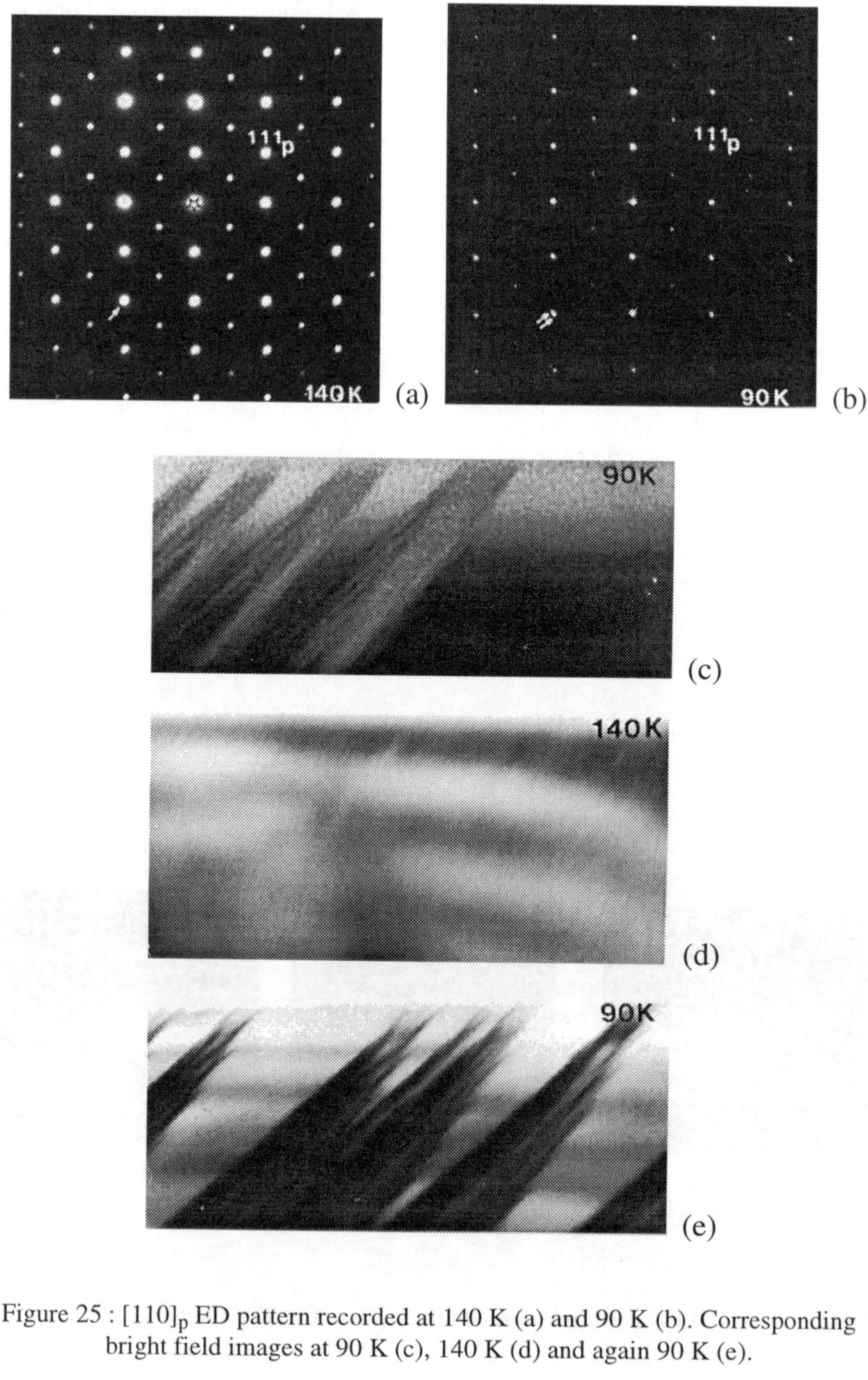

Figure 25 : $[110]_p$ ED pattern recorded at 140 K (a) and 90 K (b). Corresponding bright field images at 90 K (c), 140 K (d) and again 90 K (e).

The above results clearly show that the magnetotransport transitions are closely related to structural transitions. Nevertheless it should be emphasized that the different structures described above are not universal. For instance, $Nd_{0.5}Sr_{0.5}MnO_3$ which is characterized, like $Pr_{0.5}Sr_{0.5}MnO_3$ by AFMI-FMM-PMI transitions, exhibits fundamentally different structures [44]. Its "high temperature" structure (T > 140 K) is indeed Imma "$a_p\sqrt{2}$ x $2a_p$ x $a_p\sqrt{2}$", involving a single tiling of the almost regular MnO_6 octahedra. In this compound, the AFMI-FMM transition around 160 K - 180 K coincides also with a structural transition. But here, again, the structure of the low temperature AFMI phase is different from that of $Pr_{1/2}Sr_{1/2}MnO_3$: the cell is orthorhombic "$a_p\sqrt{2}$ x $2a_p$ x $a_p\sqrt{2}$" with Pmmm or $Pnm2_1$ as possible space groups. It results that the AFMI phase exhibits two different Mn sites for $Nd_{0.5}Sr_{0.5}MnO_3$, and consequently is charge ordered contrary to $Pr_{0.5}Sr_{0.5}MnO_3$.

It is also worth recalling that the average size of the interpolated cation is susceptible to influence the structural transition. This is for instance the case of the $Nd_{0.7}Ca_{0.3-x}Sr_xMnO_3$ series [45], which exhibits at room temperature the classical Pnma "$a_p\sqrt{2}$ x $2a_p$ x $a_p\sqrt{2}$" structure whatever x. Decreasing T, no structural transition is observed down to 1.7 K whatever x, ranging from 0.08 to 0.30, in spite of the existence of a FMM-PMI transition. In contrast, the limit phase, $Nd_{0.7}Ca_{0.3}MnO_3$ (x = 0) which is an antiferromagnetic insulator exhibits a structural transition at low temperature to a $P2_1/m$ "$a_p\sqrt{2}$ x $2a_p$ x $a_p\sqrt{2}$" structure, involving charge ordering phenomena different from those encountered in other manganites. Two kinds of Mn sites, forming rows of octahedra occupied by single Mn^{3+} species, and "mixed" Mn^{3+}/Mn^{4+} species alternately are indeed evidenced.

A tremendous work remains to be performed in order to understand the relationships between the structural transitions and the CMR properties of the manganites.

5 Doping of the manganese sites

Although the mechanisms that are at the origin of the CMR properties are still subject to discussion, it seems clear that the electronic structure of Mn(III) and Mn(IV) species plays a specific role. Consequently, the CMR properties should be considerably modified by doping the Mn sites with various elements.

5.1 The substitution of trivalent and tetravalent elements for manganese in $Pr_{1-x}(Ca, Sr)_xMnO_3$.

The systematic study of the substitution of various M elements for Mn in the CMR perovskite $Pr_{0.7}Ca_{0.2}Sr_{0.1}MnO_3$ has been performed for M = Al, Ga, In, Ti, Sn [47, 48]. It shows that whatever the element, the transition temperature T_C (or T_{max}) from the ferromagnetic metallic (FMM) to the paramagnetic insulating (PMI) state decreases dramatically as the content of the doping element increases. This is illustrated by the R (T) curves of the series $Pr_{0.7}Ca_{0.2}Sr_{0.1}Mn_{1-x}Al_xO_3$ (Fig. 26) where it can be seen that T_{max} is decreased by about 12 K per percent of Al atom introduced on the Mn site. The corresponding magnetoresistance is in fact correlated to T_C : it increases significantly as x increases, R_0/R_H reaching 10^3 at 71 K for 6 % Al per Mn in a magnetic field of 7 T.

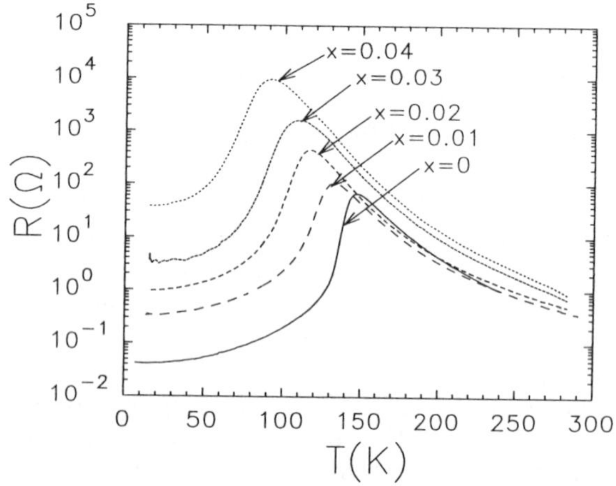

Figure 26 : R(T) curves for the series $Pr_{0.7}Ca_{0.2}Sr_{0.1}Mn_{1-x}Al_xO_3$.

A very different behaviour is observed for the phase $Pr_{0.5}Sr_{0.5}MnO_3$ [47], which exhibits two different effects according to the valence of the doping element M. Doping with trivalent elements (M = Al, Ga, In) increases the AFMI and the PMI states at the expense of the FMM state as shown from the M(T) curve for M = Al where a substantial increase of T_N is observed (Fig. 27a). On the opposite doping with tetravalent elements (M = Ti, Sn) tends to suppress the AFMI state, whereas the FMM to PMI transition still exists (Fig. 27b).

5.2 *The increase of CMR effect by iron and magnesium doping in some manganites.*

The substitution of magnesium for manganese in $Pr_{0.7}Ca_{0.2}Sr_{0.1}MnO_3$ [49] leads also to a decrease of T_C which is even larger than that observed with trivalent or tetravalent dopants. But the most remarkable feature deals with the jump of resistance near the transition temperature, which is the highest that has been observed to date in the manganites $Pr_{0.7}(Ca,Sr)_{0.3}Mn_{1-x}M_xO_3$. Resistance peaks close to four orders of magnitude are obtained at temperatures ranging from 70 K to 108 K for x ~ 0.005-0.02 (Fig. 28), suggesting that the Mg doped phases exhibit very promising CMR properties. The resistance curve registered in a magnetic field of 7 T (Fig. 29) confirms this viewpoint : a resistance ratio R_0/R_{7T} of 4.10^5 at 70 K is obtained for $Pr_{0.7}Ca_{0.2}Sr_{0.1}Mn_{0.98}Mg_{0.02}O_3$, to be compared to 230 at 150 K for the undoped phase $Pr_{0.7}Ca_{0.2}Sr_{0.1}MnO_3$.

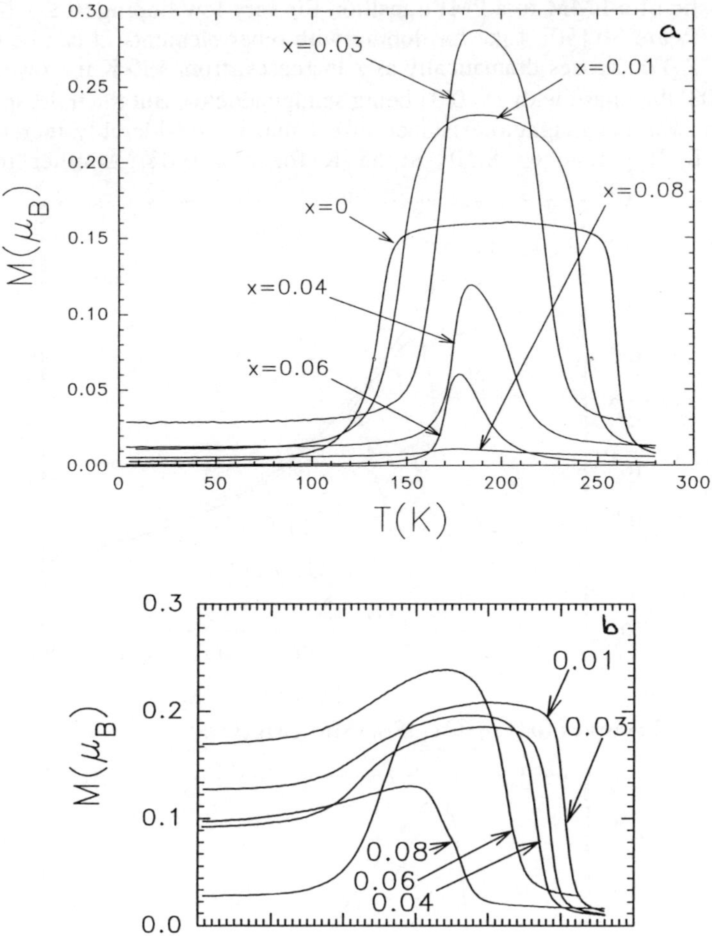

Figure 27 : M(T) curves registered under 10^{-2} T for $Pr_{0.5}Sr_{0.5}Mn_{1-x}M_xO_3$:
(a) M = Al, (b) M = Ti.

In contrast, the $Pr_{0.5}Sr_{0.5}Mn_{1-x}Mg_xO_3$ manganites are very similar to the phases $Pr_{0.5}Sr_{0.5}Mn_{1-x}M_xO_3$ with M = Al, Ga, In, differing from the homologous series doped with tetravalent elements (M = Ti, Sn) [47].

The doping of manganites by iron is also susceptible to enhance the CMR effect in the manganites. This is the case of the perovskite $Sm_{0.56}Sr_{0.44}MnO_3$. The perovskites $Sm_{0.56}Sr_{0.44}Mn_{1-x}Fe_xO_3$ exhibit indeed resistance maxima characteristic of a FMM to a PMI transition for very low doping ($0 \leq x \leq 0.03$) as shown on figure 30 [50]. Like for doping with other elements, it can be seen that T_{max} (i.e. T_C) decreases dramatically as x increases, from 130 K for x = 0 to 55 K for x = 0.03, the phase with x = 0.04 being semiconductor. But the most spectacular effect concerns the magnetoresistance effect that is considerably increased. For instance, R_0/R_{7T} reaches 8.10^5 at 55 K for x = 0.03. Another important

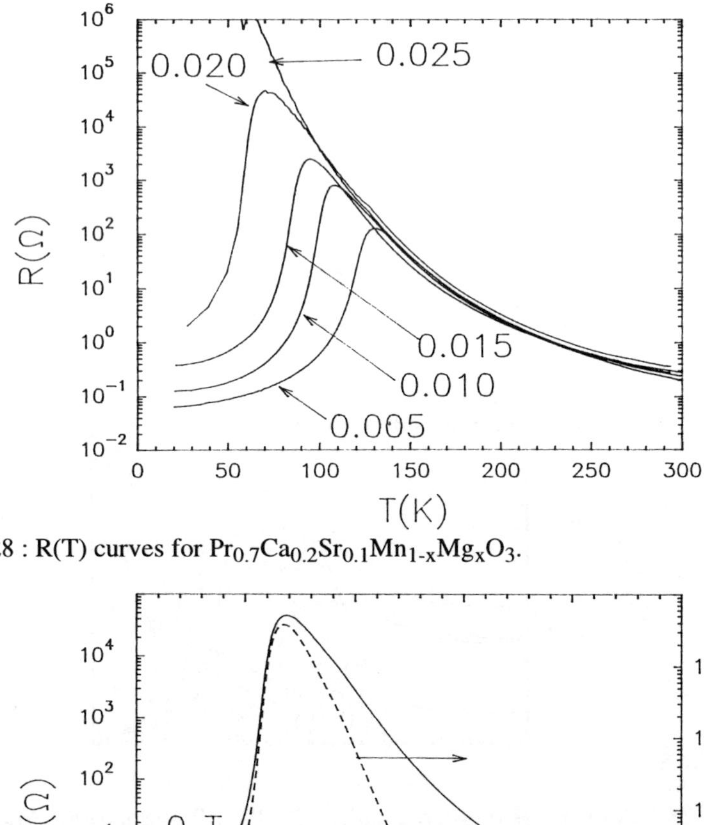

Figure 28 : R(T) curves for $Pr_{0.7}Ca_{0.2}Sr_{0.1}Mn_{1-x}Mg_xO_3$.

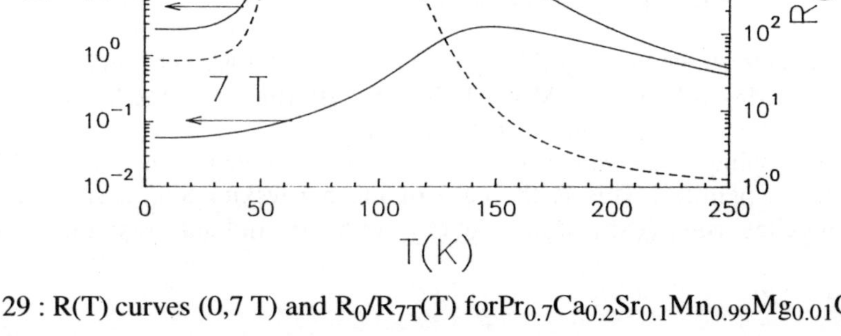

Figure 29 : R(T) curves (0,7 T) and R_0/R_{7T}(T) for$Pr_{0.7}Ca_{0.2}Sr_{0.1}Mn_{0.99}Mg_{0.01}O_3$.

characteristic deals with the evolution of the resistance versus the applied magnetic field. One observes a very abrupt decrease of the resistance for the phase $Sm_{0.56}Sr_{0.44}Mn_{0.97}Fe_{0.03}O_3$ in the range 0 - 0.02 T. Such a decrease of R has never been observed below 1 T in other manganites to date.

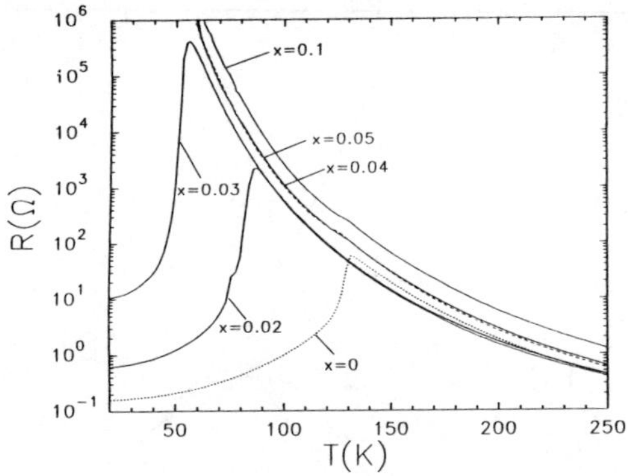

Figure 30 : R(T) curves of the samples $Sm_{0.56}Sr_{0.44}Mn_{1-x}Fe_xO_3$.

5.3 *The spectacular insulator-metal (I-M) transition induced by "Cr,Co,Ni" doping in $Pr_{0.5}Ca_{0.5}MnO_3$.*

In contrast to many other manganites with the perovskite structure, the manganite $Pr_{1-x}Ca_xMnO_3$ remains insulating whatever x [51-57]. Such a behaviour has been explained by the fact that the similar sizes of the interpolated cations Ca^{2+} and Pr^{3+} are rather small favoring a CO AFMI state.

For small x values, an I-M transition can be induced by applying a magnetic field, as shown for instance for $Pr_{0.7}Ca_{0.3}MnO_3$ [54-57], so that the latter compound exhibits CMR properties, in spite of its insulating properties. Nevertheless, for the particular value x ~ 1/2, the charge ordering between the Mn(III) and Mn(IV) species reinforces the capitulation of the I-M transition so that $Pr_{0.5}Ca_{0.5}MnO_3$ does not exhibit CMR properties, i.e. remains insulator whatever the value of the applied magnetic field up to 7T.

The study of the doping of $Pr_{0.5}Ca_{0.5}MnO_3$ with chromium, cobalt or nickel leads to a rapid disappearance of the charge ordering phenomenon [58, 59]. This is illustrated by the R(T) curves (Fig. 31a) and M(T) curves (Fig. 31b) of the series $Pr_{0.5}Ca_{0.5}Mn_{1-x}Cr_xO_3$ [58]. The fact that charge ordering vanishes by doping with a foreign element is not particular to Cr, Co or Ni : it is also observed for the manganites $Pr_{0.5}Ca_{0.5}Mn_{1-x}M_xO_3$ with M = Fe, Al, Ga, Ti [60]. But most important is the fact that this "Cr, Co or Ni" doping induces a resistance peak characteristic of an I-M transition as T decreases (Fig. 31a). This behaviour is very different from that observed for other manganites $Pr_{0.5}Ca_{0.5}Mn_{1-x}M_xO_3$, that remain insulator, whatever M = Fe, Al, Ga, Ti and whatever x. Such an I-M transition is remarkable since it has never been observed to date in manganites with a small average size of the A-site cation. Another remarkable feature concerns the transition temperature T_{max} of the Cr-doped phases (Fig. 31a) that increases significantly as the chromium content increases up to T_{max} = 150 K for x = 0.04 and then decreases for x > 0.04.

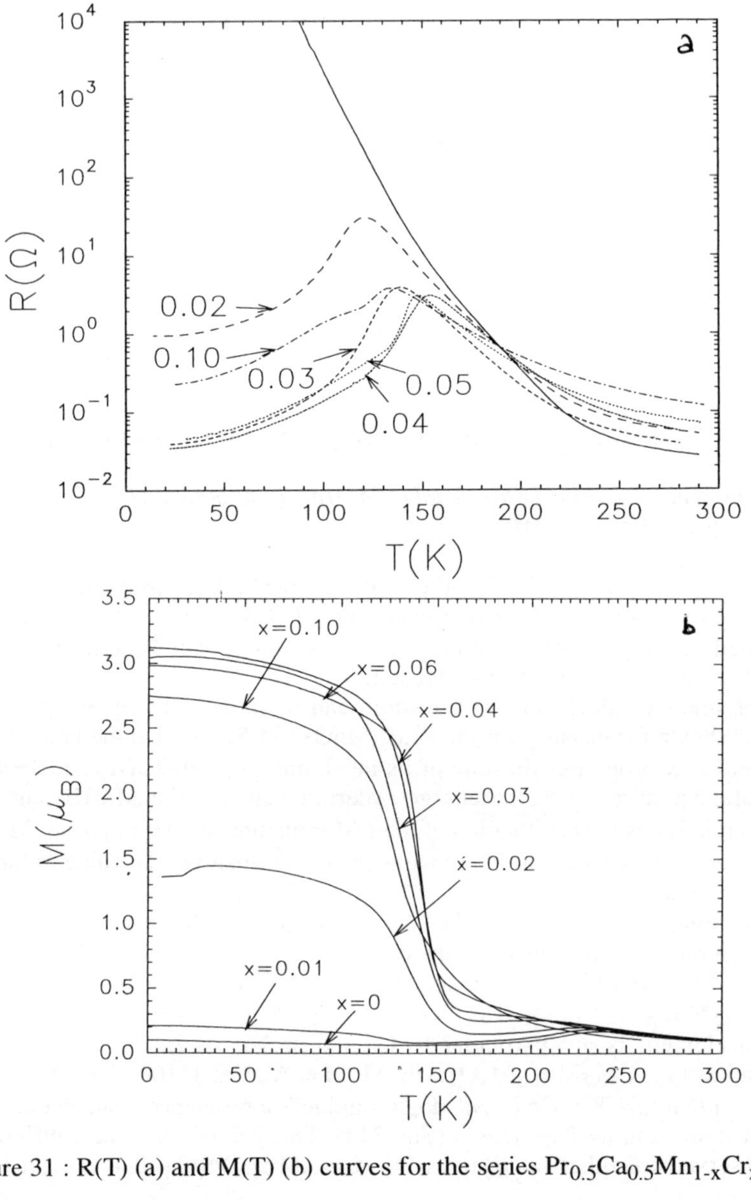

Figure 31 : R(T) (a) and M(T) (b) curves for the series $Pr_{0.5}Ca_{0.5}Mn_{1-x}Cr_xO_3$.

A similar evolution is observed for Co and Ni doped manganites. The M (T) curves (Fig. 31b) corroborate the R(T) curves. One indeed observes the appearance of a ferromagnetic contribution at low temperature by doping with Cr, Co or Ni, in contrast to the undoped phase which is antiferromagnetic at those temperatures. This is illustrated on figure 31b for the Cr-doped manganites that show a rapid increase of their magnetic moment as x increases reaching 3 - 3.10 μ_B for x ranging from 0.03 to 0.06; simultaneously, T_C increases up to 150 K as x increases up to x = 0.06. Then, T_C and ferromagnetism tend to decrease as x increases beyond x = 0.06, as shown for the x = 0.10 sample that exhibits a T_C of 125 K and a magnetic moment of 2.75 μ_B at 4.2 K. Very similar behaviours are observed for cobalt and nickel, except that the maximum value of the magnetic moment is significantly smaller than for chromium. This transition from an insulator to a ferromagnetic metal suggests that the so doped samples exhibit colossal magnetoresistance properties. The induced CMR effect is indeed spectacular as shown by comparing the undoped $Pr_{0.5}Ca_{0.5}MnO_3$ manganite that is not magnetoresistant even under 7T, with the Cr doped manganites $Pr_{0.5}Ca_{0.5}Mn_{1-x}Cr_xO_3$ (Fig. 32). For very weak doping levels, i.e. x = 0.01, that correspond to insulators in zero magnetic field, one observes high resistance ratios (RR) of 3.10^4 (Fig. 32a) at 60 K, in spite of the fact that charge ordering has not completely disappeared. It is remarkable that as soon as the doped phase exhibits an I-M transition in a zero magnetic field, the maximum RR value decreases significantly, following the T_{max} value : RR values of 30 and 40 are observed for the x = 0.05 (Fig. 32b) and 0.06 (Fig. 32c) Cr-doped phases at T_{max} values of 145 K and 140 K respectively. It is worthy of note that among these three dopants, cobalt exhibits the highest resistance ratios with RR $\sim$ 3.10^6 at 60 K under 7 T for $Pr_{0.5}Ca_{0.5}Mn_{0.99}Co_{0.01}O_3$.

The possibility to apply such an effect to other perovskites $Ln_{0.5}Ca_{0.5}MnO_3$ with a smaller lanthanide cation has been demonstrated for the series $Ln_{0.5}Ca_{0.5}Mn_{1-x}Ni_xO_3$ [59]. Nevertheless, it is observed that the ability to suppress charge ordering decreases as the size of Ln^{3+} decreases from Pr^{3+} to Sm^{3+}. In a same way, the difficulty to induce an I-M transition increases from Pr^{3+} to Sm^{3+}.

5.4 Modification of the magnetic phase diagrams by Mn site doping

The possibility to induce MI transitions by doping charge ordered manganites with small $\langle r_A \rangle$ has motivated a reinvestigation of the $T_m(\langle r_A \rangle)$ diagram for the $Ln_{0.5}Ca_{0.5}Mn_{1-x}M_xO_3$ series. Based on previous results, chromium has been found to be the most efficient cation to induce MI transitions in these perovskites. The comparison of saturated magnetization values (5 K) for different doping elements in $Pr_{0.5}Ca_{0.5}Mn_{1-x}M_xO_3$ samples illustrates the chromium superiority (Fig. 33). The highest magnetization values are obtained for the latter, reaching values close to that expected from the formula (3.5 μ_B). Moreover one can remark on the graph of figure 33 that in contrast to M = Ni or Co, the Cr effect is not limited to a narrow

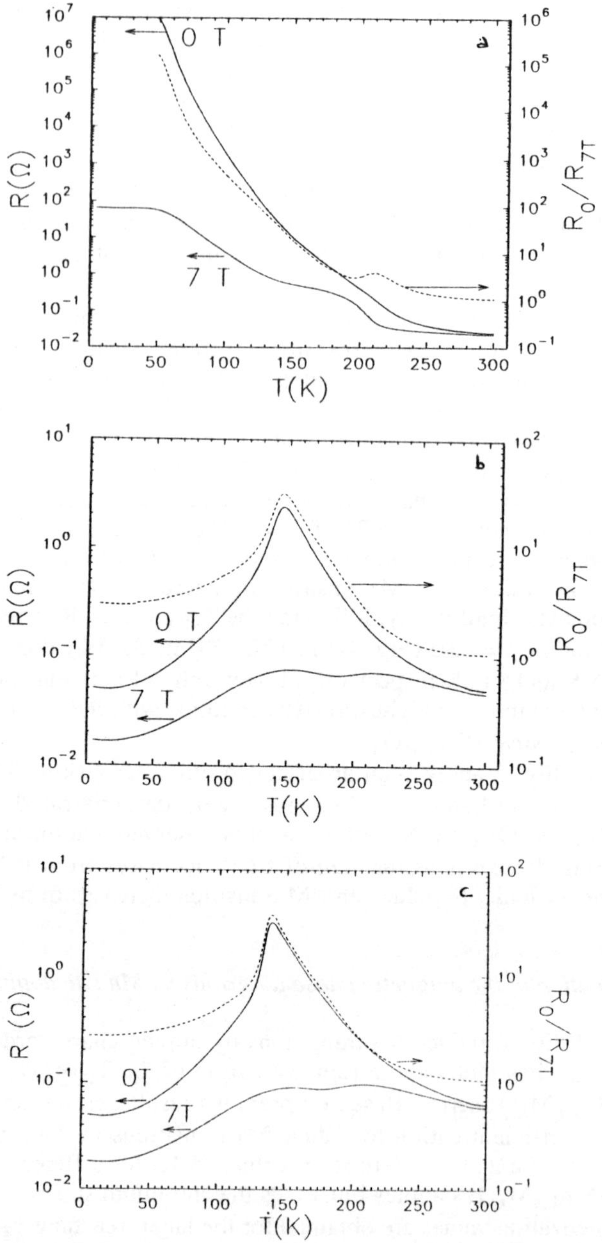

Figure 32 : $R_0(T)$, $R_{7T}(T)$ and $R_0/R_{7T}(T)$ curves for $Pr_{0.5}Ca_{0.5}Mn_{1-x}Cr_xO_3$ samples : x = 0.01 (a), x = 0.05 (b) and x = 0.06 (c).

range of compositions since from 3 to 10% of Cr the magnetic moment reaches ~ 3 μ_B.

The size effect has been investigated for M = Cr in $Ln_{0.5}Ca_{0.5}MnO_3$ charge-ordered pristine samples by varying Ln from La to Tb. Along this lanthanadic line the A-site cationic size mismatch remains very small so that the observed effects result mainly from the $\langle r_A \rangle$ variations. Finally, a new line $T_m(\langle r_A \rangle)$ has been obtained (Fig. 34) for these substituted materials. Furthermore, these data fit well with those of undoped charge ordered samples of $\langle r_A \rangle > 1.19$ Å and low σ^2. The pristine samples, with $\langle r_A \rangle < 1.19$ Å, are all AFMI compounds below T_{CO} with no CMR properties in agreement with their strong AFM state. In contrast, optimally chromium doped samples exhibit a FMM-PMI transition down to $\langle r_A \rangle = 1.15$ Å demonstrating that CMR is obtained below the size limit of ~1.19 Å common to both $Ln_{0.7}A_{0.3}MnO_3$ and $Ln_{0.5}A_{0.5}MnO_3$ samples. By killing the CO state, chromium allows to push down the $\langle r_A \rangle$ limit, and, moreover, the high value of the restored magnetization strongly suggests that Cr participates to the double-exchange mechanism i.e., is ferromagnetically coupled to its manganese neighbours, and thus that charge delocalization occurs.

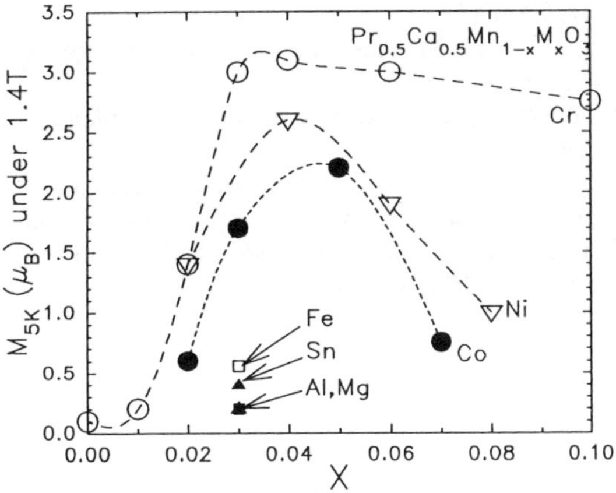

Figure 33 : Magnetization values at 5 K registered under 1.4 T versus the doping level x for the series $Pr_{0.5}Ca_{0.5}Mn_{1-x}M_xO_3$ (M = Cr, Ni, Co, Fe, Sn, Al, Mg).

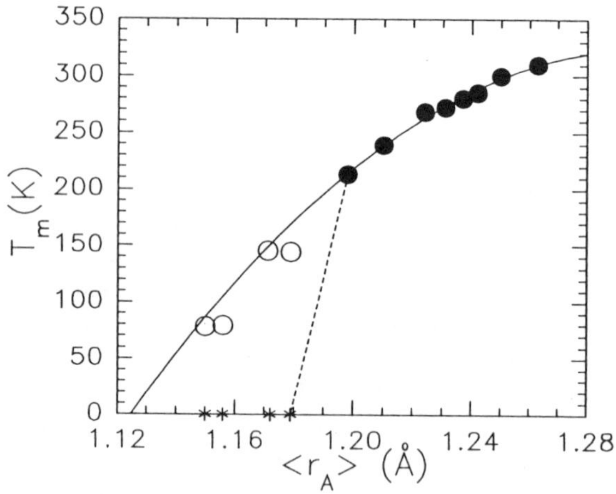

Figure 34 : Optimized M-I transition T_m versus $<r_A>$ for undoped $Ln_{0.5}A_{0.5}MnO_3$ compounds (filled circles) with low mismatch (from ref. 10) and for doped compounds (open circles) belonging to the series $Ln_{0.5}Ca_{0.5}Mn_{1-x}Cr_xO_3$. Stars represent the undoped compounds which are insulating.

References

1. R. Von Helmolt, J. Wecker, B. Holzapfel, L. Schultz and K. Sanwer, *Phys. Rev. Lett.* **71**, 2331 (1993).
2. K. Chahara, T. Ohno, M. Kasai and Y. Kozono, *Appl. Phys. Lett.* **63**, 1990 (1993).
3. H.L. Ju, C. Kwon, Qi Li, R.L. Greene and T. Venkatesan, *Appl. Phys. Lett.* **65**, 2108 (1994).
4. S.S. Manoharan, N.Y. Vasanthacharya, M.S. Hegde, K.M. Satyalakshmi, V. Prasad and S.V. Subramanyam, *J. Appl. Phys.* **76**, 3923 (1994).
5. S. Jin, T.H. Tiefiel, M. Mc Cormack, R.A. Fastnacht, R. Ramesh and L.H. Chen, *Science* **264**, 413 (1994).
6. R. Mahendiran, R. Mahesh, A.K. Raychaudhuri and C.N.R. Rao, *Solid State Commun.* **94**, 515 (1995).
7. R. Mahesh, R. Mahendiran, A.K. Raychaudhuri and C.N.R. Rao, *J. Solid State Chem.* **114**, 297 (1995) ; ibid. **120**, 204(1995).
8. J. Inoue and S. Maekawa, *Phys. Rev. Lett.* **74**, 3407 (1995).
9. Y. Tokura, A. Urushibara, Y. Moritomo, T. Arima, A. Asamitsu, G . Kido and N. Furukawa, *J. Phys. Soc. Jap.* **63**, 3931(1994).
10. H. Kuwahara, Y. Tomioka, A. Asamitsu, Y. Moritomo and Y. Tokura, *Science* **270**, 961 (1995).
11. A. Urushibara, Y. Morimoto, T. Arima, A. Asamitsu, G. Kido, Y. Tokura, *Phys. Rev.* B , 14103 (1995) .
12. Y. Tomioka, A. Asamitsu, Y. Moritomo, H. Kuwahara and Y. Tokura, *Phys. Rev. Lett.*, 56 (1995).
13. V. Caignaert, A. Maignan and B. Raveau, *Solid State Commun.* **95,** 357 (1995).

14. A. Maignan, V. Caignaert, Ch. Simon, M. Hervieu and B. Raveau, *J. Mater. Chem.*, **5**, 1089 (1995).
15. A. Maignan, Ch. Simon, V. Caignaert and B. Raveau, *Solid State Commun.* **96**, 623 (1995).
16. B. Raveau, A. Maignan and Ch. Simon, *J. Solid State Chem.* **117**, 424 (1995).
17. P.G. de Gennes, *Phys. Rev.* **118** , 141 1960).
18. J.B. Goodenough, *Phys. Rev.* **100,** 564 (1955).
19. A.J. Millis, P.B. Littlewood and B.I. Shraiman, *Phys. Rev. Lett.* **74**, 5144 (1995).
20. A. Maignan, Ch. Simon, V. Caignaert and B . Raveau, *J. Magn. Magn. Mater.* **152**, L5 (1996).
21. A. Maignan, Ch. Simon, V. Caignaert and B. Raveau, *Z. Physik* B **99**, 305 (1996).
22. J. Hejtmanek, Z. Jirak, D. Sedmidubsky, A. Maignan, Ch. Simon, V. Caignaert, C. Martin and B. Raveau, *Phys. Rev.* B **54**, 11947 (1996).
23. F. Millange, A. Maignan, V. Caignaert, Ch. Simon, and B. Raveau, *Z. Physik* B **101**, 169 (1996).
24. J. Fontcuberta, B. Martinez, A. Seffar, S. Pinol, J.L. Garcia-Munoz and X. Obradors, *Phys. Rev. Lett.* **76**, 1122 (1996).
25. P.G. Radaelli, M. Marezio, H.Y. Hwang and S.W. Cheong, *J. Solid State Chem.* **122**, 444 (1996).
26. R.D. Shannon, *Acta Cryst.* **A32**, 751 (1976).
27. A. Maignan, Ch. Simon, V. Caignaert and B. Raveau, *J. Appl. Phys.* **79**, 7891 (1996) ; R. Mahendiran, R. Mahesh, A.K. Raychaudhuri and C.N.R. Rao, *Phys. Rev.* B **53**, 12160 (1996).
28. J. Wolfman, Ch. Simon, M. Hervieu, A. Maignan and B. Raveau, *J. Solid State Chem.* **123**,) 413 (1996.
29. J. Wolfman, A. Maignan, Ch. Simon and B. Raveau, *J. Magn. Magn. Mater.* **159**, L299 (1996).
30. H.Y. Hwang, S.W. Cheong, P.G. Radaelli, M. Marezio and B. Batlogg, *Phys. Rev. Lett.* **75**, 914 (1995).
31. F. Damay, A. Maignan, C. Martin and B. Raveau, *J. Appl. Phys.* **81**, 1372 (1997).
32. L.M. Rodriguez- Martinez and J.P. Attfield, *Phys. Rev.* B **54**, R15622 (1996)
33. A. Maignan, C. Martin, F. Damay and B. Raveau, *Z. Physik* B **104**, 21 (1997).
34. A. Sundaresan, A. Maignan and B. Raveau, *Phys. Rev.* B **56** , 5092 (1997)
35. F. Damay, C. Martin, A. Maignan and B. Raveau, *J. Appl. Phys.* (in press).
36. V. Caignaert, E. Suard, A. Maignan, Ch. Simon and B. Raveau, *C.R. Acad. Sc.* **321**, 515 (1995).
37. D.N. Argyriou, J.F. Mitchell, C.D. Potter, D.G. Hinks, J.D. Jorgensen and S.D. Bader, *Phys. Rev. Lett.* **76**, 3826 (1996).

38. M. Hervieu, G. Van Tendeloo, V. Caignaert, A. Maignan and B. Raveau, *Phys. Rev.* B **53**, 14274 (1996).
39. V. Caignaert, E. Suard, A. Maignan, Ch. Simon and B. Raveau, *J. Magn. Magn. Mater.* **153**, L260 (1996).
40. D.N. Argyriou, D.G. Hinks, J.F. Mitchell, C.D. Potter, A.J. Schultz, D.M. Young, J.D. Jorgensen and S.D. Bader, *J. Solid State Chem.* **124**, 381 (1996).
41. H. Kawano, R. Kajimoto, H. Yoshizawa, Y. Tomioka, H. Kuwahara and Y. Tokura, *Phys. Rev. Lett.* **78** , 4253 (1997).
42. K. Knizek, Z. Jirak, E. Pollert, F. Zounova and S. Vratislav, *J. Solid State Chem.* **100**, 292 (1992).
43. F. Damay, C. Martin, M. Hervieu, A. Maignan, B. Raveau, G. André and F. Bourée, *J. Magn. Magn. Mater.* (submitted).
44. V. Caignaert, F. Millange, M. Hervieu, E. Suard and B. Raveau, *Solid State Commun.* **99**, 173 (1996).
45. F. Millange, V. Caignaert, G. Mather, E. Suard and B. Raveau, *J. Solid State Chem.* **127**, 131 (1996) ; V. Caignaert, F. Millange; M. Hervieu, G. Mather, B. Raveau, E. Suard, P. Laffez and G. Van Tendeloo, *Phys. Rev.* B (in press).
46. N. Kumar and CNR Rao, *J. Solid State Chem.* **129**, 362 (1997).
47. A. Maignan, C. Martin and B. Raveau, *Z. Physik* B **102**, 19 (1997).
48. C. Martin, A. Maignan and B. Raveau, *J. Mater. Chem.* **6**, 1245 (1996).
49. A. Maignan and B. Raveau, *Z. Physik B* **102**, 209 (1997).
50. F. Damay, A. Maignan, N. Nguyen and B. Raveau, *J. Solid State Chem.* **124**, 385 (1996).
51. Z. Jirak, S. Krupicka, V. Nekvasil, E. Pollert, G. Villeneuve and F. Zounova, *J. Magn. Magn. Mater.* **15**, 519 (1980).
52. E. Pollert, S. Krupicka, Z. Simsa, M. Dlouha and V. Vratislav, *J. Phys. Chem. Solids* **43**, 1137 (1982).
53. Z. Jirak, S. Krupicka, Z. Simsa, M. Dlouha and V. Vratislav, *J. Magn. Magn. Mater.* **53**, 153 (1985).
54. H. Yoshizawa, H. Kawano, Y. Tomioka and Y. Tokura, *J. Phys. Soc. Jap.* **65**, 1043 (1996).
55. H. Yoshizawa, H. Kawano, Y. Tomioka and Y. Tokura, *Phys. Rev. B* **52**, R13145 (1995).
56. Y. Tomioka, A. Asamitsu, Y. Moritomo and Y. Tokura, *J. Phys. Soc. Jap.* **64**,) 60 (1995).
57. M.R. Lees, J. Barratt, G. Balakrishnan, D. MacK. Paul and M. Yethiraj, *Phys. Rev.* B **52**, R14303 (1995).
58. B. Raveau, A. Maignan and C. Martin, *J. Solid State Chem.* **130**, 162 (1997).
59. A. Maignan, F. Damay, C. Martin and B. Raveau, *Mat. Res. Bull.* **32**, 965 (1997).
60. F. Damay, C. Martin, A. Maignan and B. Raveau, *J. Magn. Magn. Mater.* (in press).

MAGNETOTRANSPORT AND MAGNETOELASTIC EFFECTS IN MANGANESE-OXIDE PEROVSKITES

M.R. IBARRA and J.M. DE TERESA
Dpto. de Física de la Materia Condensada&ICMA, Universidad de Zaragoza-CSIC, 50009-Zaragoza, Spain

We review the relevant magnetic, structural, magnetotransport and magnetoelastic properties of some mixed-valence manganites. The main interactions found in these compounds are reported. The interplay among ferromagnetism, antiferromagnetism and charge ordering brings about a broad phenomenology. The relevance of the electron-phonon interaction is discussed. We have described three series of compounds: $La_{1-x}Ca_xMnO_3$, $(La_{1-x}Tb_x)_{2/3}Ca_{1/3}MnO_3$ and $LaMnO_{3+\delta}$. Special attention is paid to the large magnetoelastic effect in the paramagnetic phase. This was explained considering an extra volume contribution originated by the polaronic effect, which is suppressed under an applied magnetic field. The magnetic nature of the polarons, responsible for the giant magnetoresistance effect, was confirmed by using a combination of macro and microscopic experimental techniques. By using the small-angle neutron scattering technique, the magnetic correlation length and the role of the electronic phase segregation in these compounds have been determined.

1 Introduction

The magnetic and transport properties of $AMnO_3$-type compounds are closely related to the crystallographic structure. The studied compounds have crystallographic structures of perovskite type, which derive from the ideal cubic symmetry. The different structures are generated from the cubic one by three mechanisms [1, 2]:
-(i) Tilting of the MnO_6 octahedra.
-(ii) Displacements of the cations.
-(iii) Distortions of the MnO_6 octahedra.
The displacement of the cations only produces a change in the lattice parameters and a small distortion of the octahedra. Usually (ii) and (iii) are connected and are not relevant in our study. Nevertheless, the tilting of the octahedra is most relevant in order to determine the overall space-group symmetry of the particular perovskite.

The tilting of one of the MnO_6 octahedra produces the tilting of the neighbouring octahedra. To visualize how the octahedra rotate, we consider the resultant position as a linear combination of rotations about the octahedra axis of the original cubic symmetry. "abc" notation is used to indicate a rotation about the [100], [010] and [001] fourfold cubic axis (see Fig. 1).

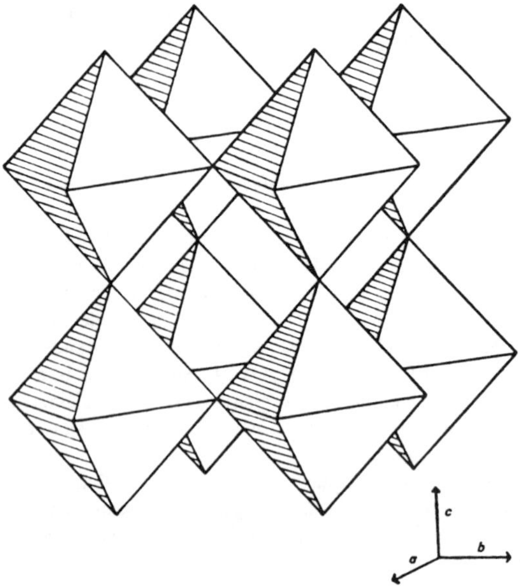

Fig. 1. Position of the Oxygen octahedra in the cubic perovskite $AMnO_3$-type. The A cation is in the center of the cube formed by the octahedra. The Mn ion is placed in the center of each octahedron.

If the rotation about two axis is of the same magnitude, they are represented by the same letter. If we consider the axis perpendicular to the basal plane of the octahedra as the rotation axis, the neighbouring octahedra in the plane are forced to rotate in opposite direction. However, the upper and lower octahedra along the axis can rotate in phase, out of phase or do not rotate. This is represented using the superscript "+", "-" or "o" respectively. For instance, $a^ob^+b^-$ is a double rotation along the [010] (in phase) and [001] (out phase) axis of the same magnitude. Any rotation produces a length reduction between the octahedra centers in the basal plane. The distances along the rotation axis keep unchanged. Under rotation of the octahedra the pseudocubic axis are given by:

$a_p = \xi \cos\alpha \cos\beta$

$b_p = \xi \cos\alpha \cos\gamma$

$c_p = \xi \cos\beta \cos\gamma$

where α, β, γ are the rotation angles about the [100], [010] and [001] axis respectively and ξ is the anion-anion distance along the octahedra centers. In Fig. 2 we show how a rotation produces a change in the distances between anions. These equations result in three different pseudocubic lattice parameters. All possible combinations of rotations give rise to a large number of different space-group symmetries [1].

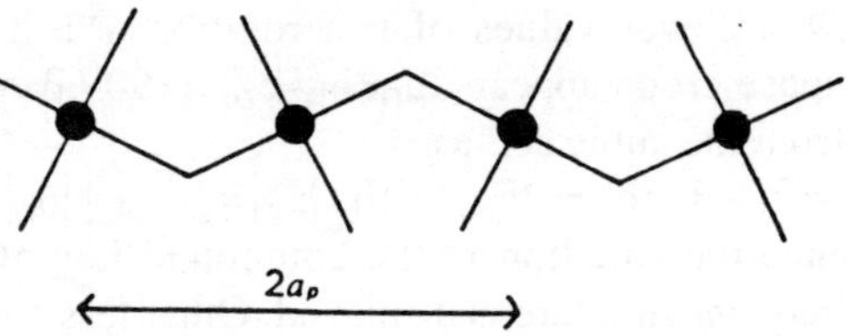

Fig. 2. Schematic diagram of the rotation of octahedra about an axis perpendicular to the plane of the figure. The B cation is represented as a black point. Distances between B cations approach in the directions perpendicular to the axis of rotation.

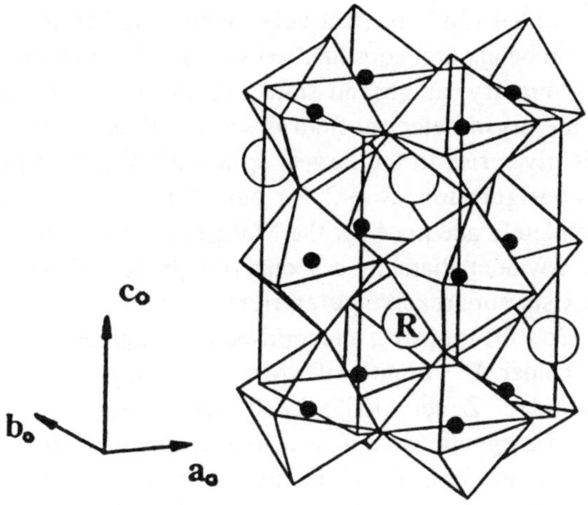

Fig.3. Crystallographic structure that results when there are two rotations of different magnitude about the axis [100] and [011]. This structure is commonly found when the tolerance factor is much lower than 1.

In the case of the manganites, the distortions are a consequence of the reduced size of the rare earth ion that occupies the A site in the $AMnO_3$ structure, and give rise to a rotation of the MnO_6 octahedra in order to fill the empty space. This distortion is characterized by the Goldsmith tolerance factor [3], defined as:

$t = d_{A\text{-}O}/\sqrt{2}\,(d_{Mn\text{-}O})$, where $d_{A\text{-}O}$ and $d_{Mn\text{-}O}$ are the average A cation-Oxygen and Manganese-Oxygen distances respectively. In the case of a cubic perovskite t=1; the smaller the A cation radius, the lower the t value. For t values close to 1, a rotation $a^-a^-a^-$ takes place and $R\bar{3}c$ (rombohedral) is the space group obtained ($a_p \approx b_p \approx c_p$, $\alpha \approx \beta \approx 90^o \neq \gamma$) . For lower values of t, a rotation $a^+b^-b^-$ occurs and the Pbnm (orthorhombic) space group appears ($a_p \neq b_p \approx c_p$, $\alpha \neq 90^o, \beta \approx \gamma$). The orthorhombic unit cell is obtained from the cubic cell as:

$\vec{a}_0 = \vec{a}_c + \vec{b}_c; \vec{b}_0 = \vec{a}_c - \vec{b}_c; \vec{c}_0 = 2\vec{c}_c$ with $|a_0| = |\sqrt{2}a_c|; |b_0| = |\sqrt{2}a_c|; |c_0| = |2a_c|$

This is for instance the situation in the compound $La_{2/3}Ca_{1/3}MnO_3$. In Fig. 3 a diagram of this frequent structure is displayed. Other less common structures can also be found [4-7].

The compounds where the large magnetoresitance effects are observed have the formula $R_{1-x}A_xMnO_3$ [R is a rare earth ion (valence=+3) and A is an alkali earth (valence=+2)]. In these compounds, already studied in the fifties [8-10], the Manganese ions show mixed valence. The ending compounds, $RMnO_3$ and $AMnO_3$, are antiferromagnetic (due to the superexchange interaction) and the Mn ions exhibit valence +3 ($3d^4$) or +4 ($3d^3$) respectively. When the Mn ion is in a crystal with cubic symmetry and octahedral coordination with negative ligands, the 3d level splits in a triplet (t_{2g} symmetry) as ground state and an excited doublet (e_g symmetry). The intensity of the crystal electric field is weaker than the intra-atomic electronic correlations. This gives rise to the largest spin multiplicity as predicted by Hund´s rule (high-spin configuration, S=3/2 for Mn^{+4} and S=2 for Mn^{+3}). This naïve picture is not completely adequate for the Manganese oxides under study as the Mn-O-Mn bond has a covalent character associated with the p-Oxygen and d-Manganese orbitals. There exists superexchange antiferromagnetic interaction through the π orbitals, $t_{2g}(Mn)\text{-}2p_\pi(O)\text{-}t_{2g}(Mn)$, and, consequently, governed by the t_π^{AF} transfer integral [11, 12]. Under A^{+2} doping of $RMnO_3$, the appearance of 3d holes with e_g symmetry takes place. Zener [13] related the existence of these holes to the mechanism for ferromagnetic exchange called double exchange. This interaction in mixed-valence manganites is provided by the covalent charge-transfer integral between e_g-Mn and σ-Oxygen orbitals (t_σ^F). The aforementioned strong Hund coupling makes this charge transfer possible between neighbouring Mn^{+3}/Mn^{+4} ions if the spin of the e_g electron of the Mn^{+3} ion is aligned with the spin of the Mn^{+4} ion, the transfer $e_g(Mn)\text{-}2p_\sigma(O)\text{-}e_g(Mn)$ being ferromagnetic in nature. Anderson and Hasegawa [14] and De Gennes [15] established that the effective transfer integral of an e_g electron to the next Mn neighbour depends on the angle between the local t_{2g} moments. If they are parallel, the electron will be easily transferred. The ferromagnetic order is consequently related to a large electronic itinerancy, i.e metallic behaviour.

In the mixed-valence manganites, where the above mentioned mechanisms are present, the spontaneous or field-induced alignment of the local magnetic moments gives rise simultaneously to a drastic reduction of the resistivity when a transition from a paramagnetic (semiconducting-like) to a ferromagnetic (metallic-like) takes place [16-18]. This effect is in the origin of the huge magnetoresistance found in these compounds, which was coined as "colossal" by Jin et al. [19].

In this section we will concentrate on the series $La_{1-x}Ca_xMnO_3$, where many interesting phenomena appear mainly because of the interplay between ferromagnetic (low resistivity) and antiferromagnetic (high resistivity) interactions. The phase diagram of this series was first described by Jonker et al. [8] and recently revised by Troyanchuk et al. [20] and Schiffer et al [21] (see Fig. 4) The ending compounds are antiferromagnetic at low temperatures and insulator in the whole range of temperatures. This behaviour persits in the high-doping region $x \geq 0.5$ as represented in the preliminar phase diagram reported in Ref. [21]. The region with $x<0.2$ was initialy reported as ferromagnetic insulator but this is not the real case as an inhomogeneous state occurs [22, 23]. The ferromagnetic region is restricted to the concentration range $0.2 \leq x \leq 0.5$.

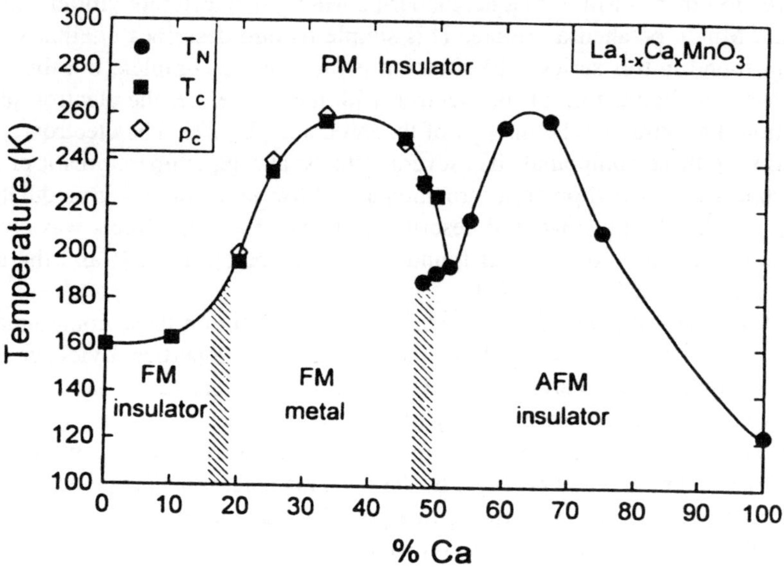

Fig. 4. Phase diagram of magnetic and transport properties of the series $La_{1-x}Ca_xMnO_3$, published in Ref. [21]. PM stands for paramagnetic; FM, ferromagnetic; and AFM, antiferromagnetic.

The maximum in the Curie temperature (T_C) takes place approximately for the concentration x=2/3, i.e. the ferromagnetic exchange intensity is optimised for such concentration. These results can not be understood from the point of view of the ferromagnetic double-exchange interaction above described. One would expect the maximum in the T_C at the concentration x=0.5, where the number of Mn^{+3}/Mn^{+4} pairs is maximun along the series. And the magnetic phase diagram should be symmetric. i.e. the compounds with concentrations x=0.3 and 0.7, with the same number of Mn^{+3}/Mn^{+4} pairs, should have the same magnetic behaviour, which is not the case. As a consequence, not only the exchange interaction is relevant in these compounds but some other mechanism must be present. J.B. Goodenough proposed the relevance of the lattice dynamics in this scenario [24, 25]. The coupling of the electronic states with the vibrational modes of the lattice gives rise to an enhancement of the ferromagnetic interaction. This interaction does not imply charge transfer but is related to the existence of a dynamic Jahn-Teller effect, being, as a consequence, proportional to the number of Mn^{+3} ions, i.e. this interaction is large for low x and negligible for high x. In Goodenough's model, the existence of two Mn^{+3} in two different vibronic states may give rise to an inversion of the ground state (dx^2-y^2 and dz^2) in the e_g energy levels (splitted by the Jahn-Teller effect) of neighbouring Mn ions. If this is the case, an electron can migrate with the same spin direction from a Mn^{+3} to a neighbouring Mn^{+3}, in a different vibrational state, when the MnO_6 octahedra vibrate. This simple picture describes qualitatively the situation found in this series. The situation is much more complex, but the idea of considering the interaction of the electron with the lattice, i.e. the electron-phonon interaction, has attracted the interest of theoreticians [26, 27]. The electron-phonon interaction in these compounds has several effects: at high temperatures it provides the mechanism for small polaron formation and at low temperatures it modulates the one-electron bandwith as we will describe. Evidence for both effects was obtained from the extra anharmonic effect found at high temperatures [28] and the strong isotopic effect on the T_C value [29].

Nowadays, the vast experimental work carried out on these compounds has established the polaronic nature of the magnetotransport properties. Evidence for the magnetic nature of the polaron has been given [30]. There exists several meanings of the polaron term and the situation may be confuse if one does not carefully describe and define this term. We consider a magnetic polaron as a lattice small polaron dressed by a magnetic cloud provided by the short-range ferromagnetic interaction.

First, we will describe the lattice polaron. Landau in 1933 introduced the concept of polaron in ionic crystals. When an electron remains for some time in the vicinity of a particular ion, the lattice will be deformed. The local change in the dielectric constant and the elastic relaxation of the system brings about a potential well where the electron can form a bound state. This phenomenon is known as

Landau trapping or self-trapping [31-33]. In a 3d metal the average life-time of an electron at an ionic position is of the order of 10^{-15} seconds. This short time makes not possible that its own electrical field can polarize the crystal. However, if the electron moves in a very narrow band, or due to some other mechanism, the electron is near the core ion for $\sim 10^{-12}$ seconds. Then, the electric field of the electron will attract the positive ions and push away the negative ions. The effective electric field at the electronic positions decreases because of the increase of the dielectric constant due to both the polarization and the ionic displacement. If the electron tries to move away, it will feel a force that tends to restore it at the initial position. The electrostatic energy plus the elastic energy add to give the potential barrier (E_{hop}) that should be overcome to release the charge from the potential well. This bound state, charge + lattice distortion, is called polaron. When the electronic wave function is localized in a radius of the order of a lattice parameter is called small-polaron. The polaron theory differs from the band theory in the relevance of the electron-phonon interaction, which is incorporated as a zero order term in the Hamiltonian and not as a perturbation.

There exist two mechanisms for the polaron to move from site to site in the lattice. One is by thermal activation at high temperatures and other by quantum tunneling at low temperatures (see figure 5). In this case the polarons form a band whose bandwith (b_p) is related to the average life-time ($b_p \Delta\tau \approx \hbar$) by Heisenberg´s incertitude relationship.

Holstein [34] proposed a model for the origin and dynamics of the small-polaron that acccounts for these two mechanisms and which we will briefly describe. This model establishes a relationship between b_p and b_t (electronic covalent transfer integral, $t_\sigma{}^F$): $b_p = b_t \exp[-\gamma \coth(\Theta_D/2T)]$, where Θ_D is the Debye temperature. This implies a modulation of the polaronic bandwith with temperature taking as the scale parameter the Debye temperature which is the charateristic parameter of the lattice vibrations. Three thermal ranges are considered in the model:

i) $T \ll \Theta_D$. Within this temperature range the polaronic and one-electron bandwith coincide ($b_p = b_t$): the only possible mechanism for polaron motion is quantum tunneling.

ii) $T < \Theta_D/2$. Under this condition $b_p \approx b_t \exp[-\gamma(2T/\Theta_D)]$. Both, tunneling and thermal activation mechanisms, are possible.

iii) $T > \Theta_D$. The polaronic band becomes very narrow (strong localization) and the only possible mechanism is thermal activation.

Summarizing, Holstein´s model proposes a narrowing of the polaronic band as the temperature increases that is exponentially modulated by the number of phonons. The thermal dependence of the resistivity in Holstein´s model in the adiabatic limit, i.e. when the carrier hops between lattice sites of identical vibrational configuration, is given by [35-38]:

$\rho=BT\exp(E_{hop}/KT)$, where B depends on the geometry and E_{hop} is the height of the energy barrier. For temperatures below $\Theta_D/2$ this model gives rise to a thermal dependence of the resistivity of the type $\rho=C\exp(T^{-1/4})$.

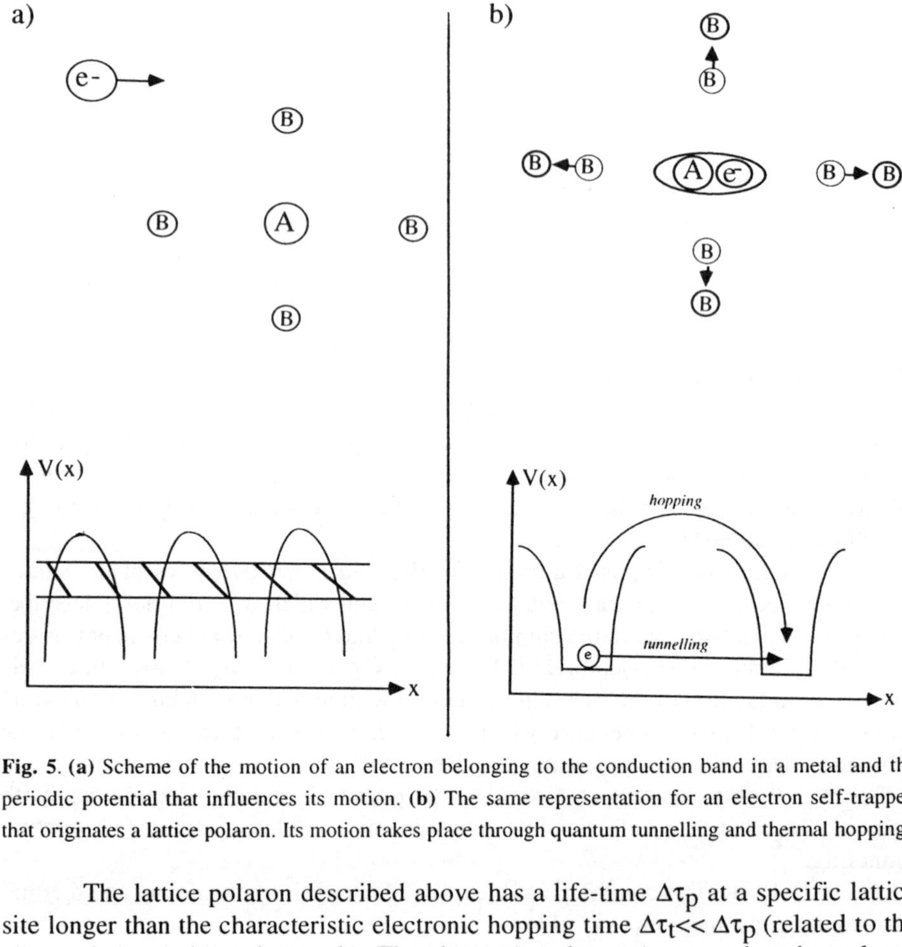

Fig. 5. **(a)** Scheme of the motion of an electron belonging to the conduction band in a metal and the periodic potential that influences its motion. **(b)** The same representation for an electron self-trapped that originates a lattice polaron. Its motion takes place through quantum tunnelling and thermal hopping.

The lattice polaron described above has a life-time $\Delta\tau_p$ at a specific lattice site longer than the characteristic electronic hopping time $\Delta\tau_t << \Delta\tau_p$ (related to the electronic bandwith as $b_t\Delta\tau_t \approx \hbar$). The charge ($e_g$ electron), trapped at the polaron time scale, visits simultaneously the neighbouring Mn^{+4} ions. This charge transfer Mn^{+4}-O^{-2}-Mn^{+3} is ferromagnetic, as mentioned above, giving rise to a short-range ferromagnetic interaction. The entity formed by the small lattice polaron plus the associated ferromagnetic cluster is referred to in this work as magnetic polaron. Röder et al. [27], using a hamiltonian which includes double-exchange interaction and electron-phonon interaction, theoretically predicted the existence of magnetic polarons which have been observed by De Teresa et al. [30] using small angle neutron scattering techniques SANS.

The concept of magnetic polaron, introduced by Kasuya and Yanase in 1968 as a giant spin molecule [39], has been used with different meanings in other fields of the Solid State Physics. For instance, in diluted magnetic semiconductors and Eu chalcogenides [40-42] the lattice does not play any role, the charge being localized at magnetic impurities sites. This type of polaron has been introduced in the manganites by Coey et al. [43] to explain the high magnetoresistance found in the ferromagnetic phase. Presently, we can certainly say that the magnetoresistance found in the ordered phase is mainly of extrinsic nature, mainly produced by the electron scattering at the grain boundaries [44, 45]. Although the model does not explain the intrinsic magnetoresistance, which is basically originated in the paramagnetic phase, it may be applied in the case of the existence of inhomogeneous ferromagnetic states, formed as a consequence of the electronic phase segregation.

Another relevant effect observed in the rare-earth mixed-valence manganites is the so-called charge ordering. This effect appears in absence of long-range ferromagnetic order, i.e. in para or antiferromagnetic phase below a characteristic temperature T_{CO} (charge ordering temperature) when the charges freeze in a lattice of Mn^{+3} and Mn^{+4} ions. This extra periodicity has been observed using microscopic techniques as electron diffraction [46] and neutron diffraction [47]. The origin for the interaction that gives rise to this phenomenon is not well known. It seems that charge ordering is produced when the mobility is low, i.e. the e_g band is very narrow and the number of carriers is a certain fraction of the ions that form the lattice. The charge lattice may be formed when the long-range coulombian interaction overcomes the electronic kinetic energy [48]. This model is similar to the prediction of the Wigner crystallization of a low density electronic gas in which the small value of the kinetic energy (low density of states at the Fermi level) is overcome by the coulombian repulsion that tends to separate the charges as far as possible [49]. These kinds of phenomena are not exclusive of the manganites and have been observed in other transition metal oxides. The archetypal system is the Fe_3O_4 compound, where Verwey observed a spatial ordering of the Fe^{+3} and Fe^{+4} ions [50]. $La_{2-x}Sr_xNiO_4$ (x=1/2 and 1/3) [51] and $La_{1-x}Sr_xMnO_4$ (x=1/2) [52] are other compounds where this effect has been observed.

This charge ordering state can be unstable under applied magnetic field, which is not surprising because the effect of the field in the manganites is to align the magnetic moments and consequently to increase the carrier mobility. If the gain in kinetic energy is enough to overcome the coulombian repulsion, the charge-ordered state can be suppressed. For a recent review on these phenomena see Ref. [53].

2 Experimental procedure

The samples were prepared either by the standard sol-gel or by solid-state reaction (for a detailed description of the procedure see references of the authors). The quality of the samples was checked by X-ray diffraction and scanning electron microscopy. The magnetization was measured using a vibrating sample magnetometer and a SQUID using high steady magnetic field up to 12 and 6 T respectively. a.c. initial susceptibilty measurements were performed using a mutual inductance bridge. The thermal expansion was measured within the temperature range 4K-1000K. Below room temperature the strain-gauge technique was used. For high temperatures the expansion was measured using a cryofurnace and the "push rod" method. The standard four-contact technique was used to measure the electrical resisitivity. The magnetostriction parallel ($\lambda_{||}$) and perpendicular ($\lambda_{\perp}$) to the field was measured using the strain-gauge technique with steady magnetic field up to 12 T and pulsed field up to 14 T. For the pressure experiments a clamp copper-berylium cell was used, reaching pressure values up to 10 kbar. The neutron diffraction experiments were performed at the high-flux reactor of the Institute Laue Langevin in Grenoble. The multidetector D1B powder diffractometer with a wavelength of 2.52 Å was used for thermal scans in the temperature range 2.5 K-600 K. Information about the structural changes and magnetic transitions and structures was obtained by refinement of the neutron spectra using the FULLPROF program [54]. The SANS experiments were performed at the D16 using a wavelength of 4.5 Å and obtaining resolution in the transfer momentum vector, q, of 5×10^{-3}Å^{-1} within the range 0.013 Å^{-1}<q<0.65 Å^{-1}. The versatility of this instrument allows applying magnetic field up to 5 T and rotating the multidetector an angle of $2\Theta=100°$, which permits also obtaining the thermal and field dependence of the intensity of Bragg peaks.

3 Giant magnetoresistance in the series $La_{1-x}Ca_xMnO_3$

We have studied several compounds of rare-earth mixed-valence manganites with different carrier concentration. As mentioned in the previous section, an increase in x causes an increase in the number of e_g holes. We will explain the behaviour found in the different concentration ranges by showing results of selected representative compounds to conclude the section with a more reliable phase diagram than the already reported in Fig.4. All the studied samples crystallize in the orthorhombic point symmetry, space group Pbnm.

3.1 x<0.2 region

The compound $LaMnO_3$ will be described in §5. Below T_N=140 K this compound is antiferromagnetic. The Mn magnetic moments are ferromagnetically ordered in the [001] planes and antiferromagnetically between adjacents planes [9]. This compound is insulating in the whole temperature range and shows a large Jahn-Teller distortion. From the point of view of the electrical resistivity, the compound $La_{0.9}Ca_{0.1}MnO_3$ has a similar behaviour. In Fig. 6 resistivity measurements at 0 and 12 T are shown, the magnetoresistance being small. The neutron diffraction results show how the antiferromagnetic peaks observed in $LaMnO_3$ disappear. The thermal dependence of the intensity of the Bragg peaks as well as the anomaly observed in the initial magnetic susceptibility (see inset of Fig. 6) indicate the existence of a ferromagnetic phase below T_C=160 K. The existence of such a ferromagnetic insulating state at low temperature is difficult to understand within the framework of the double-exchange mechanism described above. The ferromagnetic interaction is based, in the mixed-valence manganites, on the mobility of the carriers. However, a ferromagnetic insulating state takes place in this compound.

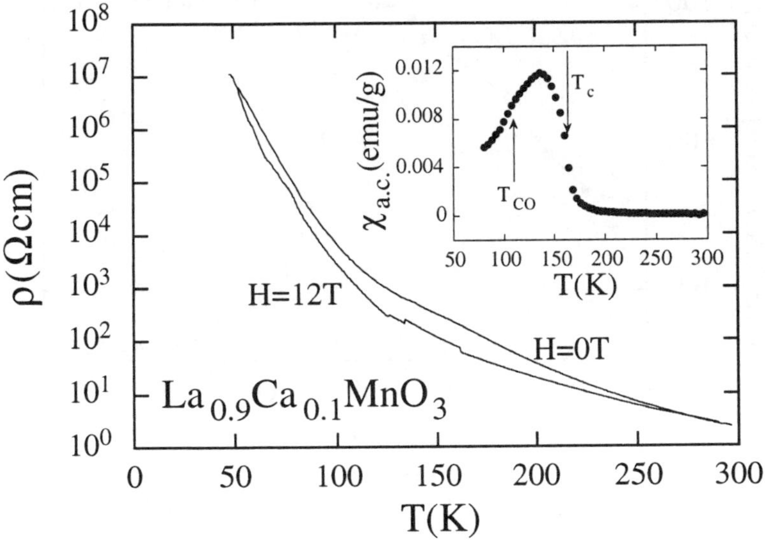

Fig. 6. Thermal dependence of resistivity at 0 and 12 T in the compound $La_{0.9}Ca_{0.1}MnO_3$. In the inset the a.c. susceptibility is shown. An anomaly connected with the onset of ferromagnetism is detected at 160 K. A second anomaly can be noticed at 110 K.

The thermal dependence of the SANS at fixed q value, [q=(4π/λ)sen2Θ], is displayed in Fig. 7. Above T_C a large increase in the intensity is observed for decreasing temperatures. This is a normal behaviour that is associated with the appearance of short-range magnetic correlation which originates ferromagnetic clusters. Surprisingly, below T_C a significant SANS contribution persists down to 2 K. The ferromagnetic correlation length (ξ) of these clusters were out of the intrumental resolution (ξ>~100Å). We propose a non-homogeneous state with the existence of large clusters at low temperatures, some of them with very large ξ, which would give contribution to the Bragg scattering. This state would not allow the electronic coherence for the existence of a metallic state, the system remaining in the insulating state. Cheong et al. [55] observed an anomaly in the a.c. initial susceptibility in the compound x=0.125 similar to our results (inset of fig. 6). They proposed the existence of charge-ordered regions below this temperature. Therefore, the low temperature ground state for x=0.1 may be a coexistence of charge-ordered and ferromagnetic regions.

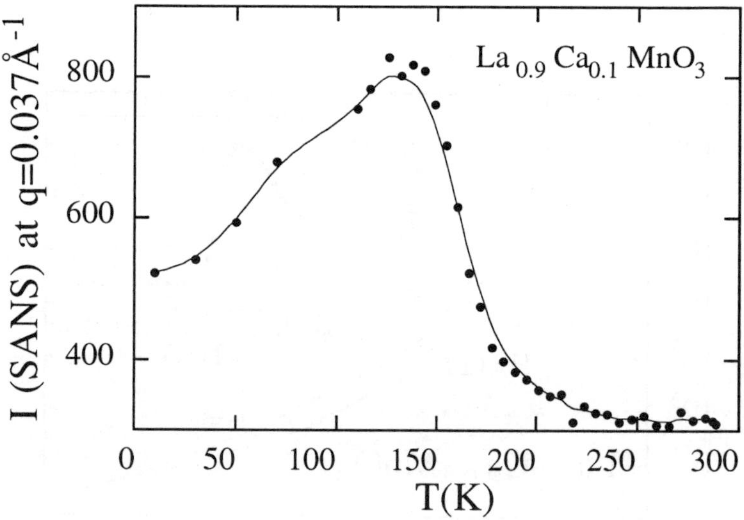

Fig. 7. SANS intensity as a function of temperature at constant q=0.03 Å^{-1} in the compound $La_{0.9}Ca_{0.1}MnO_3$. A strong magnetic contribution exists below T_C indicating that the magnetic state is non-homogeneous.

3.2 0.2≤x≤0.45 region

Four samples were studied within this concentration range: x=0.2, 0.28, 0.33, 0.45. The behaviour in all of them is very similar. We prepared the concentration x=0.33 using two different methods. The sample $La_{2/3}Ca_{1/3}MnO_3$ (#1) was obtained by using the sol-gel method as described in Ref. [28, 56] and the sample $La_{2/3}Ca_{1/3}MnO_3$ (#2) was prepared by solid state reaction (Ref. [56]). The T_C was 270 K for the sample #1 and 215 K for the sample #2. This large difference in T_C in samples prepared with the same starting nominal stoichiometric concentration is representative of the large dispersion in T_C values found in literature for this concentration. A careful characterization by using EDAX analysis, Rietveld refinement of the X-ray spectra and redox-tritation technique gives for sample #1, x=0.33±0.01 and for sample #2, x=0.28±0.01, which explains the large difference in T_C. Nevertheless, this difference in concentration is irrelevant concerning the structural, magnetic, magnetotransport and magnetostriction properties that we will report indistinctly for both compounds. In Fig. 8 we show the resistivity and magnetoresistance (inset) for sample #1.

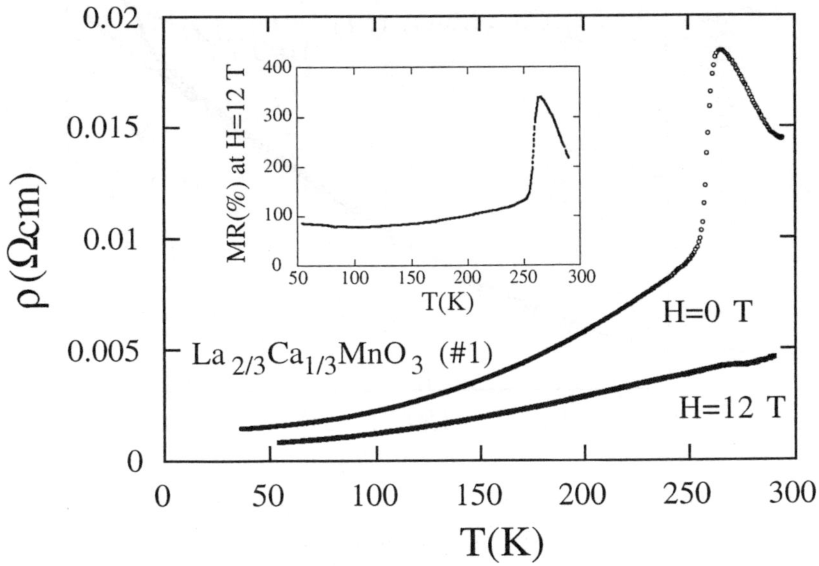

Fig. 8. Resistivity at 0 and 12 T as a function of temperature in the compound $La_{2/3}Ca_{1/3}MnO_3$ (#1). This sample shows an insulator-metal transition around 270 K. The inset shows the thermal dependence of the magnetoresistance, MR(%).

At T_C an abrupt spontaneous decrease in the resistivity takes place. This temperature clearly delimits two distinct behaviours. The semiconducting-like behaviour above T_C is associated with the existence of a polaronic regime [28]. Below T_C the long-range magnetic ordering allows the transfer of the charge, the behaviour being metallic-like. Here, the resistivity is originated by a spin-dependent scattering and tunnelling at the grain boundaries. Under an applied magnetic field of 12 T the anomaly in resistivity disappears, which gives rise to a large magnetoresistance effect in the paramagnetic phase.

Fig. 9 illustrates the volume thermal expansion for the sample #1 at 0 and 12 T. The anharmonic contribution of the lattice to the thermal expansion was calculated by using a Grüneisen law with a Θ_D=500 K (this value was obtained from the best fit at high temperatures, see also Fig. 10) [57, 28]. An extra contribution in addition to the theoretical prediction is observed in the paramagnetic phase which is spontaneously suppressed at T_C. Below this temperature the thermal expansion follows the calculated lattice contribution.

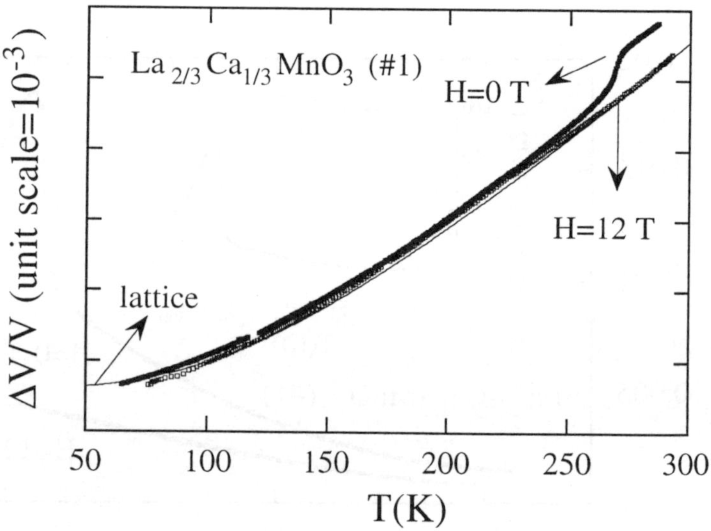

Fig. 9. Volume thermal expansion at 0 and 12 T for the compound $La_{2/3}Ca_{1/3}MnO_3$ (#1). The continuous line is an estimate of the phonon anharmonic contribution calculated with a Debye model with Θ_D=500 K. An anomaly of ≈0.66 % is observed at T_C.

This effect was interpreted by Ibarra et al. [28] as an anharmonic contribution due to polaron formation. Under an applied magnetic field of 12 T the thermal expansion is similar to that found in the ferromagnetic phase and the calculated lattice contribution, which indicates the absence of polaronic effect under such a field. This experiment clearly demonstrated the relevance of the electron-phonon interaction in these compounds and the important role of the magnetic interaction.

In Fig. 10 we report the thermal expansion at high temperatures (up to 650 K). The results indicate an anomalous contribution starting at a temperature around 2 T_C. At the highest temperature and below T_C the volume thermal expansion follows the calculated phonon contribution.

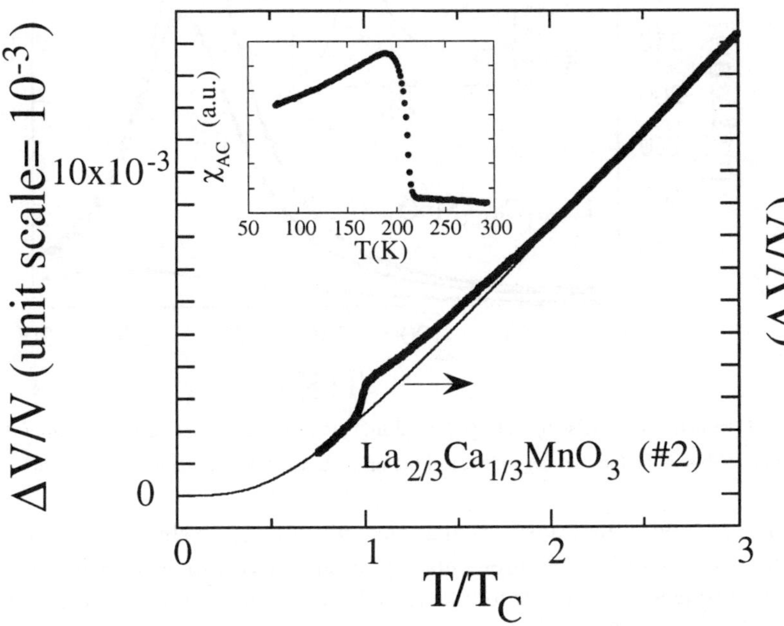

Fig. 10. Volume thermal expansion for the compound $La_{2/3}Ca_{1/3}MnO_3$ (#2) up to high temperatures. $(\Delta V/V)_0$ stands for the estimated phonon contribution. The extra anharmonic contribution disappears at T_C, which is ≈215 K as is deduced from the a.c. susceptibility measurements shown in the inset.

Considering the polaronic model described in §1, this behaviour can be explained considering a localization of the e_g electron, trapped as a small polaron, in the potential well. At very high temperatures the thermal energy allows the carrier to overcome the potential barrier. As the temperature is decreased, the localization of the carrier at specific sites is more effective and they can couple to vibrational lattice modes. This progressive localization is responsible for the extra anharmonic contribution. Below T_c this polaronic regime disappears and, consequently, the thermal expansion recovers the value that corresponds to the phonon contribution.

An important aspect of the magnetostransport properties of these compounds is the nature of the magnetoresistance found in the ferromagnetic phase. In order to give an example, we display in Fig. 11 the results found in sample #2.

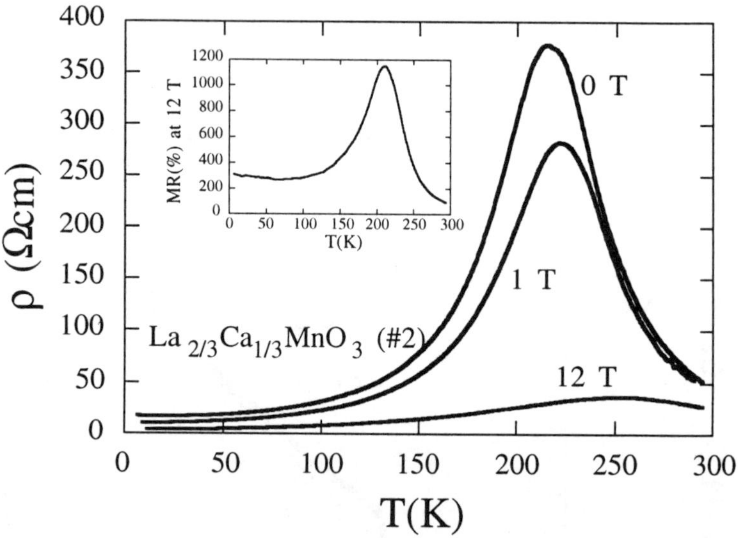

Fig. 11. Resistivity of the sample $La_{2/3}Ca_{1/3}MnO_3$ (#2) under applied magnetic field of 0, 1 and 12 T. The inset shows the thermal dependence of the magnetoresistance at the maximum field.

Even though the transition that takes place at T_c is sharp according to the a.c. initial susceptibility measurements (see inset Fig 10), the resistivity results show a smooth transition. In order to understand this phenomenon, we should consider the role of extrinsic mechanisms due to the microstructure. During the last years a large dispersion in the resistivity data on the same compound $La_{2/3}Ca_{1/3}MnO_3$ exists. The thermal dependence as well as the order of magnitude can dramatically change from sample to sample. Mahesh et al [58] and Gupta et al. [59] found a close relationship between resistivity and microstructure. The value of the resistivity increases as the grain size decreases. In single-crystal or high-quality thin films the value reported for the resistivity in the ferromagnetic phase is as low as $10^{-4}\Omega$cm [36, 59-62], several orders of magnitude lower than the value found in samples with small grain size. The magnetoresistance in the ferromagnetic phase cannot be attributed to the electronic scattering at the domain walls because samples of high quality show a negligible magnetoresistance. These results are in good agreement with the results found in single-crystals of other compositions [44, 45]. Several techniques have confirmed the extrinsic origin for the magnetoresistance in the ferromagnetic phase in polycrystalline samples [63, 64]. They found the intragrain resistivity lower than the intergrain resisitivity. In Fig. 12 we show the resistivity isotherms for the sample #2. Despite the large difference in the value of the resistivity, the field dependence qualitatively agrees with the results found in single-crystals [61]. It is worth noticing the different field dependence of the resisitivity in ferro and paramagnetic phase (see also the inset of Fig. 12).

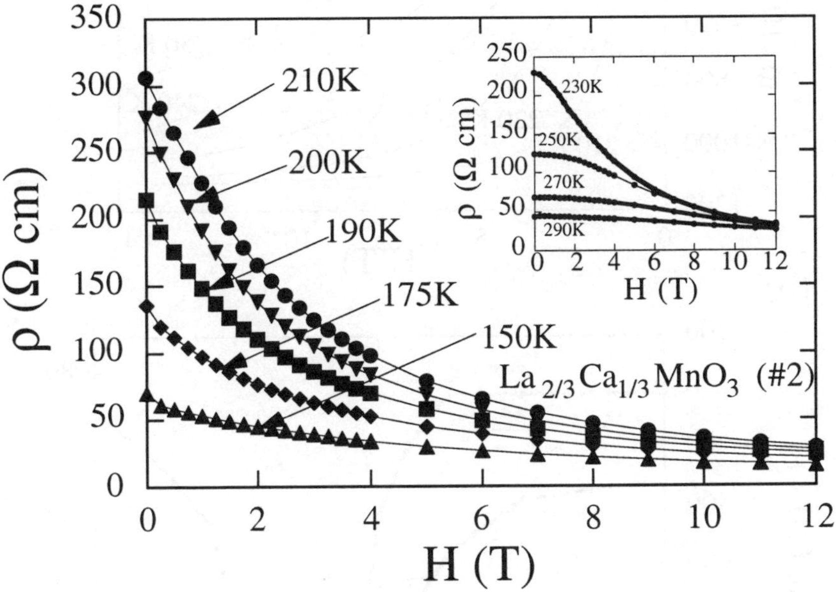

Fig. 12. Field dependence of the resistivity for the sample $La_{2/3}Ca_{1/3}MnO_3$ (#2) at temperatures above T_C. In the inset the same curves at temperatures above T_C can be seen.

The large spontaneous and field-induced volume effect suggested the relevance of magnetostriction experiments in order to investigate the ground state of this compound. The results were completely unusual and based on a new mechanism in the field of magnetoelasticity. In the ferromagnetic phase the anisotropic magnetostriction ($\lambda_t = \lambda_{||} - \lambda_{\perp}$) is small, 50x10^{-6} (see Fig. 13), this value being the typical one for 3d metals, whereas the volume magnetostriction $\omega = \lambda_{||} + 2\lambda_{\perp}$ is negligible. However, the volume magnetostriction in the paramagnetic phase was found anomalously large, about three orders of magnitude larger than in 3d metals. A new mechanism different from the usual crystal electric field and spin-orbital contribution should be considered.

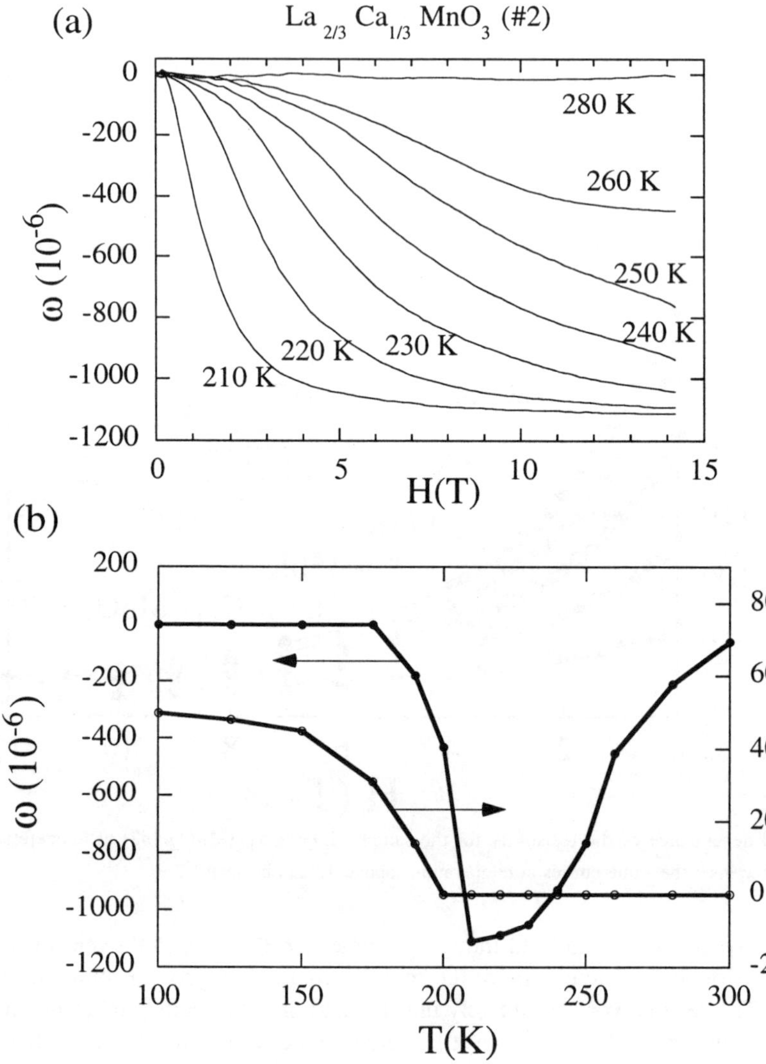

Fig. 13. **(a)** Field dependence of the volume magnetostriction at several temperatures above T_C. **(b)** Thermal dependence of the volume and anisotropic magnetostriction at the maximum field, 14 T.

The volume magnetostriction isotherms are displayed in Fig. 13 (a). At temperatures close to T_C the saturation value, $\omega \approx 1.1 \times 10^{-3}$, is reached at moderate fields. The value of ω at the maximum field is displayed in Fig. 13 (b). One can notice that at T_C ω becomes negligible and the negative value obtained above T_C is equal to the positive spontaneous anharmonic contribution due to the polaronic effect. Therefore, we associate the negative magnetostriction observed to the delocalizacion of the carriers by the applied magnetic field [28]. From these measurements a magnetic nature of the carrier is inferred.

Evidence for short-range ferromagnetic correlations exists from susceptibility measurements. Figure 14 shows the inverse of the magnetic susceptibility at high temperatures. The Curie-Weiss behaviour is followed above 1.8 T_C but below this temperature the behaviour clearly indicates the presence of short-range ferromagnetic correlations in good agreement with the neutron-scattering results.

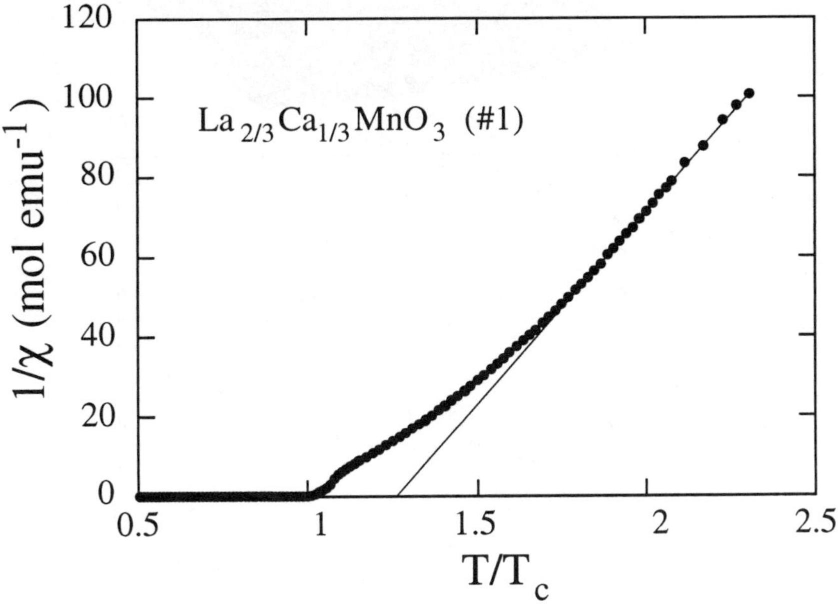

Fig. 14. Inverse of the magnetic susceptibility as a function of temperature for the compound $La_{2/3}Ca_{1/3}MnO_3$ (#1). The straight line corresponds to the Curie-Weiss behaviour.

Neutron diffraction experiments were performed at the instrument D1B at the ILL on the #1 sample. The thermal dependence of the neutron spectra is shown in Fig.

15 (a). At T_C an increase in the intensity of the Bragg peaks is observed as a consequence of the establishment of the long-range interaction. The most relevant feature is the unusual large contribution at the lowest angle. The thermal dependence of this contribution is displayed in Fig. 15 (b). Below T_C this contribution is negligible but a strong contribution appears at T_C and persists at 1.8 T_C. This is a clear indication of magnetic-cluster formation with very small correlation length.

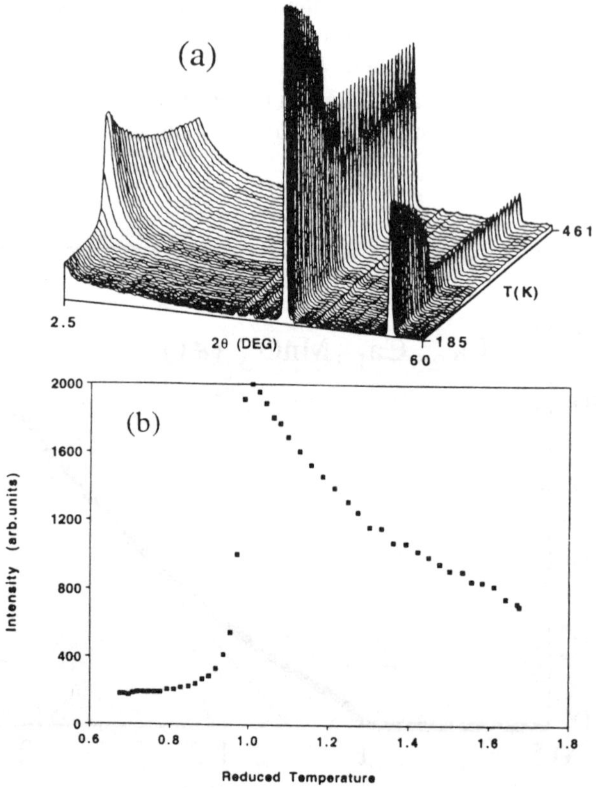

Fig. 15. **(a)** Neutron thermodiffractogram of the compound $La_{2/3}Ca_{1/3}MnO_3$ (#1) at temperatures ranging between 185 and 461 K in the angular range 2.5°≤2Θ≤60°. A strong contribution appears at low angles that disappears abruptly at T_C, when the long-range ferromagnetic order causes the appearance of magnetic contribution on some Bragg peaks. **(b)** The intensity in the angular range 2.5°-20° has been integrated and is shown as a function of the reduced temperature, T/T_C.

In order to probe into the nature of the magnetic clusters, an appropriate intrument, the SANS instrument, D16, has been used. The q range covered by this instrument (0.037 $Å^{-1}$<q<0.65 $Å^{-1}$) allows observing clusters with size smaller than 100 Å. The experiments were performed from room temperature down to 4 K and using high steady fields up to 5 T. In Fig. 16 the thermal dependence at fixed q=0.13$Å^{-1}$ is displayed. This value is very close to the value of the minimun q reached at the D1B instrument (see Fig. 14) and the obtained results are in good agreement. The SANS intensity is scaled with the extra volumen contribution observed in thermal expansion. From these results one can associate the lattice effect due to the carrier localization with the short-range ferromagnetic correlations, i.e. the lattice polarons have also magnetic character.

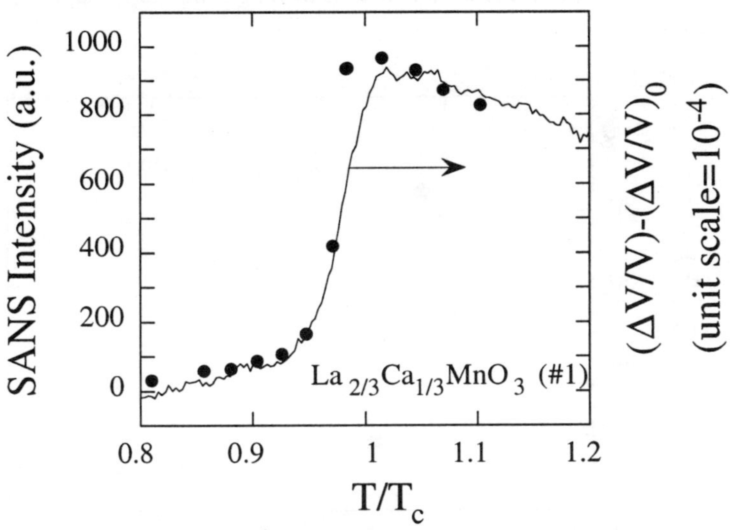

Fig. 16. SANS intensity at fixed q=0.13 $Å^{-1}$ in the neighbourhood of T_C (black dots) for the compound $La_{2/3}Ca_{1/3}MnO_3$ (#1). The continuous line stands for the extra anharmonic contribution to the volume thermal expansion due to the formation of lattice polarons.

To determine the magnetic correlation length from the analysis of the SANS intensity we should consider additional contributions. The most important is the incoherent nuclear scattering. That contribution is field and temperature independent and should be subtracted before data fitting. In the q-range measured this contribution is almost negligible but our data were corrected from it. Another source of uncertainty in SANS experiments is the contribution from grain scattering. This scattering is also field and temperature independent. Our measurements are clearly free from this contribution. First, because of the temperature and field effect observed and also because the grain size of these compounds is in the range 10^3-10^5 Å [59]. This size particle would give contribution at q values in the range 10^{-3}-10^{-5} Å^{-1}, far from the q range explored in the reported experiments. Once the data were corrected from the nuclear incoherent scattering, the magnetic scattering contribution has been analysed assuming magnetic correlations of the Ornstein-Zernike type $\langle M(0)\cdot M(r)\rangle \approx (1/r)\exp(-\kappa r)$, where $\kappa = 1/\xi$, is the inverse of the magnetic correlation length ξ [65]. This assumption gives rise to a lorentzian-type q dependence of the SANS intensity $I = I_o/(q^2+\kappa^2)$. We have identified the magnetic correlation length with the average size of the ferromagnetic clusters. In Fig. 17 the obtained thermal dependence of ξ is shown. The cluster size varies between 10-20 Å and diverges as we approach T_c due to the establishment of long-range ferromagnetic order at this temperature.

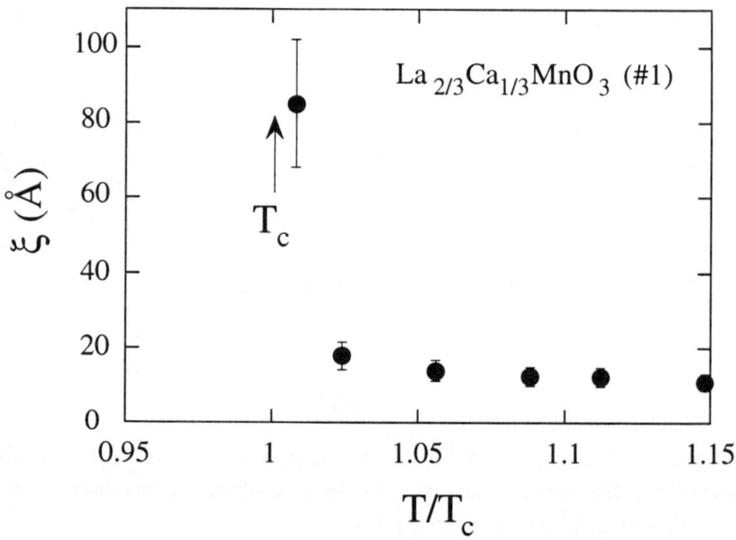

Fig. 17. Thermal dependence of the magnetic correlation length, ξ, obtained after the fit of the SANS intensity to a Lorentzian-type q dependence in the compound $La_{2/3}Ca_{1/3}MnO_3$ (#1).

The effect of the magnetic field on the extra contribution to the thermal expansion and on the intensity of the SANS intensity can be compared in Fig. 18. The observed effect is very similar in both properties. This supports the previous hypothesis that the formation of lattice polarons is linked to the existence of ferromagnetic clusters in this particular carrier concentration range.

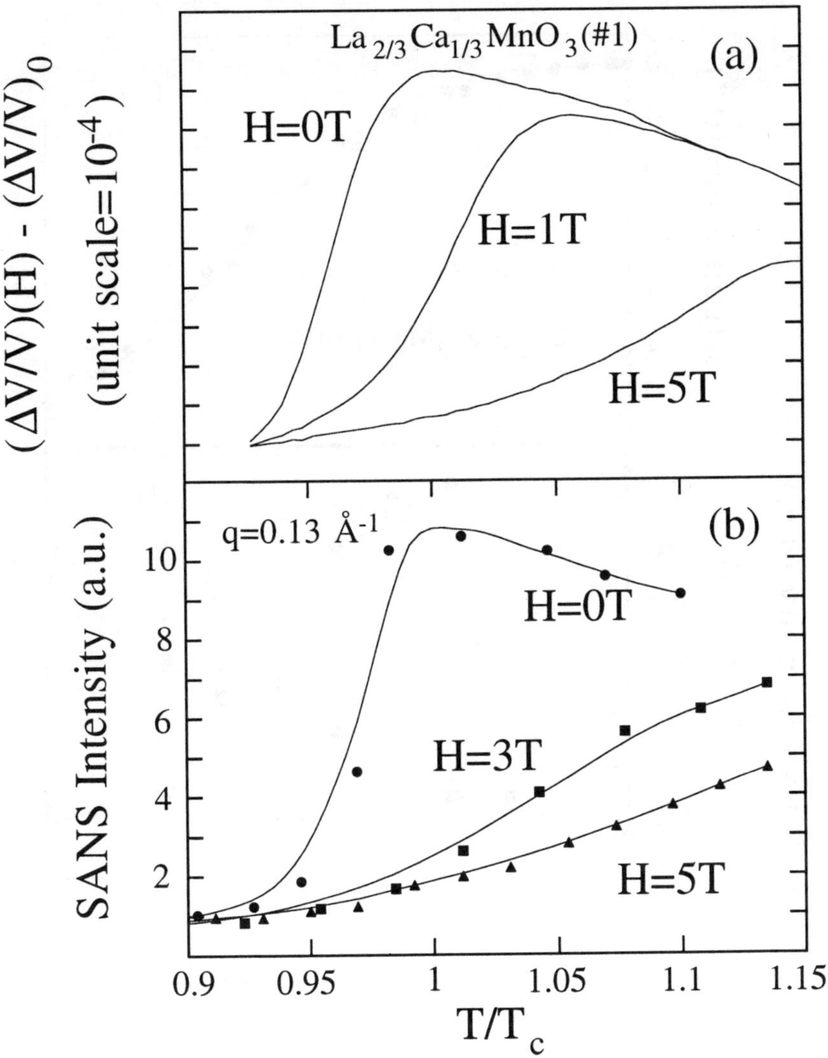

Fig. 18. Thermal dependence in the vicinity of T_C of (**a**) extra contribution to the volume thermal expansion due to the carrier localization and (**b**) SANS intensity at q=0.13 $Å^{-1}$ under several applied magnetic fields in the compound $La_{2/3}Ca_{1/3}MnO_3$ (#1).

The definitive test for the confirmation of this assumption was made by studying the isotherms of the SANS intensity and ξ up to 5 T. In figure 19 the results obtained at fixed temperature, 1.1 T_C, are shown.

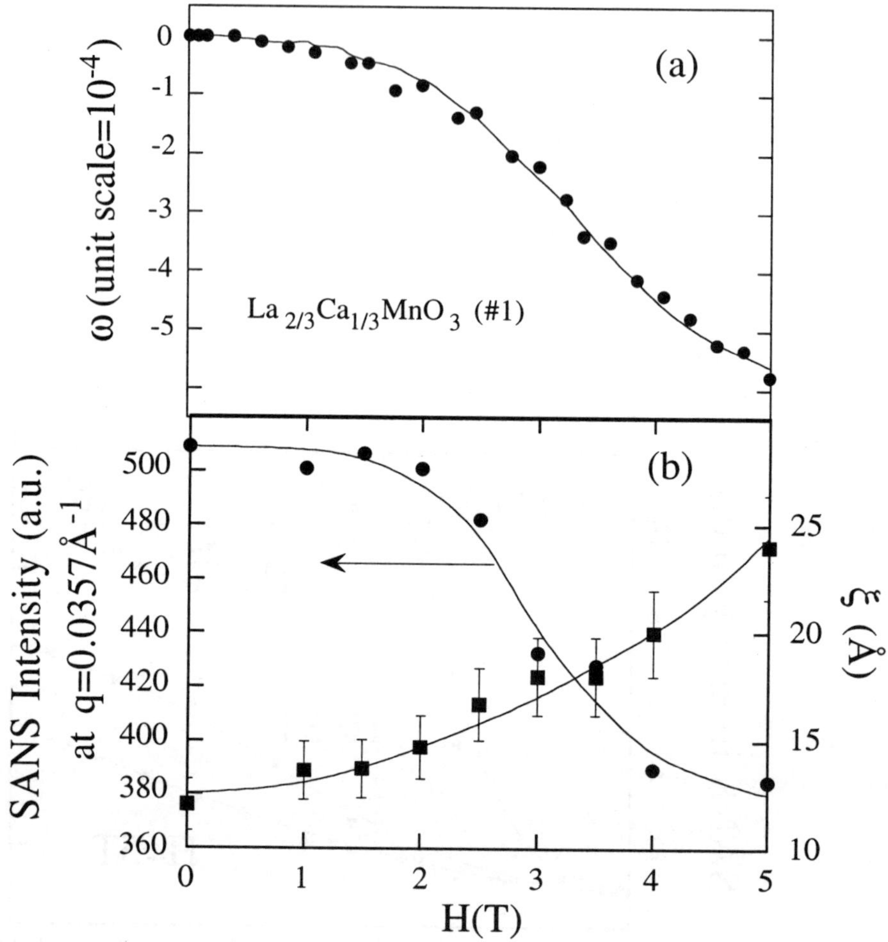

Fig. 19. Dependence with the magnetic field at T=1.1 T_C of (**a**) volume magnetostriction (**b**) SANS intensity at q=0.13 $Å^{-1}$ and the magnetic correlation length in the compound $La_{2/3}Ca_{1/3}MnO_3$ (#1).

Below 1.5 T both magnitudes as well as ω practically remain unchanged. Above this value of the field, the SANS intensity is drastically reduced as happens with ω and simultaneously the cluster size increases up to a value of 25 Å at 5 T. The drop of the SANS intensity is attributed to the reduction of the number of scattering centers, i.e. ferromagnetic clusters, whose number decreases as the size increases.

As we observe in Fig. 18 (a) and 19 (a), the volume distortion associated with the lattice polarons vanishes as the magnetic cluster grows. This can be explained as due to the delocalization of the carriers within a big cluster. If the polarons would not have magnetic character, the field should not have effect and the volume magnetostriction should be zero. This is the behaviour found in the compound $La_{0.35}Ca_{0.65}MnO_3$ [66] (see §3.4), where the ferromagnetic correlations are weak and the lattice polarons do not have magnetic character.

From the reported results we can conclude that in $La_{2/3}Ca_{1/3}MnO_3$ the lattice distortion produced by the charge localization is associated with a ferromagnetic cluster. The intrinsic magnetoresistance observed in this compound is due to the existence of magnetic polarons. The spontaneous insulator-metal transition at T_C is explained in this context considering the disappearance of the magnetic polarons when the long range is established. The effect of the magnetic field can be explained considering that the narrow polaronic band, formed because of the large electron-phonon interaction, can be widened by the increase in the electronic charge tranfer. This occurs due to the nature of the double-exchange interaction, which favours the transfer of the e_g electron between Mn^{+4} ions as the local moments are aligned by the field.

The large magnetovolume effect reported above and the close relationship of this effect to the magnetotransport properties suggested the study of volume, resistivity and susceptibility under pressure. The first hypothesis to explain the colossal magnetoresistance was given by Jin et al. [67]. These authors assumed that the huge magnetoresistance found in a La-Y (10%)-doped sample was due to the decrease of the unit-cell volume. The T_C value in this doped sample is significantly reduced ($T_C \approx 150$ K) in regard to the undoped sample ($T_C \approx 270$ K). If such a hypothesis was true, the external pressure should decrease T_C and, consequently, the intensity of the double-exchange interaction. Pressure experiments in the same Y-doped sample [68] showed an increase in T_C ($dT_C/dP = 2.2$ K/kbar) with increasing pressure. Similar results were found in the undoped sample [56, 69]. In Fig. 20 we show the thermal dependence of resistivity, susceptibility and thermal expansion at several values of hydrostatic pressure obtained on the #2 sample. The pressure enhances the double-exhange interaction and, consequently, the carriers mobility. This fact explains the decreasing value of resistivity and the increase in T_C with pressure. The thermal expansion anomaly at T_C is also reduced as pressure increases, which is in agreement with an increase of the carrier mobility.

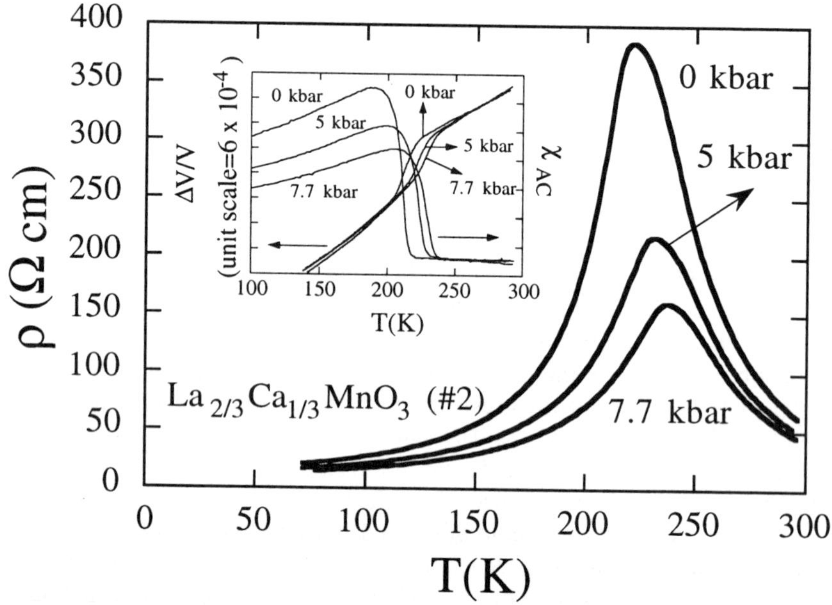

Fig. 20. Thermal dependence of the resistivity in the compound $La_{2/3}Ca_{1/3}MnO_3$ (#2) at pressures of 0, 5, and 7.7 kbar. The inset shows the a.c. susceptibility and volume thermal expansion measurements.

3.3 Region close to x=0.5

The compound $La_{0.5}Ca_{0.5}MnO_3$ shows a peculiar behaviour originated by the competition between the double-exchange, charge-ordering and antiferromagnetic superexchange interactions. A para-ferromagnetic transition takes place at 225 K [21, 70]. However, this transition in not accompanied by a metal-insulator transition or volume anomaly. Around 150 K the magnetization decreases sharply, which is associated with the establishment of the antiferromagnetic phase. This transition brings about an anisotropic structural distortion, which has recently been related to a cooperative Jahn-Teller distortion [71]. Chen and Cheong found in the ferromagnetic phase the existence of incommensurate charge ordering that becomes conmensurate at the Neel temperature. We have measured the compound x=0.47. In Fig. 21 the susceptibility results are shown. Below T_C=260 K the compound is ferromagnetic and at 160 K a decrease in the susceptibility is observed. A pronounced thermal hysteresis is observed in the thermal range 100-200 K. In Fig. 22 the spontaneous and under-field thermal dependence of the resistivity is shown. At T_C there is not any change in the behaviour of the resistivity and a smooth maximum occurs at 160 K, well below T_C. A perceptible hysteresis is also observed as in the resistivity measurements.

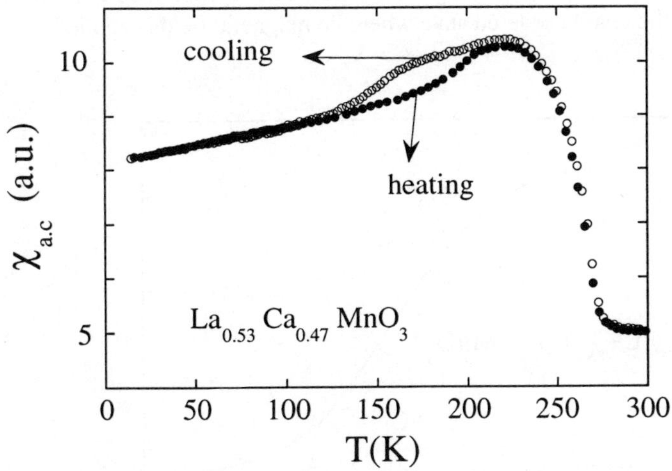

Fig. 21. a.c. susceptibility for the compound $La_{0.53}Ca_{0.47}MnO_3$ obtained cooling and heating the sample. In the thermal range 100-200 K a large hysteresis is observed.

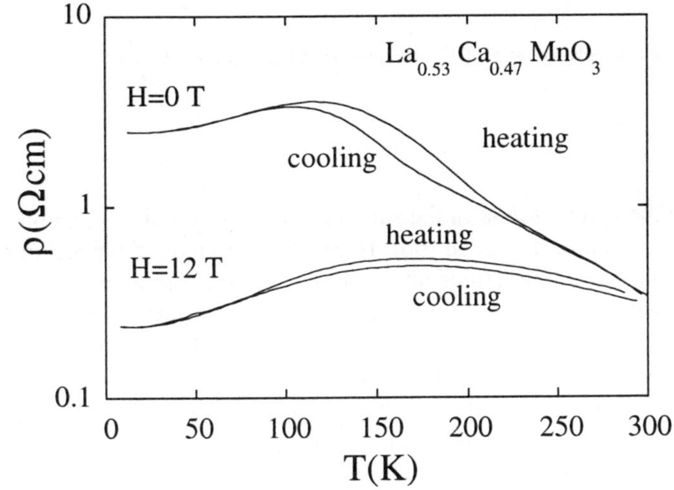

Fig. 22. Resistivity as a function of temperature at 0 and 12 T, cooling and heating, for the compound $La_{0.53}Ca_{0.47}MnO_3$.

These results can be explained as were explained in the x=0.5 compound. We consider that the low-temperature ground state is inhomogeneous, with conducting ferromagnetic regions coexisting with regions of charge ordering. In Fig. 23 we show the thermal dependence of the magnetoresistance. The value of the magnetoresistance increases below T_C. This result can be explained considering the disappearance of the charge ordered state when the magnetic field is applied.

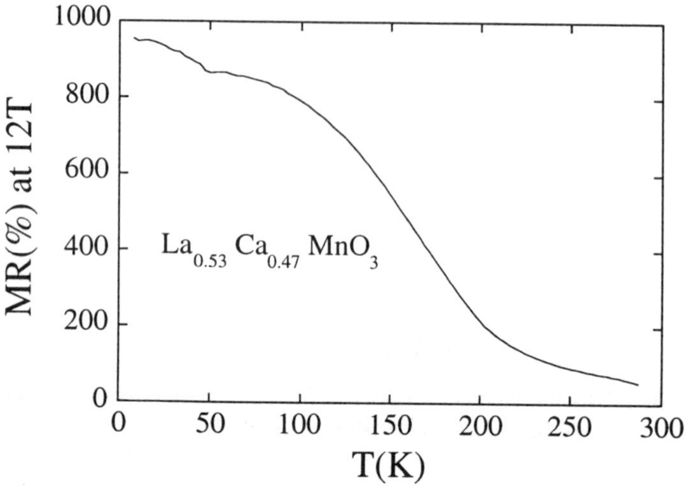

Fig. 23. Magnetoresistance, MR(%), at 12 T for the compound $La_{0.53}Ca_{0.47}MnO_3$.

3.4 x>0.5 region

We will describe the results found in the sample x=0.65 as a typical example of the behaviour in this concentration range [66]. In Fig. 24 the thermal expansion of this compound is shown. At temperatures above 400 K the thermal expansion follows the calculated phonon contribution (fine line). Below this temperature, T_p, an extra anharmonic contribution exists which is associated with the formation of lattice polarons. A distinct anomaly is observed at T_{co} related to the onset of charge ordering. This transition, as we will also describe in the case of $Pr_{2/3}Ca_{1/3}MnO_3$ [72], is associated with a transition from a non-coherent charge-localized (polaronic-type) regime to a coherent charge-localized (charge-ordered-type) state, where the e_g electrons are fixed at specific lattice sites, producing a determined Mn^{+3}/Mn^{+4}

pattern. The inset of Fig. 24 shows the hysteresis associated with the crossover from the polaronic to the charge-ordered state.

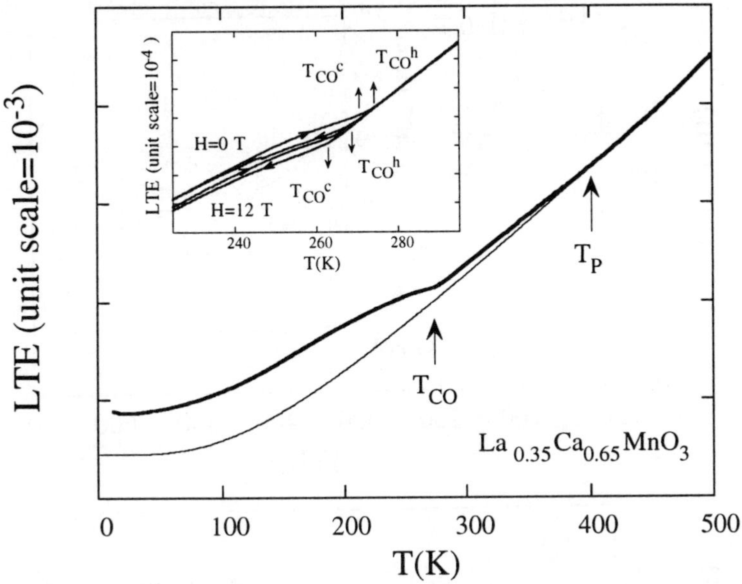

Fig. 24. Linear thermal expansion (LTE) of the compound $La_{0.35}Ca_{0.65}MnO_3$ (thick line). At T_p the experimental curve and the estimated phonon contribution (fine line) separate. A clear anomaly appears at T_{co}. In the inset the LTE curves at 0 and 12 T in the vicinity of T_{co} when cooling and heating are shown. Hysteresis in the thermal range 220-270 K is seen. The arrows indicate the charge-ordering temperatures when cooling ($T_{co}{}^c$) and heating ($T_{co}{}^h$).

Experiments of resistivity and neutron diffraction support this assumption. The resistivity measurements are displayed in Fig. 25. Above room temperature the resistivity reaches a value of around 0.01 Ωcm. At T_{co} an abrupt increase in the resistivity was found [see Fig. 25 (a)], indicating a change in the conduction mechanism at T_{co}. The high-temperature region seems to be rather anomalous. An inflexion in the resistivity is observed [see inset of Fig. 25 (a)]. It seems that a semiconductor-metal transition takes place, as was proposed by Taguchi and Shimada [73]. We consider that this is not the case but the change in the slope is due to a typical polaronic behaviour $\rho=BT\exp(E_{hop}/KT)$. This is confirmed by the fitting of the experimental data, shown in Fig 25 (b). A linear relationship appears when $\ln(\rho/T)$ is represented versus 1/T. The resistivity measured under an applied field of 12 T is also shown in Fig. 25 (a), a negligible effect on the resistivity being observed. This is an indication of the high stability of the charge-ordered state.

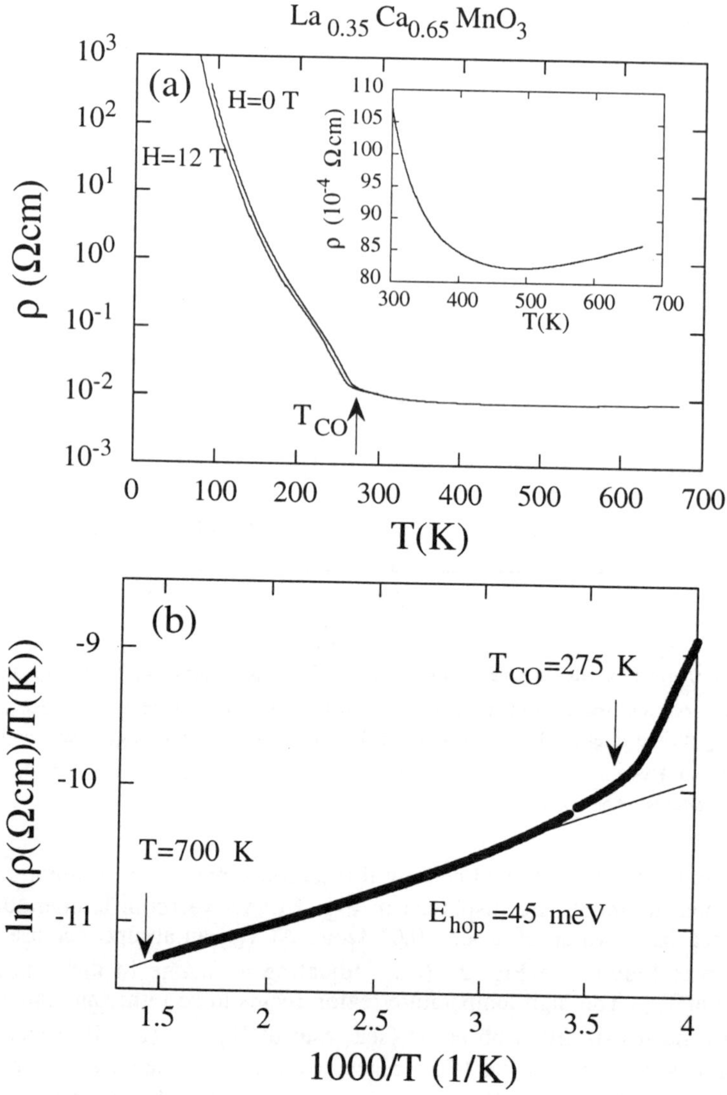

Fig. 25. **(a)** Resistivity as a function of temperature in the compound $La_{0.35}Ca_{0.65}MnO_3$ under magnetic fields of 0 and 12 T. In the inset the high-temperature behaviour is shown in detail. **(b)** Same data in the form ln(ρ/T) vs 1/T. The linear relationship that exists above room temperature corresponds in Holstein´s model to a hopping energy of 45 meV.

Neutron diffraction experiments were performed in the temperature range 1.5-320 K (see Fig. 26). Below T_N=160±3 K a set of antiferromagnetic peaks that correspond to the CE magnetic structure [9] are visible. At T_{CO} a splitting of some Bragg peaks is observed.

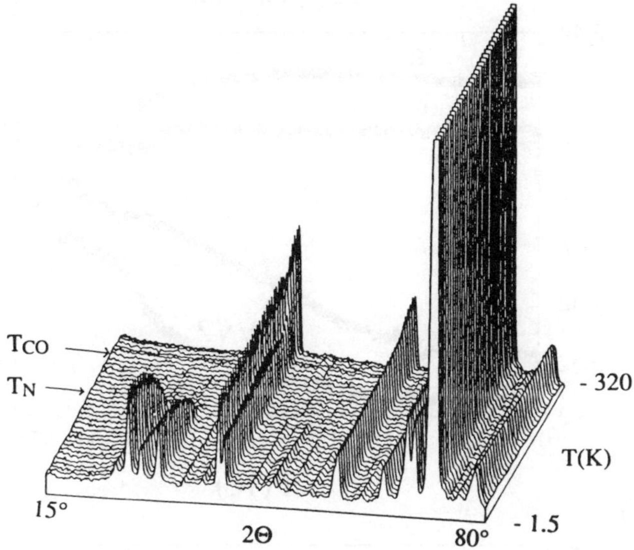

Fig. 26. Neutron thermodiffractogram of the compound $La_{0.35}Ca_{0.65}MnO_3$ between 1.5 and 320 K and angular range 25-80°. At T_{CO}=275 K a structural distorsion is reflected in a splitting of some Bragg peaks. At T_N=160 K, a set of antiferromagnetic-CE-type peaks become visible. T_{CO} and T_N do not coincide.

The spectra were refined with the Pbnm space group at all temperatures. From the lattice parameters obtained in the refinements, one can notice that a large distortion occurs at T_{CO} [see Fig. 27 (a)]. The Mn-O1 and Mn-O2 bond distances were also obtained from the refinement of the neutron diffraction data and are shown in Fig. 27 (b). O1 and O2 stand for the Oxygen atoms at the apical and basal plane positions respectively in the octahedra. Assuming that 2/3 of the octahedra (Mn^{+4}) are not distorted and 1/3 (Mn^{+3}) are Jahn-Teller distorted as in $LaMnO_3$, from the average octahedral distortion one can calculate the distortion in one Mn^{+3} ion. The obtained value in $La_{0.35}Ca_{0.65}MnO_3$, ~0.2 Å, is the same as in $LaMnO_3$ [74]. From neutron diffraction and attenuation of ultrasonic waves experiments, Ramirez et al. [75] suggested the existence of a cooperative Jahn-Teller effect in

$La_{0.35}Ca_{0.65}MnO_3$. These kind of transitions can be found in compounds with high concentration of Jahn-Teller ions [46, 76]. Our results seem to confirm this asumption.

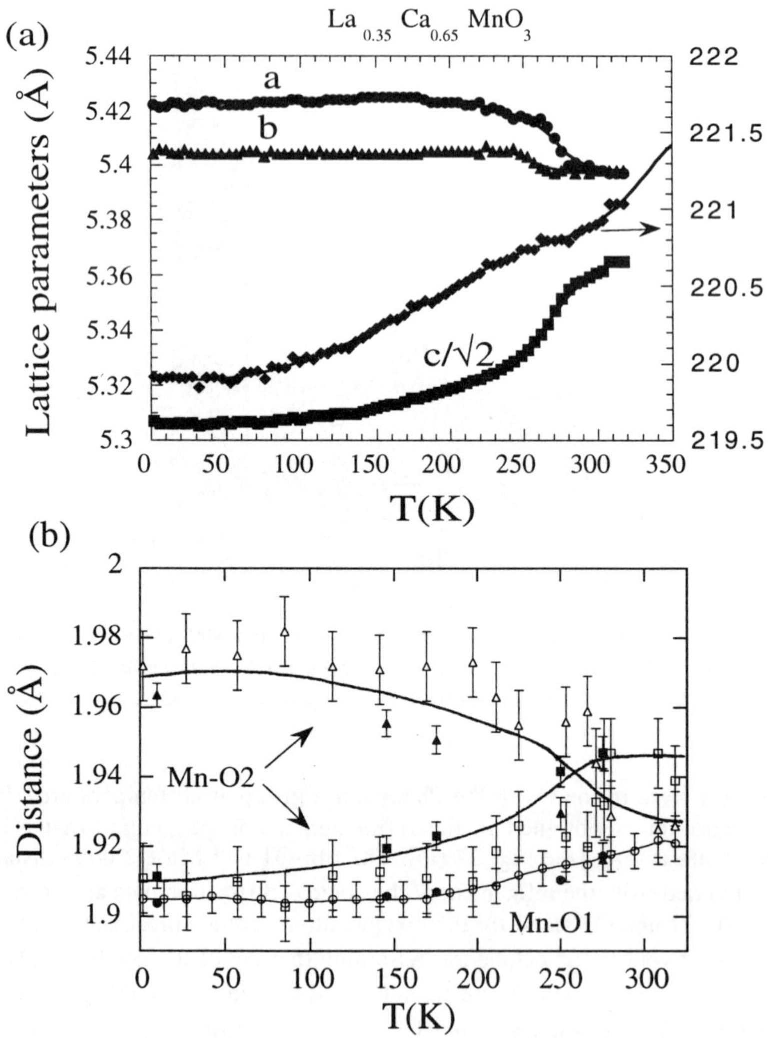

Fig. 27. **(a)** Thermal dependence of the lattice parameters and the cell volume in the compound $La_{0.35}Ca_{0.65}MnO_3$ as obtained from the fits of the neutron difracction patterns. A distinct anomaly appears at T_{co}=275 K. **(b)** Mn-O distances obtained from the same refinements.

3.5 Phase diagram of the $La_{1-x}Ca_xMnO_3$ series.

Along this section we have proved that the earlier phase diagram proposed for this series in not completely correct and should be modified. Cheong et al. [55] have carefully studied this series obtaining practically the same conclusions. The proposed phase diagram is displayed in Fig. 28.

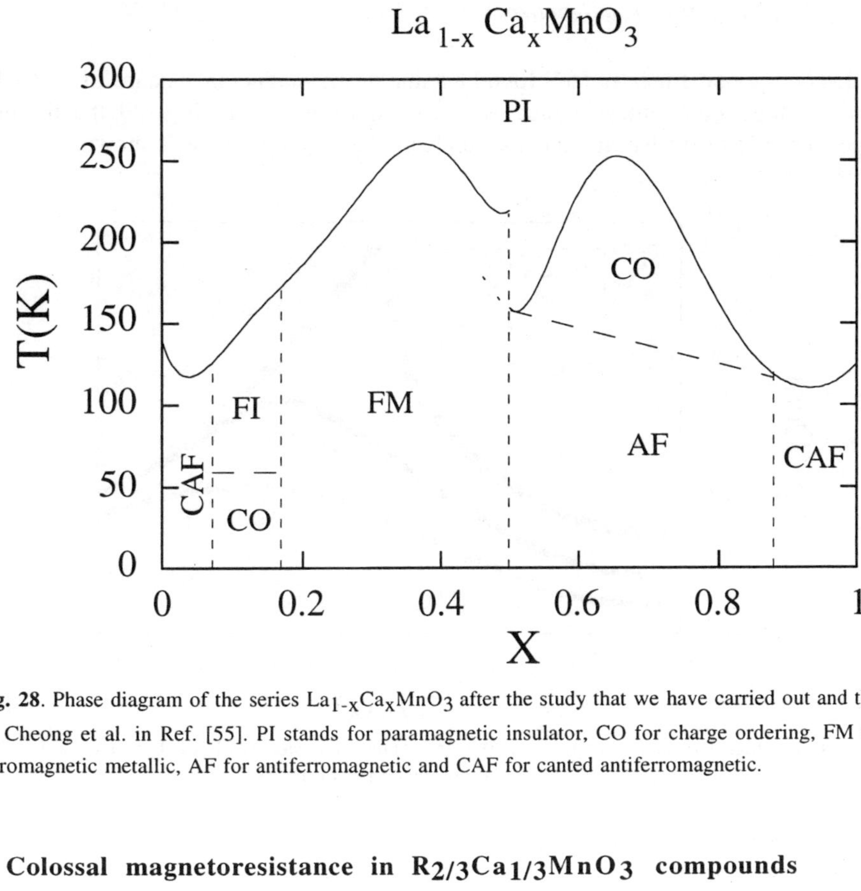

Fig. 28. Phase diagram of the series $La_{1-x}Ca_xMnO_3$ after the study that we have carried out and that by Cheong et al. in Ref. [55]. PI stands for paramagnetic insulator, CO for charge ordering, FM for ferromagnetic metallic, AF for antiferromagnetic and CAF for canted antiferromagnetic.

4 Colossal magnetoresistance in $R_{2/3}Ca_{1/3}MnO_3$ compounds

In this section we will report the results obtained for several compounds with this formula, characterized by having the same carrier concentration, $Mn^{+4}/Mn^{+3}=1/2$ (R=La-Y, Pr, La-Tb). The change of the rare-earth ion produces a distortion of the perovskite structure in the following way: the smaller the ions, the smaller the

Mn-O-Mn bond angle. This fact brings about a narrowing of the e_g electronic bandwidth and, consequently, the double-exchange interaction is weakened. We have selected the compounds with R=$La_{0.6}Y_{0.07}$ (where the colossal magnetoresistance was found), R=Pr (in which the double-exchange interaction is weak and charge-ordering effects are relevant) and the series corresponding to R=La-Tb (where a cluster-glass state appears at high Tb concentration).

4.1 R=$La_{0.6}Y_{0.07}$: ferromagnetism

In this compound Jin et al. [67] found a value of the magnetoresistance of 10^4-10^5. Such a huge magnetoresistance was coined *colossal*. In Fig. 29 the thermal dependence of the resistivity under several applied magnetic fields is shown.

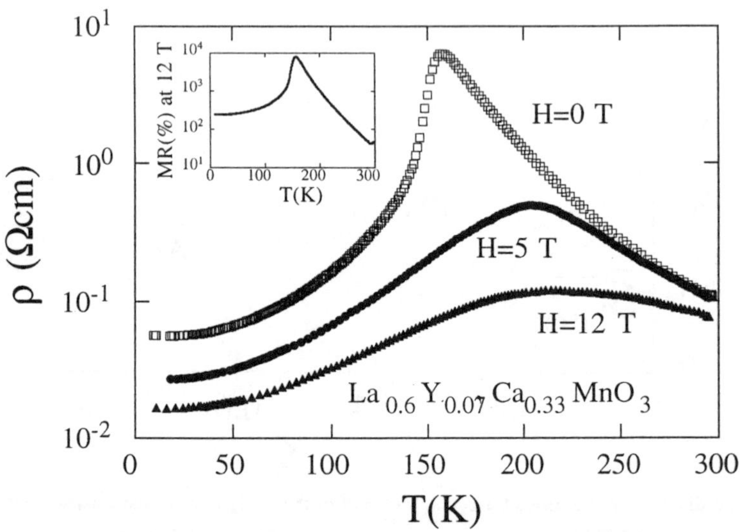

Fig. 29. Thermal dependence of the resistivity at applied magnetic fields of 0, 5, and 12 T in the compound $La_{0.6}Y_{0.07}Ca_{1/3}MnO_3$. The inset shows the MR(%) at 12 T.

At 160 K an insulator-metal transition occurs and, simultaneously, a para-ferromagnetic transition takes place. The drastic decrease of T_c in respect of the undoped compound is explained in terms of the narrowing of the e_g band as a consequence of the lattice distortion produced by the substitution of La by a smaller ion. In the inset of Fig. 29 the thermal dependence of the magnetoresistance can be seen. The large value of the residual magnetoresistance at low temperatures is due to the contribution from the tunneling through the grain boundaries. An additional contribution in this compound in the ferromagnetic phase can be attributed to the non-collinear alignment of the magnetic moments below T_C even at the lowest temperatures [12, 77]. This is in good agreement with our measurements of the field dependence of the intensity of a magnetic Bragg peak (see inset of Fig. 30). In Fig. 30 we display the thermal expansion without and under field. The calculated phonon contribution is also shown. As in the case of the undoped compound, an extra anharmonic contribution below 400 K is observed. The volume anomaly at T_C is $\approx 1.2 \times 10^{-3}$, which is about twice the value observed in the undoped compound. The anomaly is almost supressed at high fields.

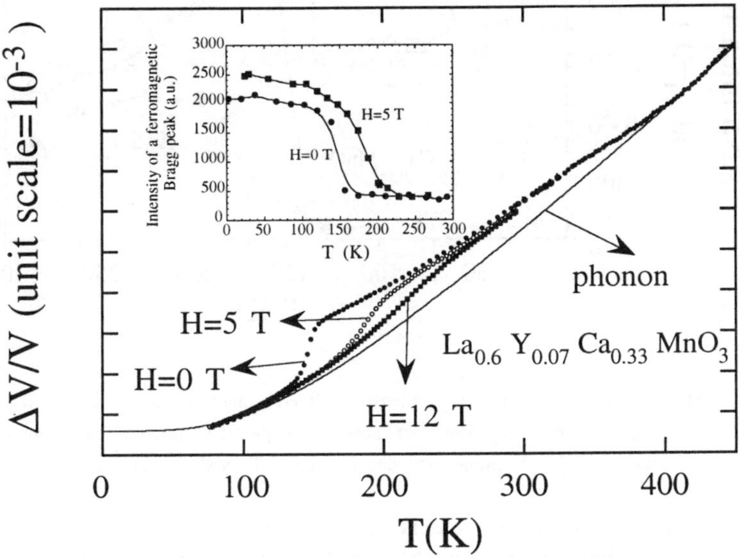

Fig. 30. Volume thermal expansion under several applied magnetic fields (0, 5 and 12 T) and calculated phonon Grüneissen-type contribution for the compound $La_{0.6}Y_{0.07}Ca_{1/3}MnO_3$. The inset shows the intensity of a ferromagnetic Bragg peak under magnetic fields of 0 and 5 T.

In order to investigate the presence of magnetic polarons we have performed SANS experiments in this compound using magnetic fields up to 5 T. In Fig. 31 the thermal dependence of the SANS intensity at fixed q=0.1 $Å^{-1}$, without and under an applied magnetic field of 5 T, is shown. The behaviour is close to that observed in the undoped compound: a large contribution to the SANS intensity in the paramagnetic phase persists up to room temperature. This contribution disappears sharply at T_C. The effect of the magnetic field below room temperature is to reduce significantly the SANS intensity, i.e. the number of magnetic polarons decreases while its size increases so that many of them can percolate giving rise to regions of long-range magnetic order. This explanation is supported by the thermal dependence of the magnetic correlation length (ξ), showed in the inset of Fig. 31. At around 210 K, ξ=10-12 Å when the field is off. However, at 5 T ξ tends to diverge because magnetic clusters larger than 30 Å start to form.

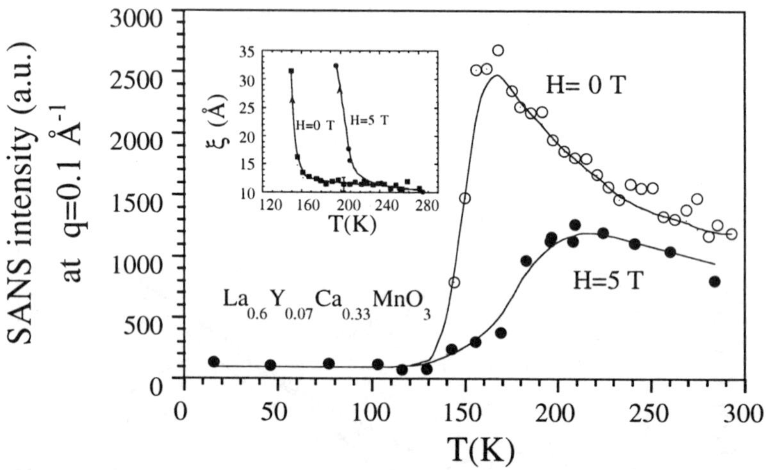

Fig. 31. Thermal dependence of the SANS intensity at fixed q under magnetic fields of 0 and 5 T in the compound $La_{0.6}Y_{0.07}Ca_{1/3}MnO_3$. The inset shows the magnetic correlation length obtained from the fits at each temperature of the lorentzian-type dependence of the SANS intensity with q.

A more clear evidence for the existence of the magnetic polarons in paramagnetic phase and how they are responsible for the intrinsic magnetoresistance in this compound can be reached from the results in Fig. 32. In this figure the isotherms of several relevant physical magnitudes in the paramagnetic phase are displayed. We have selected three temperatures at which the field dependence of volume magnetostriction, resistivity, intensity of the magnetic Bragg scattering, magnetic correlation length, and SANS intensity was measured. The large negative value and the field dependence of ω in Fig 32 (a) indicates the disappearance of the extra positive spontaneous volume contribution coming from the carrier trapping. As the carrier are released, the volume is reduced. We can observe this effect in the field dependence of the resistivity as well [see Fig. 32 (b)]. The carriers increase their mobility as the volume decreases.

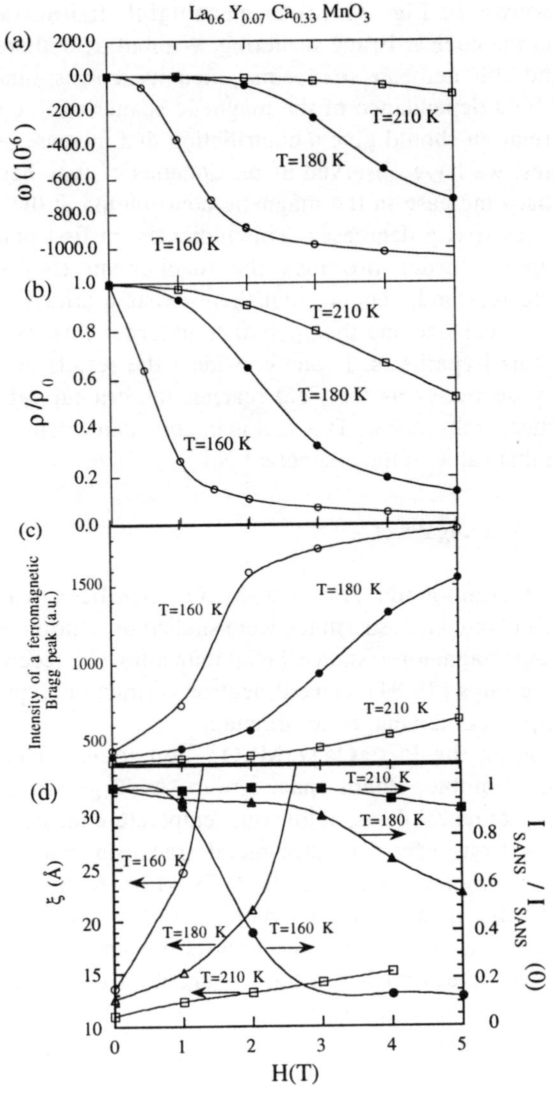

Fig. 32. Results obtained at fixed temperatures (160, 180 and 210 K) in paramagnetic phase and under magnetic fields up to 5 T of several relevant magnitudes in the compound $La_{0.6}Y_{0.07}Ca_{1/3}MnO_3$: **(a)** volume magnetostriction; **(b)** resistivity; **(c)** intensity of a ferromagnetic Bragg peak; **(d)** magnetic correlation length, ξ, and normalized SANS intensity at fixed q.

The result shown in Fig. 31 (c) is meaningful. It illustrates the magnetic contribution over the nuclear Bragg scattering. Without field, there is not long-range contribution and the neutron scattering intensity corresponds to the nuclear scattering. The field dependence of the magnetic scattering due to the presence of short-range correlation should give a contribution that increases smoothly with the field. Nevertheles, we have observed a "metamagnetic"-type behaviour. At 160 K there exists a sharp increase in the magnetic contribution at the field in which the volume and the resistivity decreases. This indicates, in first place, that the short-range ferromagnetic order provides the mechanism to form *well defined* ferromagnetic clusters and, second, that there exists a critical field at which the clusters reach a critical size and then percolate in larger clusters. The results in Fig 32 (d) confirm this hypothesis. If one considers the results at 160 K, the SANS intensity sharply decreases as the field reaches the critical value of 1 T, i.e. the ferromagnetic clusters disappear. This fact is in good agreement with the divergence of ξ observed at this value of the magnetic field.

4.2 R=Pr: charge-ordering effect

The series of compounds $Pr_{1-x}Ca_xMnO_3$ has been widely studied. The crystallographic and magnetic structures were studied by neutron diffraction [46, 78]. The magnetic and magnetoresistance behaviour attracted recently the interest of several research groups [79-81]. Recent neutron diffraction experiments [82] have clarified some aspects of the magnetic structure.

We will focus on the $Pr_{2/3}Ca_{1/3}MnO_3$ compound. The crystallographic structure is orthorhombic, Pbnm space group. At $T_{CO}\approx 210$ K a transition to a charge-ordering state takes place. Below this temperature the compound is insulator down to 4 K. A first-order insulator-metal transition can be induced by the application of a magnetic field. Below $T_N\approx 150$ K this compound orders antiferromagnetically in a CE-type structure. Below ≈ 100 K a canting of the magnetic moment takes place giving rise to a ferromagnetic component [83]. Much effort was devoted to establish the tight connection between the magnetic and transport properties in this compound. We have studied the lattice effects trying to get an insight into the underlying spin-charge-lattice coupling.

In Fig. 33 (a) the thermal expansion results of $Pr_{2/3}Ca_{1/3}MnO_3$ under applied magnetic fields of 0, 6 and 12 T are shown. A comparison of the experimental results and the estimated lattice contribution (using a Θ_D=500K) gives three different temperature ranges. Below $\approx$400 K there exists an extra contribution, as observed in the La-Ca compound, which we associated with the charge localization of small polarons. At $T_{CO}\approx 210$ K we observe a smooth anomaly that coincides with the temperature at which the charge ordering was observed in neutron-diffraction experiments [83]. We propose that at this temperature a transition from an incoherent charge-localization regime above T_{CO} (small-polaron-type) to a coherent charge-localization regime in which a charge lattice is formed [72]. We would like to underline the lack of volume effect (isotropic distortion) when this transition occurs even though the lattice is strongly distorted (anisotropic change in the lattice parameters). Under an applied magnetic field of 6 T, at ≈80 K a change in volume is observed (≈0.14%). At 12 T the field-induced transition is shifted up to 210 K. It seems that at this value of the magnetic field, the charge ordering is completely suppressed in the whole temperature range. These results can be understood considering that the volume distortion associated with the charge localization disappears as the charge is released. When the whole charge is released, the thermal expansion recovers the Grüneisen thermal dependence due to the phonon contribution, which is the situation of the metallic state in $La_{2/3}Ca_{1/3}MnO_3$. In order to make more clear this effect, the volume effect associated to the extra anharmonicity is displayed in the inset of Fig. 33.

In this compound, as happens in the previously reported $La_{2/3}Ca_{1/3}MnO_3$ compound, there exists a close relation of the volume with the resistivity. As can be observed in Fig. 33 (b), the spontaneous behaviour is characterised by a semiconductor-like dependence of the resistivity with a smooth anomaly at 210 K. Under a field of 6 T, an insulator-metal transition takes place at ≈80 K and at 12 T the charge-ordered state is not reached in good agreement with the thermal-expansion measurements. In the inset of Fig. 33 (b) the extremely huge values of magnetoresistance are shown.

The magnetic properties of this compound were investigated using neutron-diffraction techniques. The D16 diffractometer at ILL was used. We placed the multidetector at two fixed angular positions, at q=0.58 $Å^{-1}$, where the first antiferromagnetic peak of the CE strucutre appears and at q=2.3 $Å^{-1}$, which coincides with the position of a Bragg peak. The results found at zero and 5 T are shown in Fig 34. In absence of applied field the para-antiferromagnetic transition occurs at T_N=160 K, where the intensity of the antiferromagnetic peak starts to increase [see Fig. 34 (a)]. At T=50 K the intensity of this peak decreases as a consequence of the antiferro-canted structure. This brings about the appearance of a ferromagnetic contribution on the Bragg peak as observed in Fig. 34 (b). Under an applied magnetic field of 5 T, at around 80 K the intensity of the antiferromagnetic peak starts to decrease to disappear at the lowest temperature. Simultaneously, the ferromagnetic contribution increases below 80 K. This contribution saturates at the lowest temperature, when the antiferromagnetic phase is completely suppresed. The antiferro-ferromagnetic transition is in good agreement with the resistivity and thermal expansion experiments.

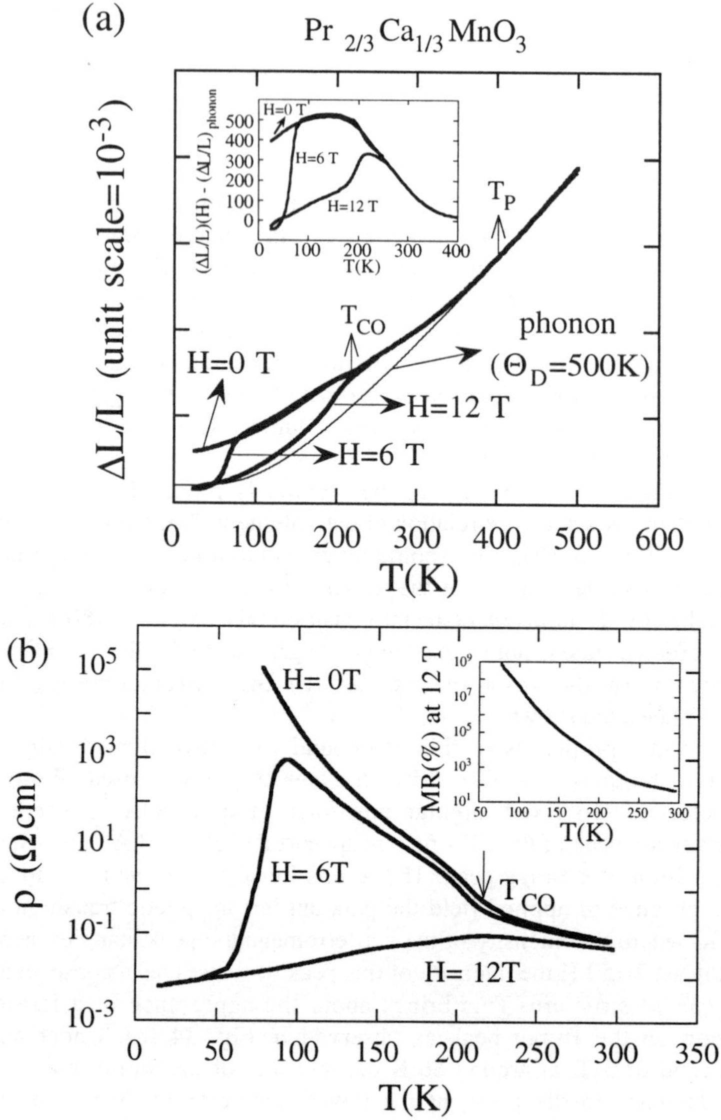

Fig. 33. **(a)** Linear thermal expansion under applied magnetic fields of 0, 5, and 12 T (thick lines) and the estimated phonon contribution (fine line) in the compound $Pr_{2/3}Ca_{1/3}MnO_3$. The arrows mark the temperatures at which the experimental and theoretical curves separate (T_p) and the charge-ordering temperature (T_{co}). The inset shows the extra volume at different magnetic fields. **(b)** Resistivity as a function of temperature under the same magnetic fields. The inset shows the MR(%) at 12 T.

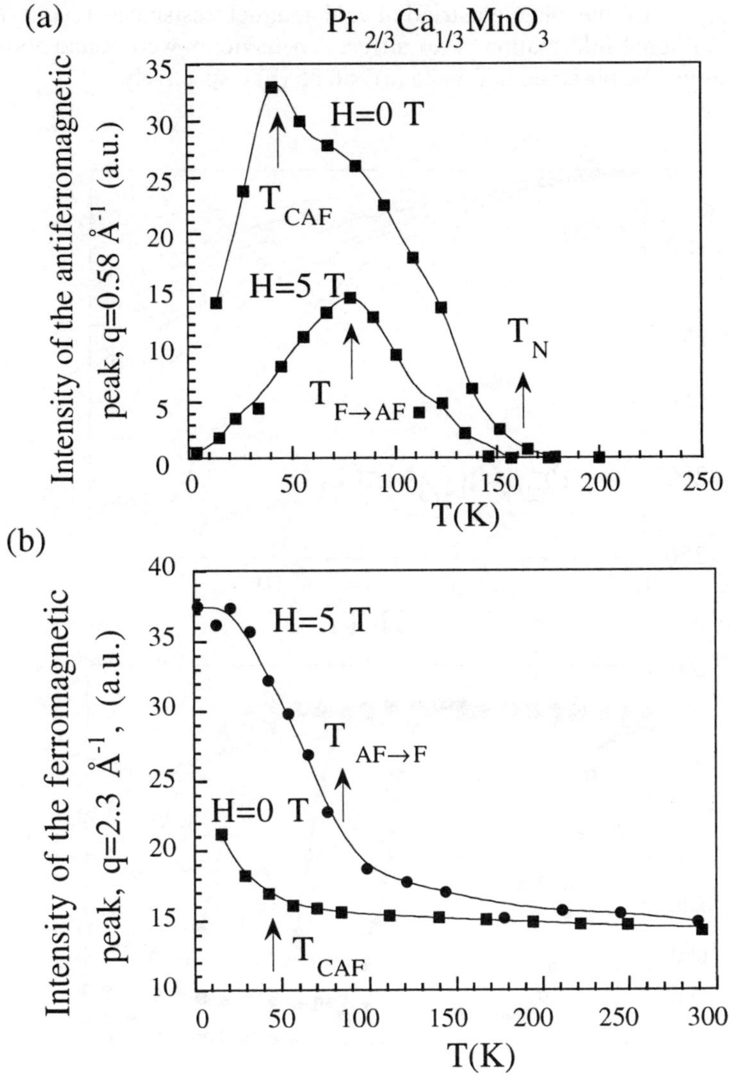

Fig. 34. Thermal dependence of the integrated intensity of two magnetic diffraction peaks in the compound $Pr_{2/3}Ca_{1/3}MnO_3$ as measured in the instrument D16.

The analysis of the magnetostriction and magnetoresistance results allows obtaining additional information. Two different behaviours were found above and below T_{CO} as can be observed in Fig. 35 (a) and 35 (b) respectively.

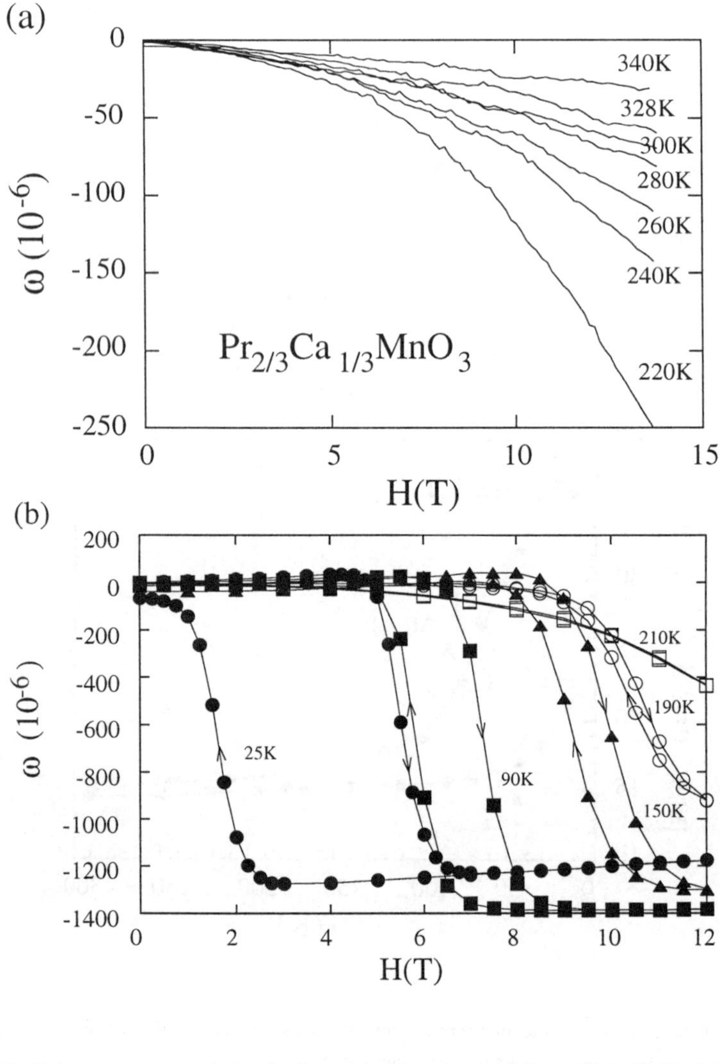

Fig. 35. Volume magnetostriction isotherms in the compound $Pr_{2/3}Ca_{1/3}MnO_3$ **(a)** above T_{CO} **(b)** below T_{CO}.

The volume magnetostriction above T_{CO} shows a continuous increase of volume with field and no hysteresis is observed. This behaviour is similar to the previously reported for the $La_{2/3}Ca_{1/3}MnO_3$ compound in the paramagnetic phase. Thus, we associate this magnetostriction with the progressive delocalization of the carriers. The small value of this magnetostriction and the lack of saturation may be related to a weak magnetic character of the polaron. Fields higher than 14 T would be neccesary in order to release the charge completely. The behaviour below T_{CO} is markedly different. Below 150 K the volume magnetostriction isotherms show a clear saturation with a very large hysteresis. This constitutes an indication of the complete disappearance of the charge-ordered state with the field and also the first-order nature of the insulator-metal transition. The change in volume at the field-induced transition has the same value as the spontaneous extra volume contribution and opposite sign. A comparison of Fig. 35 (a) and (b) with the magnetoresistance isotherm of Fig. 36 and inset, confirms the close relation between the magnetoelastic and magnetotransport properties in this compound.

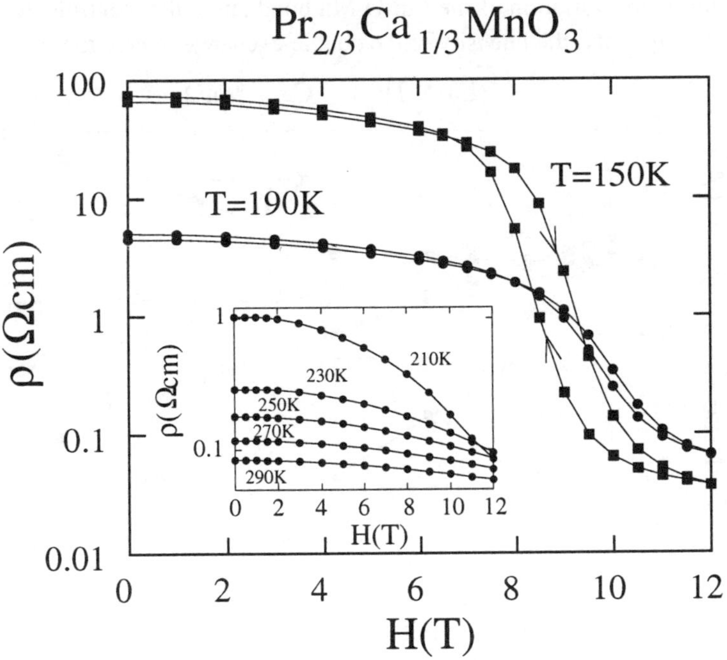

Fig. 36. Change in the resistivity with the magnetic field at temperatures below and above (inset) T_{CO}.

All the reported results on this compound reveal the different nature of all the interactions that play a relevant role in mixed-valence manganites. Above T_{CO} the Weiss temperature is positive [81] and the presence of lattice polarons with a weak magnetic character support the presence of short-range ferromagnetic correlations of double-exchange origin. The existence of small-polarons itself is an indication of the strong electron-phonon interaction. Below T_{CO} the progressive localization of the carriers favours the ordering of the charges probably by long-range coulombian interaction. Below T_N the compound orders antiferromagnetically due to the superexchange interaction. It is not surprising a large field effect given the different ground states available, almost degenerated in energy.

4.3 R=(La-Tb): appearance of spin-glass behaviour

Several samples of the series $(La_{1-x}Tb_x)_{2/3}Ca_{1/3}MnO_3$ have been studied. The Tb^{+3} ion is smaller than the La^{+3} ion. This allows changing the tolerance factor between 0.885 for x=1 and 0.915 for x=0. The main effect of this change on the structure is the variation of the Mn-O-Mn bond angle that controls the e_g bandwith and, consequently, the intensity of the double-exchange interaction.

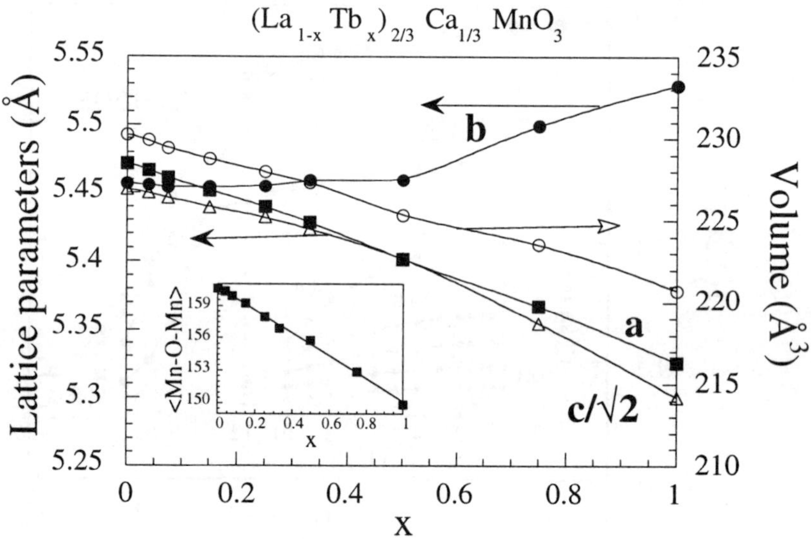

Fig. 37. Evolution with the Tb concentration of the lattice parameters and volume at room temperature in the series $(La_{1-x}Tb_x)_{2/3}Ca_{1/3}MnO_3$. The inset shows the dependence of the average angle Mn-O-Mn, obtained from the X-ray diffraction data, with the Tb content.

The X-ray analysis of the samples studied were refined in the Pbnm space group. This is the common orthorhombic structure ($GdFeO_3$ type) in rare-earth perovskites. Details about the structural parameters can be found in reference [84]. The results are summarized in Fig. 37, where the lattice parameters and volume at room temperature as a function of the concentration is shown. The decreasing values of the Mn-O-Mn bond angle with x (as volume decreases) can be observed in the inset of Fig. 37.

The magnetic behaviour of these compounds was studied by initial magnetic susceptibility. The results are shown in Fig. 38.

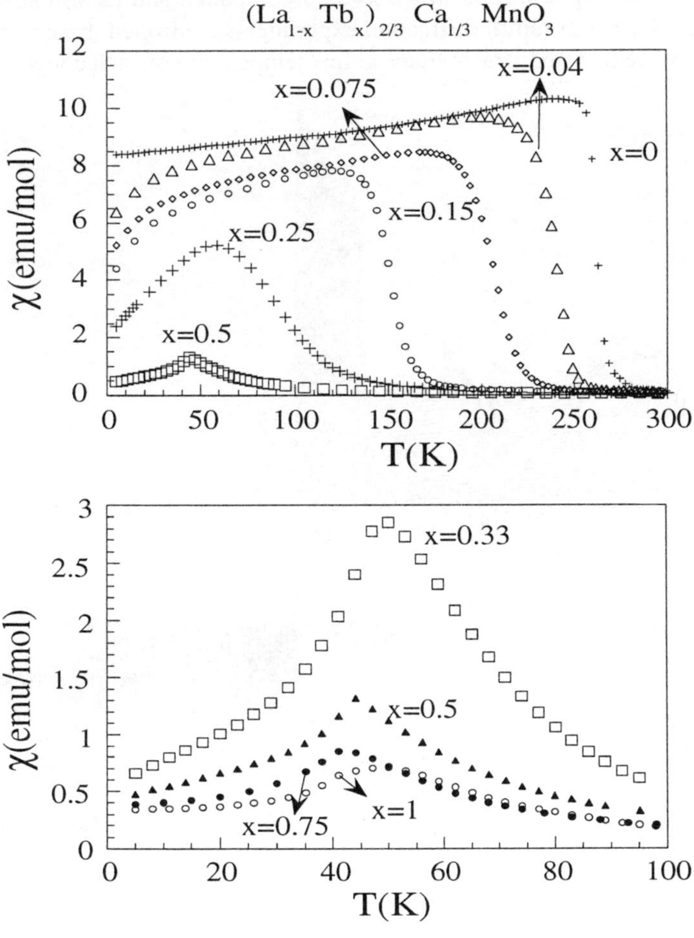

Fig. 38. a.c. susceptibility for all the studied compounds of the series $(La_{1-x}Tb_x)_{2/3}Ca_{1/3}MnO_3$.

In the low-doping range ($x \leq 0.15$) the susceptibility increases sharply, indicating a para-ferromagnetic transition. At the other end of the series ($x \geq 0.5$), the magnetic susceptibility shows a low value, showing a cusp-like anomaly at around 50 K. This anomaly, as will be discussed, is related to the freezing temperature of a cluster-glass state at low temperatures. The behaviour of the intermediate concentrations is quite difficult to understand only with the a.c. susceptibility results. We have selected the concentrations x=0.25 and x=0.33 to be carefully described.

From the Arrot´s plot of the magnetization isotherms we observed a small value for the spontaneous magnetization in the x=0.25 compound and its absence in the case of x=0.33 [84, 85]. Neutron diffraction experiments confirmed these results. In Fig. 39 we show the thermodiffractograms at low temperature of such compounds.

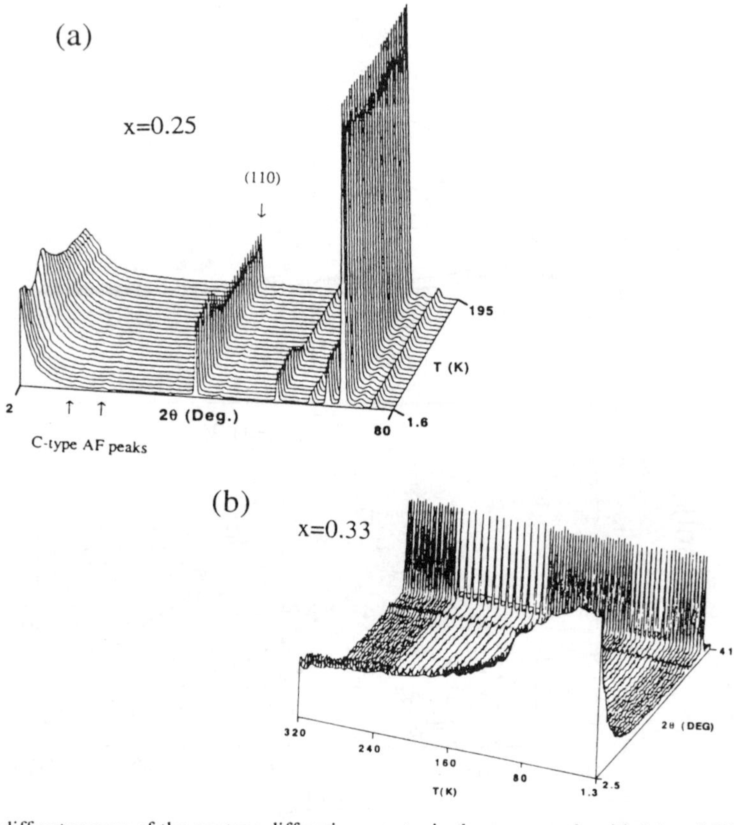

Fig. 39. Thermodiffractograms of the neutron diffraction spectra in the compounds with **(a)** x=0.25 and **(b)** x=0.33. While for the x=0.25 compound there appear long-range magnetic peaks, in the x=0.33 compound there exists only magnetic neutron scattering at small angle.

In Fig. 39 (a) one can notice that below T_C=70 K a long-range ferromagnetic contribution appears on the (110) and (020/112) Bragg peaks for the x=0.25 compound. Simultaneously, at T_C small antiferromagnetic peaks appear. These are very weak so that it was not possible to refine the structure. They appear at angular positions that correspond to the C-type structure, the same as was also observed in $Pr_{0.7}Ca_{0.25}Sr_{0.05}MnO_3$ [86]. From the refinement of the neutron spectra at 1.5 K, a ferromagnetic moment for the Mn ions of 2.6 μ_B is obtained. We made an estimate of the total magnetic moment considering the small antiferromagnetic contribution, giving the result of 2.9 μ_B. This value is below that expected for the whole polarization of the Mn magnetic moments, 3.7 μ_B. We will discuss further on this and also on the peculiar SANS region. In Fig. 39 (b) the neutron thermodiffractogram of the x=0.33 compound is shown. There is not any long-range magnetic contribution. However, at the low-angle scattering, below T≈95 K one can observe a magnetic contribution, which is not expected in a paramagnet. Both characteristics suggested the existence in this system of a cluster-glass state at low temperatures [87]. This state consists of ferromagnetic clusters which are randomly oriented. Other magnetic measurements were performed in order to confirm the existence of this cluster-glass state. In Fig. 40 we show the thermal dependence of the a.c. susceptibility at different values of the excitation frequency.

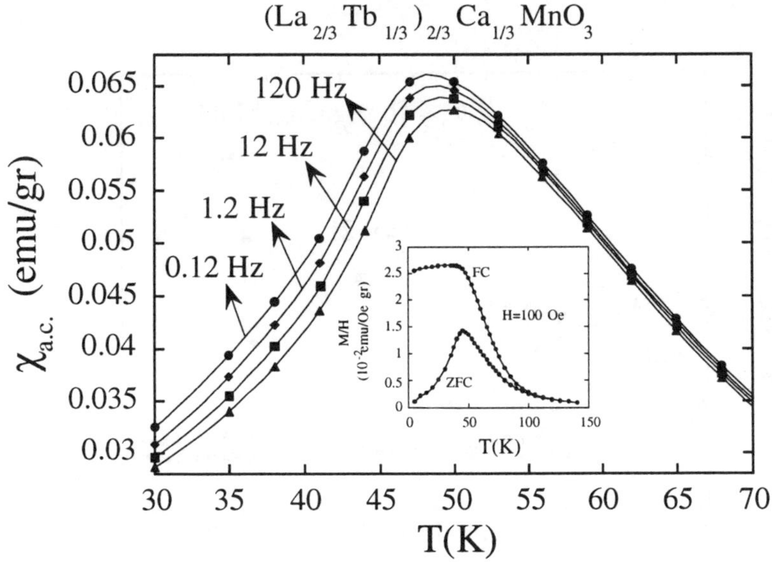

Fig. 40. a.c. susceptibility at different frequencies in the compound $(La_{2/3}Tb_{1/3})_{2/3}Ca_{1/3}MnO_3$. A shift in the maximum occurs with the change in the frequency. The inset shows the field-cooling (FC) and zero-field-cooling (ZFC) magnetization under an applied magnetic field of 100 Oe.

The shift of the maximum as well as the hysteresis observed in the field-cooling and zero-field-cooling magnetization curves (see inset of Fig. 40) suggest a random system of ferromagnetic clusters with a freezing temperature of 45 K.

In Fig. 39 one can see an anomalous behaviour in the SANS region. Some information can be obtained from the study of this region. In Fig. 41 the thermal dependence of the scattering intensity at the lowest diffraction angle is shown. The data were scaled to the background signal, which basically depends on the counting time and the sample mass. The SANS contribution for the undoped sample in the paramagnetic phase was explained in §3, being originated by the formation of magnetic polarons. This contribution disappears at T_C. The behaviour observed for x=0.25 is very similar to that of the undoped compound and should be interpreted in the same manner. However, this compound shows a new increase of the SANS below 50 K. This result can be explained considering the existence of magnetic phase segregation in the sample at low temperature, i.e. the para-ferromagnetic transition does not extend over the whole sample. The long-range ordered part of the sample contributes to the Bragg scattering, whereas the short-range region gives contribution to the SANS. The compounds with x=0.33, 0.5 and 0.75 show a similar behaviour, characteristic of a spin-glass system. In the case of x=0.5 and 0.75 the maximum occurs at the freezing temperature ≈40 K.

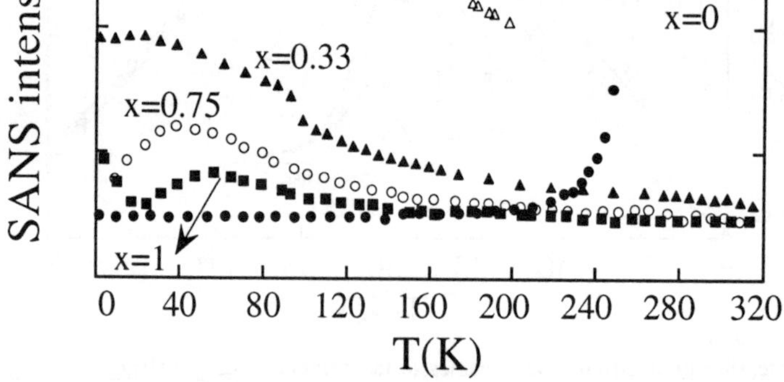

Fig. 41. SANS intensity obtained at the D1B instrument of ILL for selected compounds of the $(La_{2/3}Tb_{1/3})_{2/3}Ca_{1/3}MnO_3$ series.

We have investigated the SANS of the compounds x=0.25 and 0.33 using the D16 instrument .

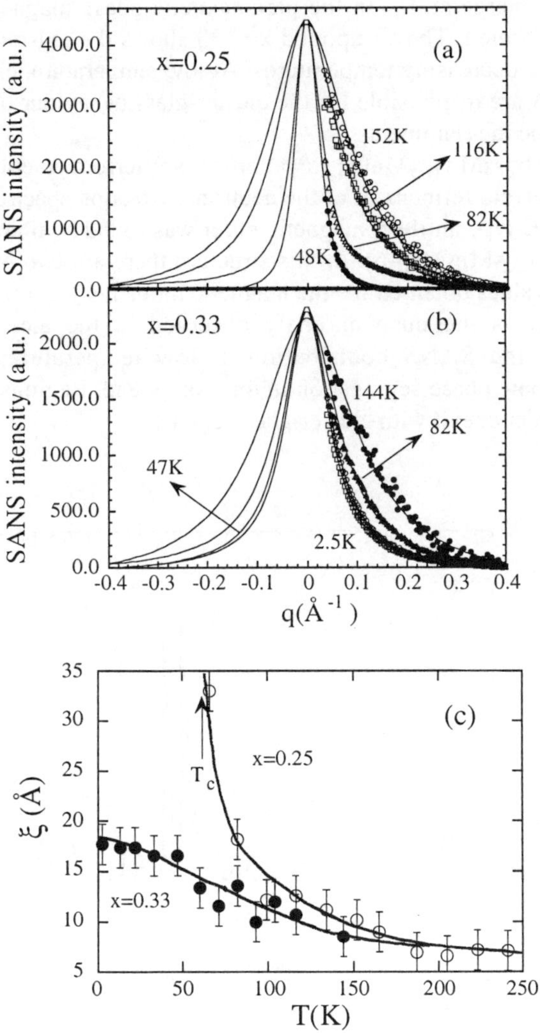

Fig. 42. SANS intensity as a function of q in arbitrary units at different temperatures and the corresponding fits to a iorentzian-type q dependence for the **(a)** x=0.25 and **(b)** 0.33 compounds of the $(La_{1-x}Tb_x)_{2/3}Ca_{1/3}MnO_3$ series. In **(c)** the thermal dependence of the magnetic correlation length obtained from such fits is shown.

As explained in §3, we have performed a lorentzian fitting of the SANS intensity. In Fig. 42 (a) and (b) such fits are shown as well as the magnetic correlation length obtained from them, in Fig. 42 (c). In the case of the compound x=0.25, ξ tends to diverge at T_C. In the paramagnetic phase magnetic clusters of about 10-20 Å are formed. The compound x=0.33 shows the existence of clusters whose size grows for decreasing temperatures. At low temperature, clusters with an average size of 18 Å are responsible for the cluster-glass behaviour observed in the macroscopic magnetic measurements.

The compound $Tb_{2/3}Mn_{1/3}MnO_3$ (x=1) orders antiferromagneticaly at T_N=75 K. In Fig. 43 the Rietveld refinement of the neutron diffraction spectrum obtained at 10 K is shown. A CE-type antiferromagnetic order was found similar to the already observed in $Pr_{0.5}Ca_{0.5}MnO_3$ [46]. In this structure there are two Mn sites. From the refinement, the values obtained for the magnetic moments were: 1.7 and 0.8 μ_B. This result as well as the cusp anomaly observed in the a.c. susceptibility measurements and the SANS contribution at low temperatures support the hypothesis of electronic phase segregation in this compound. Regions of long-range antiferromagnetic order coexit with cluster-glass regions.

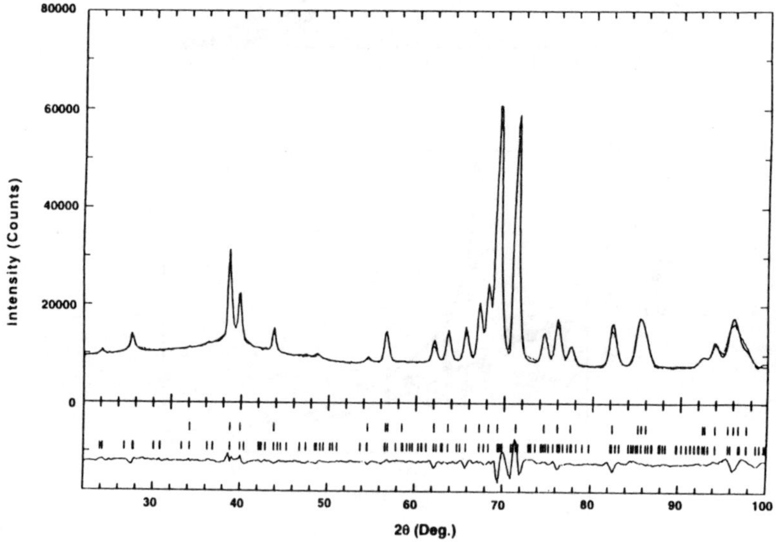

Fig. 43. Rietveld refinement of the diffraction data of the $Tb_{2/3}Ca_{1/3}MnO_3$ compound at 10 K. There appear two magnetic peaks that have been refined in a CE-type structure.

The thermal dependence of the resisitvity for selected compounds of the series is shown in Fig. 44. The values were normalized to the room temperature value. In the paramagnetic phase all the compounds show a semiconductor-like behaviour. In the concentration region x≤0.25 a transition to the metallic state occurs at the same temperature as the para-ferromagnetic transition takes place. The residual resistivity at low temperature grows with the Tb content. This in agreement with the increase in the magnetic frustration and the existence of phase segregation in the case of x=0.25. For x≥0.33 the resistivity shows a semiconducting behaviour in the whole temperature range. A tight relationship is deduced from these results between the magnetic and transport properties across the series. The ferromagnetic state is always related to a metallic state and the paramagnetic, antiferromagnetic, and spin-glass states to an insulating behaviour.

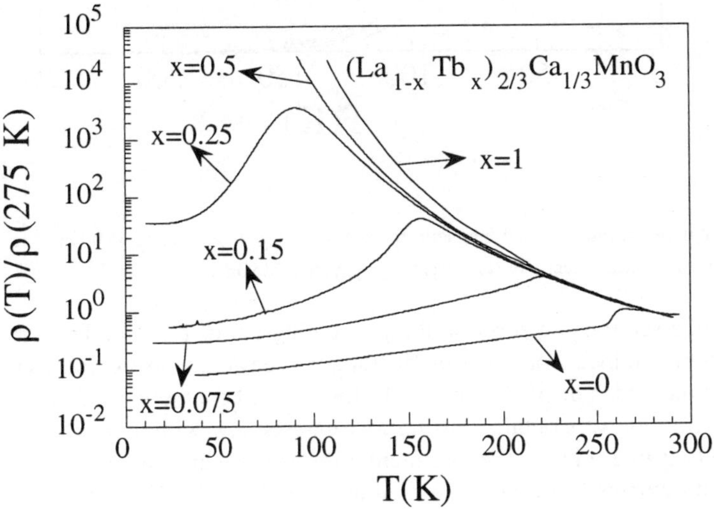

Fig. 44. Thermal dependence of the normalized resistivity for selected compounds of the $(La_{2/3}Tb_{1/3})_{2/3}Ca_{1/3}MnO_3$ series.

In Fig. 45 the thermal dependence of the resistivity under several values of the magnetic field for the compound with x=0.33 is displayed. At fields higher than 5 T the resistivity is low enough to be measured with our experimental set-up at low temperature. The role of the field is to align the clusters and increase their size to produce the percolation. A similar explanation is valid for the x=0.25 compound, where the part of the sample without magnetic long-range order becomes aligned by the field.

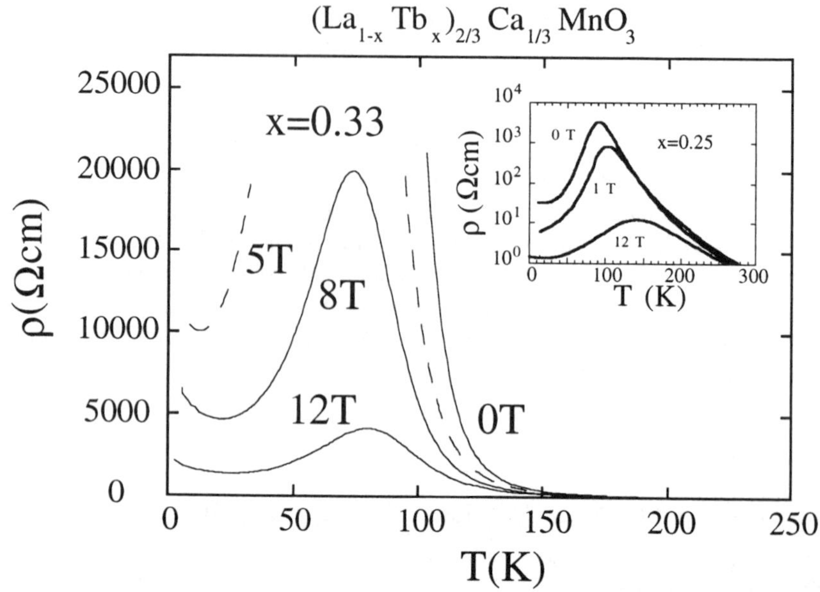

Fig. 45. Resistivity under several applied magnetic fields for $(La_{2/3}Tb_{1/3})_{2/3}Ca_{1/3}MnO_3$. The inset shows the same kind of measurements for $(La_{0.75}Tb_{0.25})_{2/3}Ca_{1/3}MnO_3$.

To sum up this section, we propose the phase diagram of Fig. 46. The volume thermal expansion obtained from neutron diffraction measurements as well as using the strain-gauge method is displayed in Fig. 47 for selected compounds of the series. The lattice contribution was obtained using a Grüneisen law with a Θ_D=500 K. In the semiconducting-like region all experimental thermal expansion curves collapse into a curve that deviates from the calculated at around 400 K. The compounds in which a spontaneous insulator-metal transition occurs, like x=0, at the transition temperature the thermal expansion curve collapses into the calculated curve. It seems that there exist two volume states associated with the localized or delocalized character of the carrier [88]. This assumption is confirmed by the peculiar character of the x=0.25 compound. In this compound, where the sample at low temperature is formed by coexisting insulator and metallic regions, the thermal expansion curves deviate from both curves reaching and intermediate volume state at low temperature.

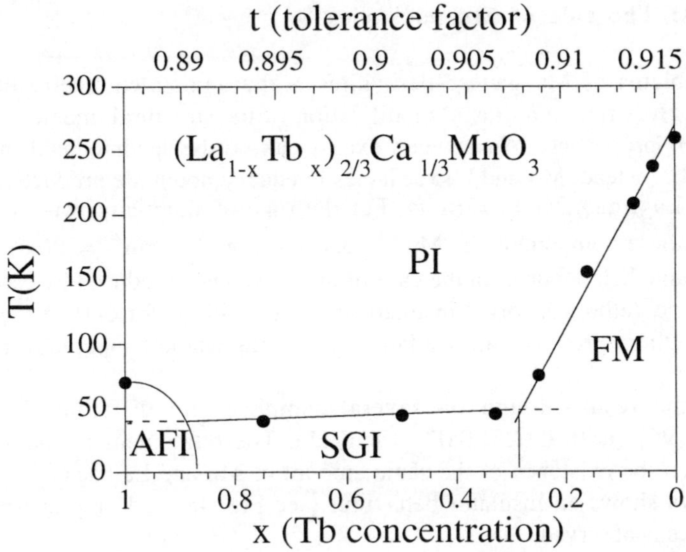

Fig. 46. Phase diagram of the $(La_{1-x}Tb_x)_{2/3}Ca_{1/3}MnO_3$ series. PI stands for paramagnetic insulator; FM, for ferromagnetic metallic; SGI, for spin-glass insulator, and AFI, for antiferromagnetic insulator. The broken line indicates phase segregation (see text).

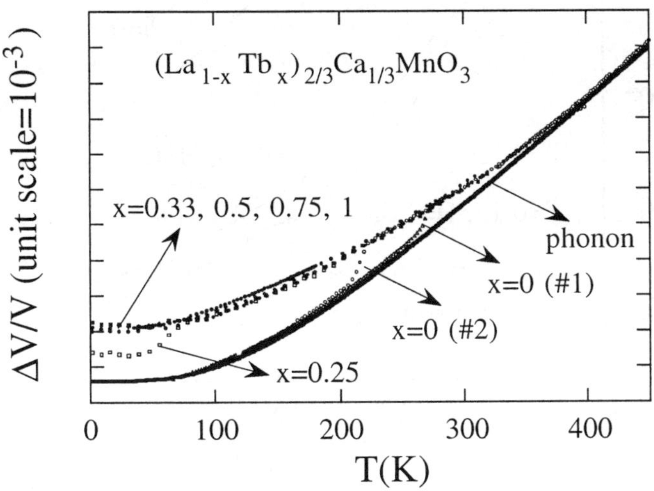

Fig. 47. Volume thermal expansion curves for selected compounds of the series $(La_{1-x}Tb_x)_{2/3}Ca_{1/3}MnO_3$ (symbols). The phonon contribution is represented by a continuous line.

5 $LaMnO_{3+\delta}$: The role of Mn vacancies

One of the problems of Manganese perovkites is the non-stoichiometry of the Oxygen, which gives rise to a drastic modification of the structural, magnetic and magnetotransport properties. An Oxygen excess can not be accomodated in this strucutre [89, 90]. Instead, Mn and La vacancies in equal amount are produced. The real formula is $La_{3/(3+\delta)}Mn_{3/(3+\delta)}O_3$. For the sake of simplicity, we will use $LaMnO_{3+\delta}$. In these compounds the Mn^{+4} concentration is $\%Mn^{+4}=200x\delta$. One can expect the same behaviour as in the case of mixed valence produced by cationic substitution. Some authors reported insulator-metal transitions for $\delta \geq 0.15$ [91-93] whereas other authors proposed an insulator state in the whole temperature range [94-96].

We report the results found on several samples with different Oxygen concentrations [96] (δ=0, 0.025, 0.07, 0.1, 0.15). The refinement of the X-ray spectra confirmed the existence of the same amount of Mn and La vacancies. The resistivity always shows an insulator behaviour (see Fig. 48) and only at δ=0.15 magnetoresistance is observed.

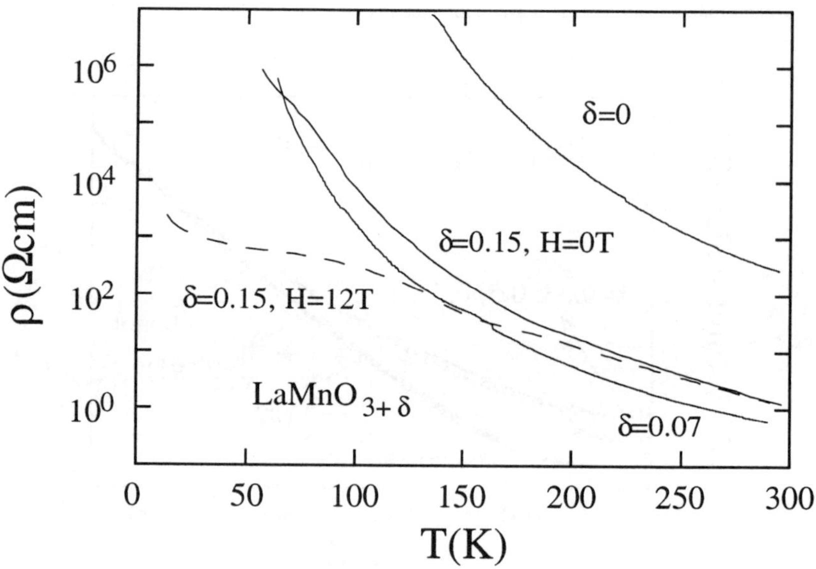

Fig. 48. Thermal dependence of the resistivity in selected compounds of the series $LaMnO_{3+\delta}$. The only compound showing giant magnetoresistance is that with δ=0.15.

The thermal dependence of the a.c. magnetic susceptibility is displayed in Fig. 49. The magnetic behaviour is quite different along the series and should be described carefully. The δ=0 compound is antiferromagnetic below T_N=140 K, temperature at which a weak but well-defined peak is observed. The magnetic structure of this compound was determined by Wollan and Koehler [9]. They proposed an in-plane ferromagnetic ordering, with an antiferromagnetic coupling between (001) planes. As δ increases, also the number of Mn^{+3}-Mn^{+4} pairs does. That means an enhancement of the ferrromagnetic double-exchange interaction. For all compounds an increase of the a.c. susceptibility is observed at temperatures around ≈160 K, with a decrease at ≈110 K. These results recall the behaviour found in the La-Ca compounds, where a charge-ordering transition takes place in the ferromagnetic phase (see Fig. 21) [55].

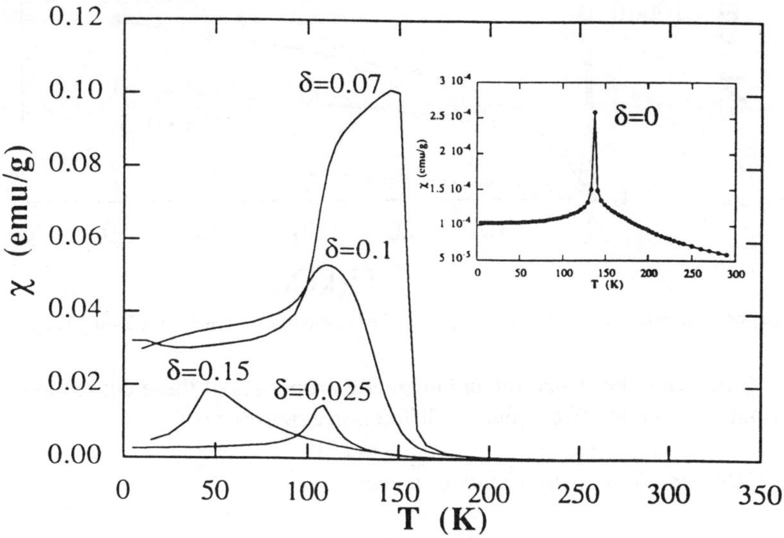

Fig. 49. a.c. susceptibility as a function of temperature for all the compounds of the series $LaMnO_{3+\delta}$ (in the inset the δ=0 compound is shown).

The isotherms at 5 K of these compounds can clarify the magnetic behaviour (see Fig. 50). The compounds δ=0.07 and 0.1 are ferromagnetic and the value of the magnetic moment found for the Mn ions are 3.43 μB and 3.05 μ_B respectively. The low value of the magnetization for δ=0.1 indicates the effect of the Mn vacancies. This effect is even more evident in the case of δ=0.15, in which the magnetic moment determined per Mn ion is 0.9 μ_B. This result is surprising because the compound with the largest number of Mn^{+3}-Mn^{+4} pairs shows weak ferromagnetic behavior. The explanation is found in the great amount of Mn vacancies (≈5%), which prevent the establishment of the long-range order in the whole sample. The cusp-like anomaly found at 50 K for this compound is associated with the freezing temperature of the spin-glass [96, 97]. The magnetic properties of the studied compounds are summarized in Table I (at the end of the section).

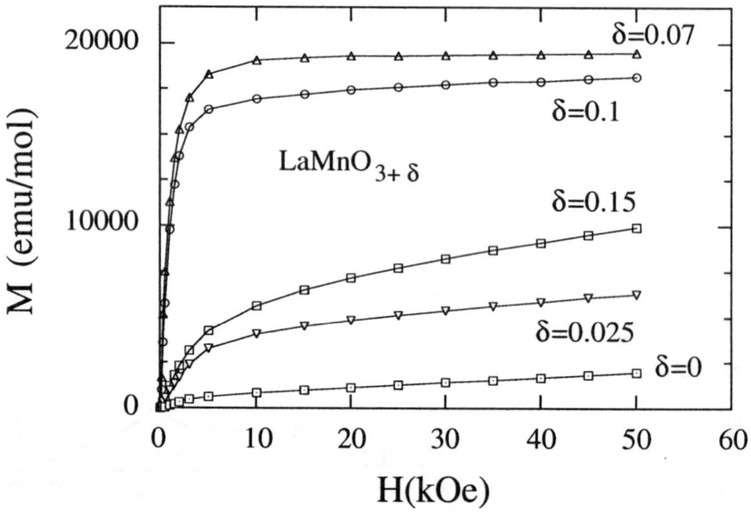

Fig. 50. Magnetization isotherms at 4 K of all the studied compounds of the series $LaMnO_{3+\delta}$.

We will describe the structural and magnetic properties of these compounds using the information obtained from neutron-diffraction experiments [96].

5.1 δ=0, 0.025 and 0.07: orthorhombic structure.

These compounds show an orthorhombic structure in the whole temperature range measured (1.5 K-300 K). The thermal scan performed at the D1B diffractometer is shown in Fig. 51. The antiferromagnetic structure observed for δ=0 (A-type [9]) evolves towards canted antiferromagnetic at δ=0.025. A long-range ferromagnetic contribution appears on the (110) Bragg peak. The refinement of the data at 1.5 K gives values of 2.52 μ_B for the antiferromagnetic component and 1.48 μ_B for the ferromagnetic component (see table I). In the compound δ=0.07, the antiferromagnetic contribution is negligible (0.25 μ_B). A large ferromagnetic contribution is observed below T_C=160 K, which gives a magnetization value of 3.25 μ_B/Mn at 1.5 K. This value is in good agreement with the value obtained from magnetization measurements (Table I).

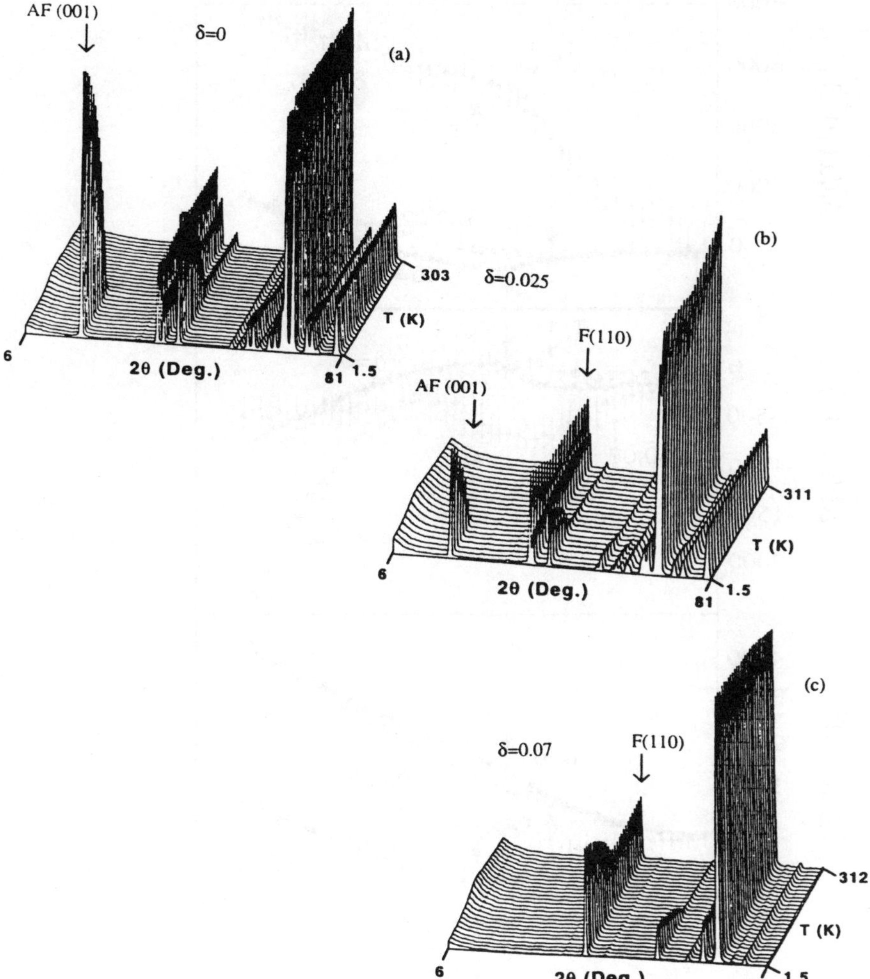

Fig. 51. Neutron thermodiffractograms obtained at the instrument D1B of the compounds with δ=0, 0.025 and 0.07 of the series $LaMnO_{3+\delta}$. Some magnetic peaks are marked by arrows.

From the analysis of the neutron-diffraction spectra, the thermal dependence of the lattice parameters, shown in Fig. 52, was obtained.

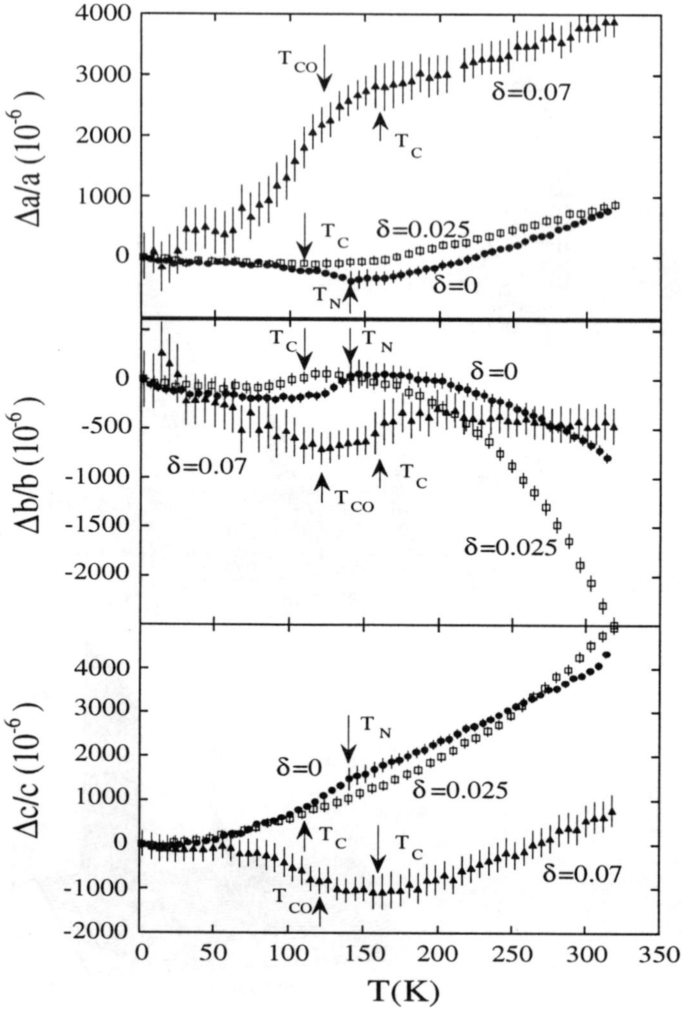

Fig. 52. Thermal dependence of the lattice parameters in the compounds with δ=0, 0.025 and 0.07 belonging to the series $LaMnO_{3+\delta}$. The magnetic transition temperatures and the charge-ordering temperatures are marked with arrows.

At the magnetic transition temperatures (T_N and T_c) smooth anomalies are observed. The compound δ=0.07 shows a very large anisotropic distortion at 110 K, temperature at which a pronounced decreasing of the a.c. susceptibility was observed. As has been discussed in §3 for the compound $La_{0.35}Ca_{0.65}MnO_3$, this huge anisotropic effect can be related to the establishment of a charge-ordered state. The coexistence of a ferromagnetic state and a charge-ordered has been proposed [96] as the ground state for this compund. This assumption was supported by SANS experiments. Once again we can verify the relevance of (in this case static) electronic phase segregation in mixed-valence manganites. Other relevant structural parameters in these compounds are the Mn-O distances. In Fig. 53 these distances for all the studied compounds at 1.5 K and 300 K are shown.

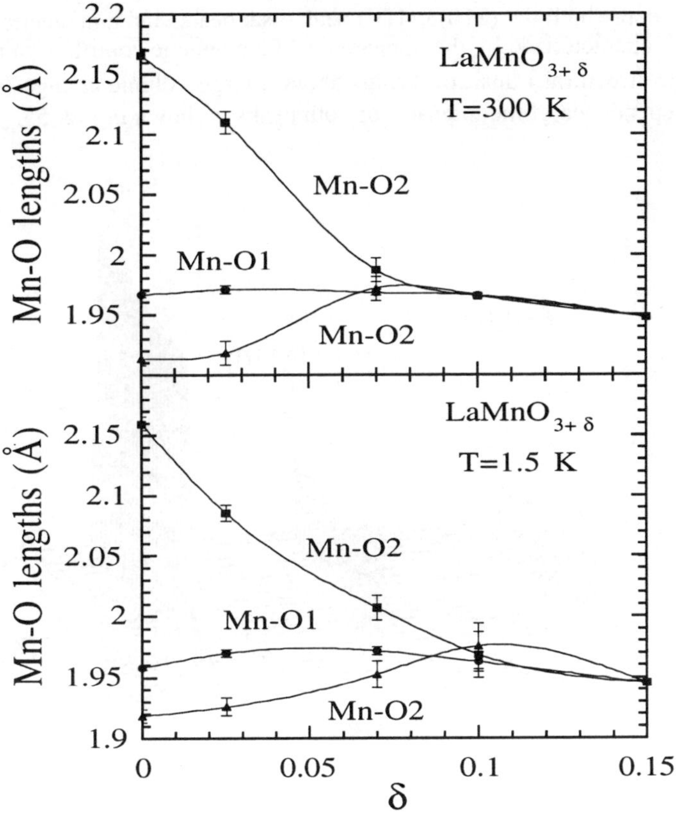

Fig. 53. Mn-O distances in the MnO_6 octahedra for the compounds with δ=0, 0.025 and 0.07 belonging to the series $LaMnO_{3+\delta}$ at temperatures of **(a)** 300 K and **(b)** 1.5 K.

The compounds δ=0 and 0.025 show a clear Jahn-Teller distortion of the MnO_6 octahedra (O1 and O2 are the apical and in-plane Oxygen atoms). As can be noticed, this distortion is suppressed as δ increases. In fact, the compound δ=0.1 shows at room temperature similar values for the three distances, which corresponds to the rombohedral symmetry. In Table II (at the end of the section) we summarize the main structural parameters [96].

5.2 δ=0.1: rombohedral-orthorhombic transition

The X-ray and neutron diffraction spectra were refined in the $R\bar{3}c$ space group [96]. Whereas for the low-temperature data, the Pbnm space group was used. The neutron thermodiffractogram (see Fig. 54) shows a structural transition at T_S=300 K, temperature at which the (210)&(120) peak disappears, and a magnetic transition at T_C=150 K, characterized by the appearance of a magnetic contribution on the Bragg peaks. The structural transition brings about a large volume change (≈0.8%). The thermal dependence of the intensity of both peaks is shown in Fig. 55.

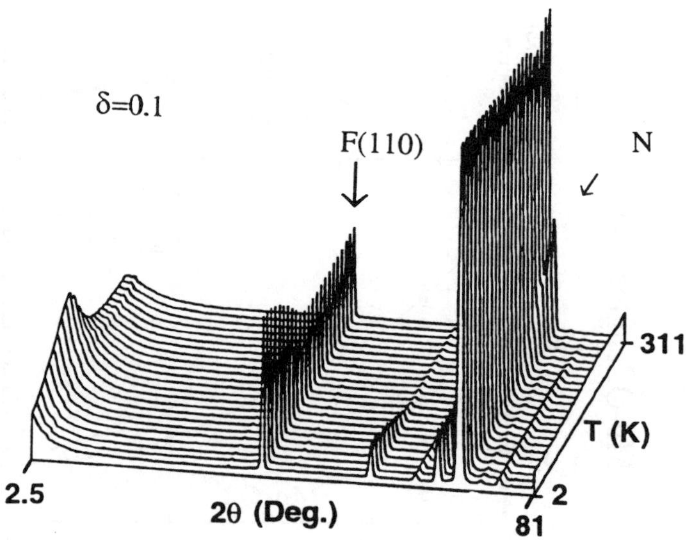

Fig. 54. Neutron thermodiffractogram of the compound $LaMnO_{3.1}$. The structural orthorhombic-rombohedral transition that takes place at 300 K is marked with an arrow. The ferromagnetic (110) peak is also marked with an arrow.

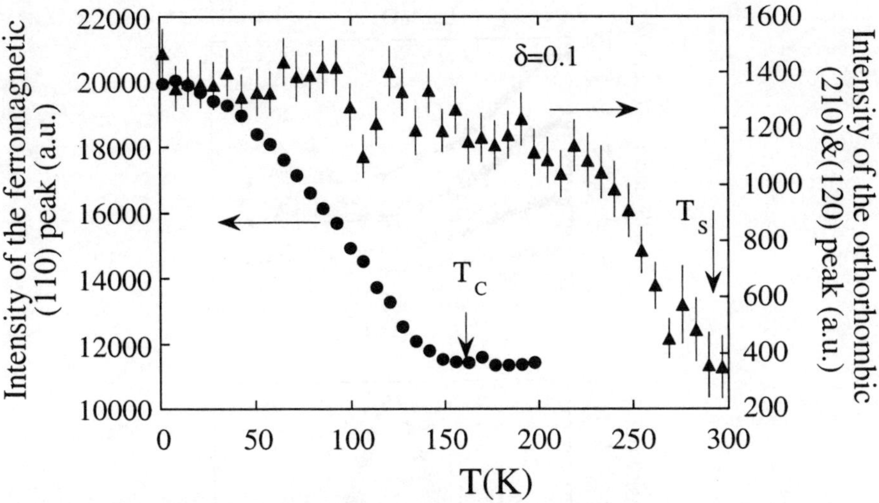

Fig. 55. Thermal dependence of the intensity of the structural orthorhombic (210)&(120) peak and the ferromagnetic (110) peak of the compound $LaMnO_{3.1}$.

5.3 δ=0.15: rhombohedral structure.

The δ=0.15 compound has about 5% of Mn vacancies and the largest number of Mn^{+4}-Mn^{+3} pairs (30%). However, the neutron-diffraction results [96-98] revealed a weak ferromagnetic component of 0.97 μ_B/Mn at 1.5 K. The anomaly at 50 K found in the a.c. susceptibility and the hysteresis in the magnetization suggest the existence of a cluster-glass state at low temperature. The appearance of this state can be understood considering the large number of Mn vacancies, which break the coherence of the long-range magnetic interaction. The results of SANS experiments are shown in Fig. 56. The thermal dependence of the SANS intensity at several values of q confirms the lack of a para-ferromagnetic transition. From the fit of the SANS intensity to Lorentzian functions [Fig. 56 (b)] we have obtained the ξ values shown in Fig. 56 (c). These values represent average sizes of the clusters, some of them should be large enough to give a small contribution to the Bragg scattering.

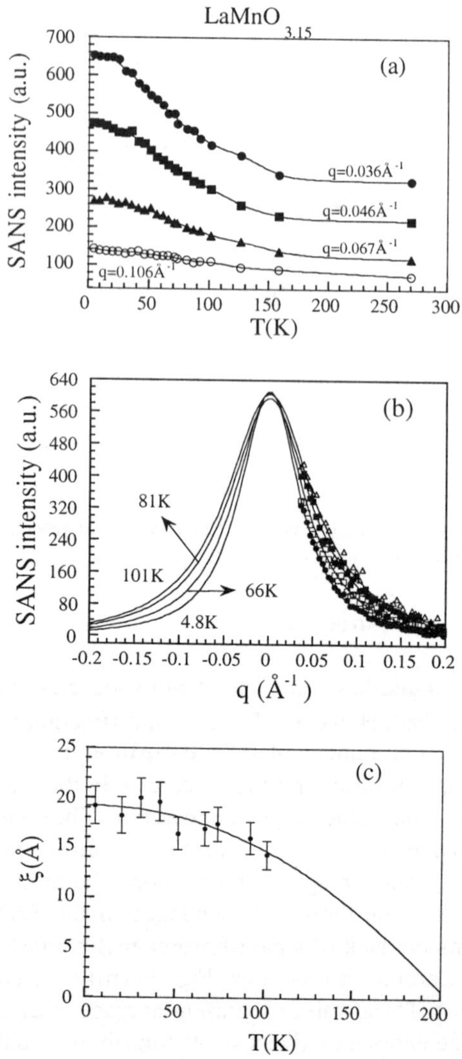

Fig. 56. Results obtained in the SANS instrument D16 in the compound $LaMnO_{3.15}$ **(a)** SANS intensity at several values of q. **(b)** Fits of the SANS intensity to Lorentzian-type q dependence at several temperatures. **(c)** Magnetic correlation length as a function of temperature obtained from such fits.

The results reported in this section may help to clarify some aspects of the magnetic, structural and magnetotransport properties of the non-stoichiometric compounds $LaMnO_{3+\delta}$. The main results can be summarized as follows:

-The increase of δ tends to stabilize the rhombohedral structure.

-The static Jahn-Teller effect observed at δ=0 and 0.025 disappears as δ increses as occurs in the case of the series $La_{1-x}Ca_xMnO_3$ as x increases.

-There is not any correlation between the crystallographic structure and the magnetic ordering.

-Increasing values of δ tend to weaken the antiferromagnetic interaction in favour of the ferromagnetic interaction as a consequence of the increasing number of Mn^{+4} ions.

-The role of the Mn vacancies is very important for $x \geq 0.07$, giving rise to a low-temperature non-homogeneous state. In the same sample (same crystallographic phase), regions of different magnetic and transport properties (paramagnetic, charge ordering, long-range ferromagnetic and cluster-glass) coexist.

-All compounds are insulating with negligible magnetoresistance except for the δ=0.15 compound.

The existence of spontaneous metal-insulator transtion in this series is still a matter of controversy. We have not observed this transition in our samples. Experiments performed in La_xMnO_3, where the La vacancies introduce Mn^{+4} ions, reveal a metal-insulator transition as in the case of the Ca-substituted compounds, exhibiting giant magnetoresistance as well [99, 100]. We suggest that samples with metal-insulator transition could have La instead of Mn vacancies, keeping the Mn lattice unchanged.

Table I. Data of the magnetic behaviour for all the studied compounds

δ	0	0.025	0.07	0.1	0.15
Mn^{4+} (%)	0	5	14	20	30
Formula unit	$LaMnO_3$	$La_{0.992}Mn_{0.992}O_3$	$La_{0.977}Mn_{0.977}O_3$	$La_{0.968}Mn_{0.968}O_3$	$La_{0.952}Mn_{0.952}O_3$
Magnetic order	AF T_N=140 K	F + AF T_C=110 K	F + AF T_C=160 K	F T_C=150 K	F T_C=160 K
μ (μ_B)	3.49	2.92 F 1.48 AF 2.52	3.26 F 3.25 AF 0.25	2.86	0.78
Paramagnetic region					
C (emu K/mol)	3.48	5.38	4.34	4.45	4.66
θ_P (K)	46	126	171	170	129
μ_{eff}	5.3	6.6	5.9	6.0	6.1
Ferromagnetic region					
M_S (5 K,H=0) (emu/mol)	624.4	3919.5	19181	17046	5421.2
ΔM/ΔH (5 K) (emu/mol Oe)	2.7×10^{-2}	4.86×10^{-2}	6.9×10^{-3}	2.3×10^{-2}	0.97×10^{-2}
μ_{eff}	0.11	0.70	3.43	3.05	0.97

Table II. Structural data for all the studied compounds.

δ	0	0.025	0.07	0.1	0.15
Mn^{4+} (%)	0	5	14	20	30
Formula unit	$LaMnO_3$	$La_{0.992}Mn_{0.992}O_3$	$La_{0.977}Mn_{0.977}O_3$	$La_{0.968}Mn_{0.968}O_3$	$La_{0.952}Mn_{0.952}O_3$
Space group 1 K	Pbnm	Pbnm	Pbnm	Pbnm	$R\bar{3}c$
Space group 300 K	Pbnm	Pbnm	Pbnm	$R\bar{3}c$	$R\bar{3}c$
1 K a (Å)	5.5330	5.5351	5.5069	5.5109	5.4759
1 K b (Å)	5.7273	5.6188	5.4948	5.4714	5.4759
1 K c (Å)	7.6681	7.7157	7.7638	7.7642	13.222
1 K Vol/z ($Å^3$)	60.75	60.00	58.73	58.53	57.22 *
300 K a (Å)	5.5350	5.5394	5.5278	5.5322	5.4814
300 K b (Å)	5.7230	5.6072	5.4922	5.5322	5.4814
300 K c (Å)	7.6970	7.7507	7.7681	13.3422	13.261
300 K Vol/z ($Å^3$)	60.96	60.18	58.96	58.94	57.51
1 K $Mn-O_I$	1.9579	1.9697	1.9715	1.9737	1.9449
1 K $Mn-O_{II}$	2.1597	2.0857	2.0070	1.9684	1.9449
1 K $Mn-O_{II}$	1.9181	1.9256	1.9523	1.9756	1.9449
300 K $Mn-O_I$	1.9662	1.9708	1.9683	1.9737	1.9480
300 K $Mn-O_{II}$	2.1658	2.1100	1.9870	1.9737	1.9480
300 K $Mn-O_{II}$	1.9134	1.9182	1.9718	1.9737	1.9480
$Mn-O_I-Mn$ (1 K)	168.27	168.32	169.89	171.52	172.20
$Mn-O_I-Mn$ (300 K)	168.16	169.40	170.63	170.76	172.27
$Mn-O_{II}-Mn$ (1 K)	168.22	169.87	169.39	169.89	172.20
$Mn-O_{II}-Mn$ (300 K)	168.16	168.64	169.84	170.66	172.27

* z=6 for $R\bar{3}c$

6 Conclusions

The recent effort in the understanding of the magnetoresistance phenomena in mixed-valence perovskites has allowed the understanding of many interesting extrinsic and intrinsic features unknown during the last forty years. One of the most relevant contributions has been the consideration of the electron-phonon interaction as most important in driving the conduction mechanisms in these materials. The relevance of this interaction in manifested by thermal expansion measurements and the isotopic effect.

The interplay between magnetic interactions (modulated by structural parameters and the lattice dynamics) and charge ordering gives rise to the existence of electronic phase segregation in which conducting ferromagnetic regions coexist with antiferromagnetic or disordered insulating regions.

The mechanism for conduction is metallic, i.e. e_g electrons moving in a narrow band, when ferromagnetic long-range order is established and polaronic in the antiferromagnetic or disordered state. Due to the strong electron-phonon coupling the carriers are "small polarons" in the paramagnetic state. When the ferromagnetic interaction is large enough to form short-range ferromagnetic clusters, the existence of magnetic polarons is possible as occurs in many of the reported systems. Under these conditions, phenomena of giant magnetoresistance occurs. Other source of huge magnetoresitance effects is the disappearance of a charge-ordered state under applied magnetic field.

Thermal expansion and magnetostriction measurements have revealed a positive isotropic volume effect associated with the polaronic regime. The transition from polaronic to metallic regime at the spontaneous or field-induced insulator-metal transition brings about unusual magnetoelastic effects. On the other hand, at the charge-ordering temperature volume effects are not observed but anisotropic distortion of the lattice.

From all the reported results for fixed carrier concentration, $Mn^{+4}/Mn^{+3}=1/2$, we have found a universal behaviour [88], which is illustrated in Fig. 57. All the thermal expansion curves which correspond to an insulating regime (Pr, LaTb) collapse onto a high-volume universal curve. This curve is also followed by the La and LaY compounds in the paramagnetic insulator regime. When these compounds undergo an insulator-metal transition, a change in volume takes place which corresponds to the change from the high-volume thermal expansion curve into the low-volume thermal expansion curve (phonon Grüneissen contribution). The existence of two volumes, high volume (associated with localized carrier) and low volume (delocalized carrier), recalls the model of two volumes proposed by Weiss [101] for invar metallic alloys (recently confirmed from *ab initio* calculations [102]). Two thermally or under-field accesible states may be in the origin for the metal-insulator transition in manganites. This would prove the first-order nature for this transition and also would allow drawing general conclusions on the nature of the invar effect in metallic alloys and electronic transitions in oxides [88].

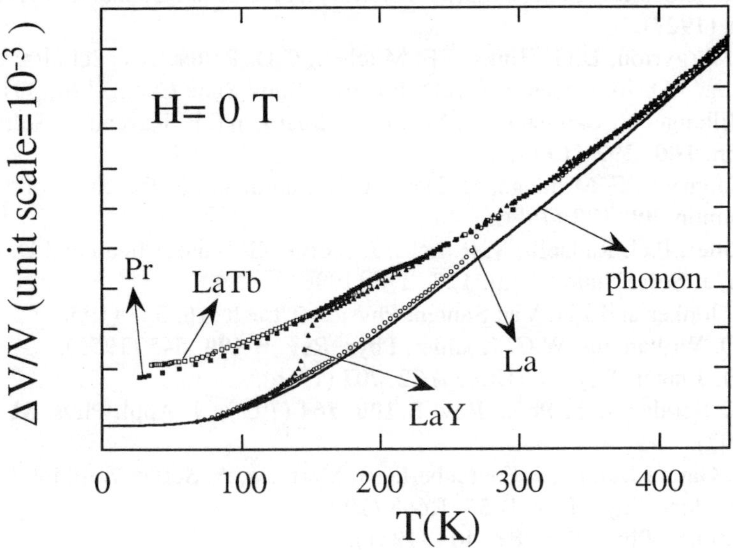

Fig. 57. Volume thermal expansion versus temperature for $La_{2/3}Ca_{1/3}MnO_3$ (La), $La_{0.6}Y_{0.07}Ca_{0.33}MnO_3$ (LaY), $La_{1/3}Tb_{1/3}Ca_{1/3}MnO_3$ (LaTb), and $Pr_{2/3}Ca_{1/3}Mn_3O_3$ (Pr). A large change in volume occurs at the insulator-metal transition.

Acknowledgements

We acknowledge P.A. Algarabel, C. Marquina, J. Blasco, J. García, C. Ritter, Z. Arnold, B. García-Landa and A. Del Moral for the collaboration in this field. We also acknowledge financial support from the Spanish CICYT.

References

1. A.M. Glazer, Acta Cryst. **B28**, 3384 (1972).
2. A.M. Glazer, Acta Cryst. **A31**, 756 (1975).
3. V.M. Goldsmith, Geochemische Verteilungsgesetze der Elemente **VII-VIII**, 1928 (1927).
4. D.N. Argyriou, D.G. Hinks, J.F. Mitchell, C.D. Potter, A.J. Schultz, D.M. Young, J.D. Jorgensen, and S.D. Bader, J. Solid State Chem. **124**, 381 (1996).
5. F. Millange, V. Caignaert, G. Mather, E. Suard, and B. Raveau, J. Solid State Chem. **100**, 292 (1996).
6. V. Caignaert, F. Millange, M. Hervieu, E. Suard, and B. Raveau, Solid State Commun. **99**, 173 (1996).
7. C. Ritter, P.G. Radaelli, M.R. Lees, J. Barrak, G. Balakrishnan and D. Mck Paul, J. Solid State Chem. **127**, 276 (1996).
8. G.H. Jonker and J.H. Van Santen, Physica (Utrech) **16**, 337 (1950).
9. J.E.O. Wollan and W.C. Koehler, Phys. Rev. B **100**, 545 (1955).
10. G.H. Jonker, Physica (Utrech) **22**, 707 (1956).
11. J.B. Goodenough, Phys. Rev. B **100**, 564 (1955); J. Appl. Phys. **81**, 5330 (1997).
12. J.L. García-Muñoz, J. Fontcuberta, B. Martínez, A. Seffar, S. Piñol and X. Obradors, Phys. Rev. B **55**, R668 (1997).
13. C. Zener, Phys. Rev. **82**, 403 (1951).
14. P.W. Anderson and H. Hasegawa, Phys. Rev. **100**, 675 (1955).
15. P.G. De Gennes, Phys. Rev. B **118**, 141 (1960).
16. R.M. Kusters, J. Singleton, D.A. Keen, R. McGreevy and W. Hayes, Physica B **155**, 362 (1989).
17. R. Von Helmolt, J. Wecker, B. Holzapfel, L. Schultz and K. Sammwer, Phys. Rev. Lett. **71**, 2331 (1993).
18. K. Chahara, T. Ohno, M. Kasai and Y. Kozono, Appl. Phys. Lett. **63**, 1990 (1993).
19 S. Jin, T.H. Tiefel, M. McCormack, R.A. Fastnacht, R. Ramesh and L.H. Chen, Science **264**, 413 (1994).
20. I.O. Troyanchuk, Sov. Phys. JETP **75**, 132 (1992).
21. P. Schiffer, A.P. Ramirez, W. Bao, and S.-W. Cheong, Phys. Rev.Lett. **75**, 3336 (1995).
22. E. L. Nagaev, Uspekhi Fizicheskikh Nauk **39**, 781 (1996) *englished translated.*
23. P.A. Algarabel, J.M. De Teresa, C. Marquina, B. García-Landa, C. Ritter and M.R. Ibarra, submitted.
24. J.S. Zhou, W. Archibald and J.B. Goodenough, Nature (London) **381**, 770 (1996).

25. G.F. Dionne, J. Appl. Phys. **79**, 5172 (1996).
26. A.J. Millis, P.B. Litlewood and B.I. Shraiman, Phys. Rev. Lett. **74**, 5144 (1995).
27. H. Röder, Jun Zhang and A. Bishop, Phys. Rev. Lett. **76**, 1356 (1996).
28. M.R. Ibarra, P.A. Algarabel, C. Marquina, J. Blasco and J. García, Phys. Rev. Lett. **75**, 3541 (1995).
29. G. Zhao, K. Conder, H. Keller and K.A. Müller, Nature **381**, 676 (1996).
30. J.M. De Teresa, M.R. Ibarra, P.A. Algarabel, C. Ritter, C. Marquina, J. Blasco, J. García, A. del Moral and Z. Arnold, Nature (London) **386**, 256 (1997).
31. N.F. Mott and E.A. Davies, "Electronic processes in non-crystalline materials", (Oxford University Press), (1971).
32. D. Adler in Solid State Physics, ed. F. Seitz, D. Turnbull, H. Ehrenreich, volume 21, page 1 (1968).
33. J. Appel, ibidem, page 193.
34. T. Holstein, Annals of Physics **8**, 343, (1959).
35. D. Emin and T. Holstein, Annals of Physics **53**, 439 (1969).
36. D.C. Worledge, G. Jeffrey Snyder, M.R. Beasley, and T.H. Geballe, J. Appl. Phys. **80**, 5158 (1996).
37. M. Jaime, M.B. Salamon, M. Rubinstein, R.E. Treece, J.S. Horwitz and D.B. Chrisey, Phys. Rev. B **54**, 11914 (1996).
38. M. Jaime, H.T. Hardner, m.B. Salamon, M. Rubinstein, P. Dorsey and D. Emin, Phys. Rev. Lett. **78**, 951 (1997).
39. T. Kasuya and A. Yanase, Reviews of Modern Physics **40**, 684 (1968).
40. S. Von Molnar, A. Briggs, J. Flouquet and G. Remenyi, Phys. Rev. Lett. **51**, 706, (1983).
41. P.A. Wolff, *Semiconductors and semimetals* **25**, 413, edited by J.K. Furoyna & J. Kossut, Academic press Inc. (1988)
42. T. Kasuya, Y. Haga and T. Suzuki, J. Alloys and Compounds **219**, 290 (1995).
43. J.M.D. Coey, M. Viret, L. Ranno and K. Ounadjela, Phys. Rev. Lett **75**, 3910 (1995).
44. H.Y. Hwang, S.W. Cheong, N.P. Ong and B. Batlogg, Phys. Rev. Lett **77**, 2041 (1996).
45. E.P. Svirina, L.P. Shlyakhina, and F.F. Shakirova, Sov. Phys. Solid State **30**, 2129 (1988).
46. C.H. Chen and S.W. Cheong, Phys. Rev. Lett **76**, 4042 (1996).
47. Z. Jirák, S. Krupicka, Z. Simsa, M. Dlouhá and S. Vratislav, J. Magn. Magn. Mater. **53**, 153 (1985).
48. Y. Tomioka, A. Asamitsu, Y. Moritomo, H. Kuwahara and Y. Tokura, Phys. Rev. Lett. **74**, 5108 (1995).
49. E.L. Wigner, Trans. Faraday Soc. **34**, 678 (1938).

50. E.J.W. Verwey, P.W. Haayman and F.C. Romejin, J. Chem. Phys. **15**, 181 (1947).
51. C.H. Chen, S.W. Cheong and A.S. Cooper, Phys. Rev. Lett. **71**, 2461 (1993); S.W. Cheong, H.Y. Hwang, C.H. Chen, B. Batlogg, L.W. Rupp, Jr. and S.A. Carter, Phys. Rev. B **49**, 7088 (1994).
52. Y. Moritomo, Y. Tomioka, A. Asamitsu, and Y. Tokura, Phys. Rev. B **51**, 3297 (1995).
53. C.N.R. Rao, R. Mahesh, A.K. Raychaudhuri, and R. Mahesh, J. Phys. Chem. Solids (in press).
54. J.L Rodríguez-Carvajal and M.A. Parnetier, ILL report 87TR014T (1987).
55. S.W. Cheong, C.M. López and H.Y. Hwang, submitted
56. J.M. De Teresa, M.R. Ibarra, J. Blasco, J. García, C. Marquina, P.A. Algarabel, Z. Arnold, K. Kamenev, C. Ritter and R. von Helmolt, Phys. Rev. B **54**, 1187 (1996).
57. T.H.K. Barron, Advances in Physics **29**, 609 (1980).
58. R. Mahesh, R. Mahendiran, A.K. Raychaudhur and C.N.R. Rao, Appl. Phys. Lett. **68**, 2291 (1996).
59. A. Gupta, G.Q. Gong, Gang Xiao, P.R. Duncombe, P. Lecoeur, P. Trouilloud, Y.Y. Wang, V.P. Dravid, and J.Z. Sun, Phys. Rev. B **54**, R15629 (1996).
60. M.F. Hundley, M. Hawley, R. H. Heffner, Q.X. Jia, J.J. Neumeier, J. Tesmer, J.D. Thompson, and X.D. Wu, Appl. Phys. Lett. **67**, 860 (1995).
61. J.Z. Liu , I.C. Chang, S. Irons, P. Klavins, R.N. Shelton, K. Song, and S.R. Wasserman, Appl. Phys. Lett. **66**, 3218 (1995).
62. G. Snyder, R. Hiskes, S. KiCaroli, M.R. Beasley and T.H. Geballe, Phys. Rev. B **53**, 14434 (1996).
63. J.S. Ramachandran, S.M. Bhagat, J.L. Peng and M. Rubinstein, Solid State Commun. **96**, 127 (1995).
64. K.H. Kim, J.Y. Gu, H.S. Choi, D.J. Eom, J.H. Jung, and T.W. Noh , Phys. Rev. B **55**, 4023 (1997).
65. S.K. Burke, Cywinski and B. Rainford, J. Appl. Cryst. **11**, 644 (1978).
66. M.R. Ibarra, J.M. De Teresa, J. Blasco. P.A. Algarabel, C. Marquina, J. García and Y. Stankiewicz, Phys. Rev. B, issue of 1st October, (1997).
67. S. Jin, H.M. O'Brian, T.H. Tiefel, R.M. Flemi and J.M. Philips, Appl. Phys. Lett. **66**, 382 (1995).
68. A. Arnold, K. Kamenev, M.R. Ibarra, P.A. Algarabel, C. Marquina, J. Blasco and J. García, Appl. Phys. Lett. **67**, 2875 (1995).
69. J.J. Neumeier, M.F. Hunkley, J.D. Thompson and R.H. Heffner, Phys. Rev. B **52**, R7006 (1995).
70. P.G. Radaelli, D.E. Cox, M. Marezzio, S.-W. Cheong, P.E. Schiffer, and A.P. Ramirez, Phys. Rev. Lett. **75**, 4488 (1995).

71. P.G. Radaelli, D.E. Cox, M. Marezio and S.W. Cheong, Phys. Rev. B **55**, 3015 (1997).
72. J.M. De Teresa, M.R. Ibarra, C. Marquina, P.A. Algarabel and S. Oseroff, Phys. Rev. B **54**, R12689 (1996).
73. H. Taguchi and M. Shimada, J. Sol. State Chem. **67**, 37 (1986)
74. S. Sapathy, Z.S. Popovic, F.R. Vukajlovic, Phys. Rev. Lett. **76**, 960 (1996).
75. A.P. Ramírez, P. Schiffer, S.-W. Cheong, C.H. Chen, W. Bao, T.T.M. Palstra, B. Zegarski, P.L. Gammel, and D.J. Bishop, Phys. Rev. Lett. **76**, 3188 (1996).
76. S. Gota, J. García, J. Chaboy and J. Bartolomé, Phys. Rev. B **44**, 11632 (1991).
77. J. Fontcuberta, B. Martínez, A. Seffar, S. Piñol, J.L. García-Muñoz, and X. Obradors, Phys. Rev. Lett. **76**, 1122 (1996).
78. E. Pollert , S. Krupicka, and E. Kuzmicova, J. Phys. Chem. Solids **43**, 1137 (1982).
79. Y. Tomioka, A. Asamitsu, H. Kuwahara, Y. Moritomo, and y. Tokura, Phys. Rev. B **53**, R1689 (1996).
80. M.R. Lees, J. Barratt, G. Balakrishnan, D.McK. Paul, and M. Yethiraj, Phys. Rev. B **52**, R14303 (1995).
81. M.R. Lees, J. Barratt, G. Balakrishnan, D.McK. Paul, and C.D. Dewhurst, J.Phys.: Condens. Matter **8**, 2967 (1996).
82. H. Yoshizawa, H. Kawano, Y. Tomioka and Y. Tokura, J. Phys. Soc. Jpn. **65**, 1043 (1996)
83. H. Yoshizawa, H. Kawano, Y. Tomioka and Y. Tokura, Phys. Rev. B **52**, R13145 (1995).
84. J. Blasco, J. García, J.M. De Teresa, M.R. Ibarra, P.A. Algarabel and C. Marquina, J. Phys.:Condens. Matter **8**, 7427 (1996).
85. J.M. De Teresa, M.R. Ibarra, J. García, J. Blasco, C. Ritter, P.A. Algarabel, C. Marquina and A. del Moral, Phys. Rev. Lett. **76**, 3392 (1996).
86. V. Caignaert, E. Suard, A. Maignan, Ch. Simon, and B. Raveau, J. Magn. Magn. Mater. **153**, L260 (1996).
87. K. Moorjani and J.M.D. Coey, in *Magnetic Glasses*, ed.Wolsky S.P. and Czanderna A.W., Methods and Phenomena **6** (Elsevier, The Netherlands), pages 181 and 137, (1984).
88. M.R. Ibarra, J.M. De Teresa, P.A. Algarabel, C. Marquina, and B. García-Landa, submitted.
89. B.C. Tofield and W.R. Scott, J. Solid State Chem. **10**, 183 (1974)
90. J.A.M. Roosmalen and E.H.P. Cordjunke, Solid State Chem. **110**, 109 (1994).
91. M. Verelst, N. Rangavittal, C.N.R. Rao, and A. Rousset, J. Solid State Chem. **104**, 74 (1993).
92. R. Mahendiran, R. Mahesh, N. Rangavittal, S.K. Tewari, A.K. Raychaudhuri,

T.V. Ramakrishnan, and C.N.R. Rao, Phys. Rev. B **53**, 3348 (1996).
93. G. Zhao, M.B. Hunt, and H. Keller, Phys. Rev. Lett. **78**, 955 (1997).
94. J. Töpfer, J-P.Doumerc, and J.C. Grenier, J. Mater. Chem. **6**, 1511 (1996).
95. L. Ranno, M. Viret, A. Mari, R.M. Thomas, and J.M.D. Coey, J.Phys.: Condens. Matter **8**, L33 (1996).
96. C. Ritter, M.R. Ibarra, J.M. De Teresa, P.A. Algarabel, C. Marquina, J. Blasco, J. García, S. Oseroff and S.W. Cheong, Phys.Rev.B (issue of 1st October) (1997).
97. B.C. Hauback, F. Helmer, and N. Sakai, J.Solid State Chem. **124**, 43 (1996).
98. J.A. Alonso, M.J. Martínez-López, M.T. Casáis, and A. Muñoz, Sol. State Commun. **102**, 7 (1997).
99. A. Gupta et al., Appl. Phys. Lett. **67**, 3494 (1995).
100. A. Arulraj, R. Mahesh, G.N. Subbanna, R. Mahendiran, A.K. Raychaudhuri, and C.N.R. Rao, J. Solid State Chem. **127**, 87 (1996).
101. R.J. Weiss, Proc. Phys. Soc. **82**, 281 (1963).
102. V.L. Moruzzi and P.M. Marcus, *Handbook of Magnetic Materials*, ed. K.H.J. Buschow, North-Holland **7**, 97 (1993).

COLOSSAL MAGNETORESISTIVE MANGANITES : THE PUSH TOWARDS LOW FIELD MAGNETORESISTANCE

R. Ramesh, S. B. Ogale, M. Rajeswari, R.L. Greene, and T.Venkatesan
Center for Superconductivity Research and Materials Research Science and Engineering Center, University of Maryland, College Park, MD 20742.

ABSTRACT : We present a review of our progress towards achieving large, low field, room temperature magnetoresistance values in substituted rare earth manganites. The re-emergence of research and scientific interest in these materials has been triggered by the observation of large magnetoresistance values, albeit at large magnetic fields. Over the past 3-4 years research effort in this field has focused on : (i) models to understand the origin of the large magnetoresistance that is observed ; (ii) approaches to reduce the field dependence and increase the temperature range in which large magnetoresistance is observed. In our own research program, we continue to focus on both these directions with parallel efforts on thin films and bulk materials. This review summarizes a few of the highlights of our progress to date.

Perovskite materials continue to be focal points for materials research and development effort around the world. Perovskites such as barium titanate were extensively studied and used in the forties as active elements in communications systems. More recently, the confluence of novel thin film deposition techniques and the observation of superconductivity at temperatures above that of liquid nitrogen, has triggered off a flurry of activity in a broad class of perovskite and related materials. Another parallel area of considerable technological and scientific interest is ferroelectric and dielectric perovskites. Thin films of these materials, integrated on Si/GaAs are strong candidates for solid state memories, sensors, actuators, monolithic

$$\underline{A_{1-x}B_xMnO_3}$$

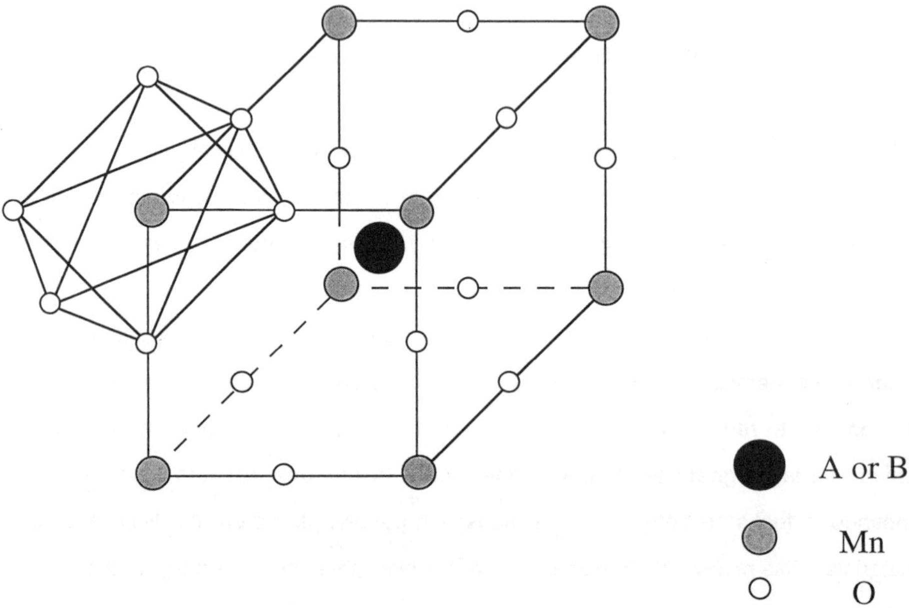

$$A = La, Nd, Pr$$

$$B = Ba, Ca, Sr$$

$$x = 1/3$$

Figure 1 : A schematic of the perovskite crystal structure of substituted rare earth manganites.

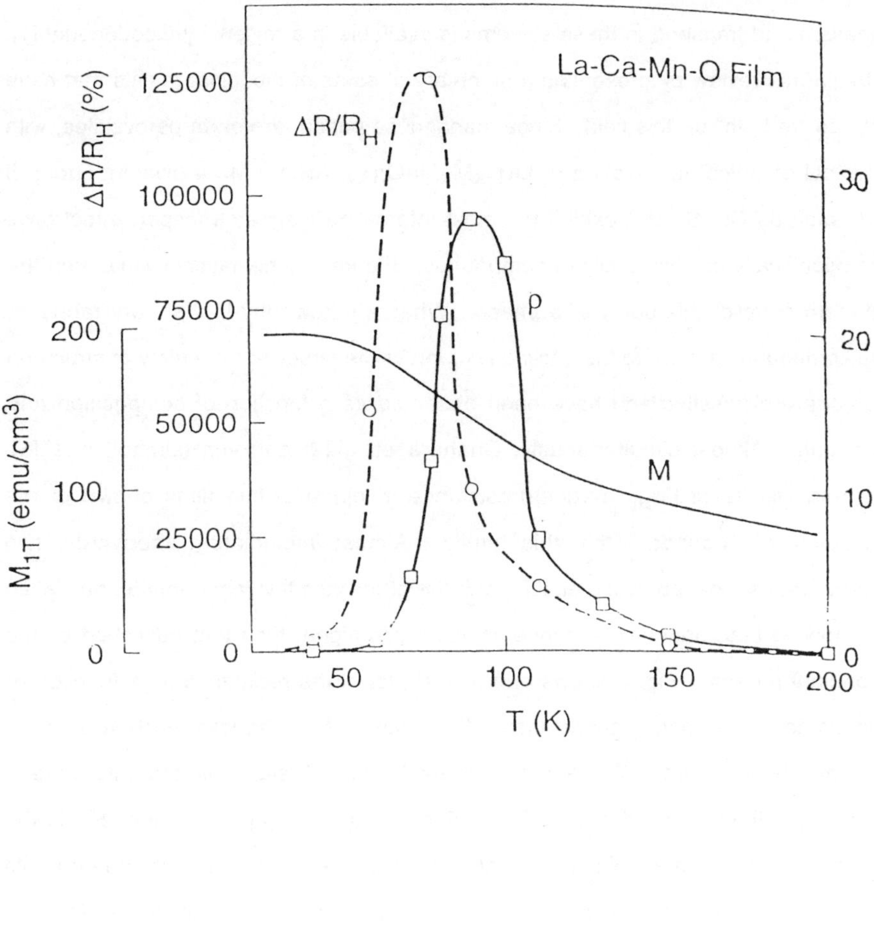

Figure 2 : Temperature dependence of saturation magnetization, resistivity and magnetoresistance for an epitaxial La-Ca-Mn-O film.

microwave elements and infrared sensors, etc. Focusing specifically on the magnetic perovskites, considerable work has already been carried out on magnetic perovskites based on Fe, Co, Ni and Mn. A very detailed review of the structural chemistry, magnetism and transport in these systems is available in a review by Goodenough[1]. We begin this review by presenting a summary of some of the initial results that have led to the "re-birth" of this field. Since manganese based rare earth perovskites, with the typical chemical formulation of $La_{1-x}M_xMnO_{3\pm y}$, (where M= a divalent Group II cation such as Ca, Sr, Ba) exhibit the most interesting magnetotransport effects, we focus specifically on this family of compounds. Figure 1 schematically illustrates the crystal structure of this perovskite phase. What is noteworthy is that very rarely do these compounds adhere to the basic cubic perovskite structure ; a variety of structural distortions are possible and have been observed as a function of composition and processing. Almost simultaneously, Chahara et. al.[2] and vonHelmholt et.al.[3], presented results of large magnetoresistance in epitaxial thin films of two of the prototypical compounds within this family. Almost immediately afterwards, two different groups, one led by Jin et.al.[4] and the other from this laboratory[5], published data showing that magnetoresistance values much higher than that published by the two original papers. Fig. 2 shows one typical plot of the resistance as a function of magnetic field ; the magnetoresistance (MR) value in this case was much larger than 10^5 % at a temperature of 77K and a magnetic field of 5Tesla. Note that this value of MR is larger than 100% due to its definition as : $(R_0 - R_H)/R_H$, where R_0 is the resistance at zero field and R_H is the value in an applied field of "H". When one puts this into a technological perspective, the field values needed to get useful MR are much too high ! This was the trigger point in our program, to undertake a very detailed study of the various factors that impact the temperature and field dependence of magnetoresistance. Among them, the effect of lattice strain, oxygen stoichiometry and the effects of layering and heterostructuring are being studied. The thin films and

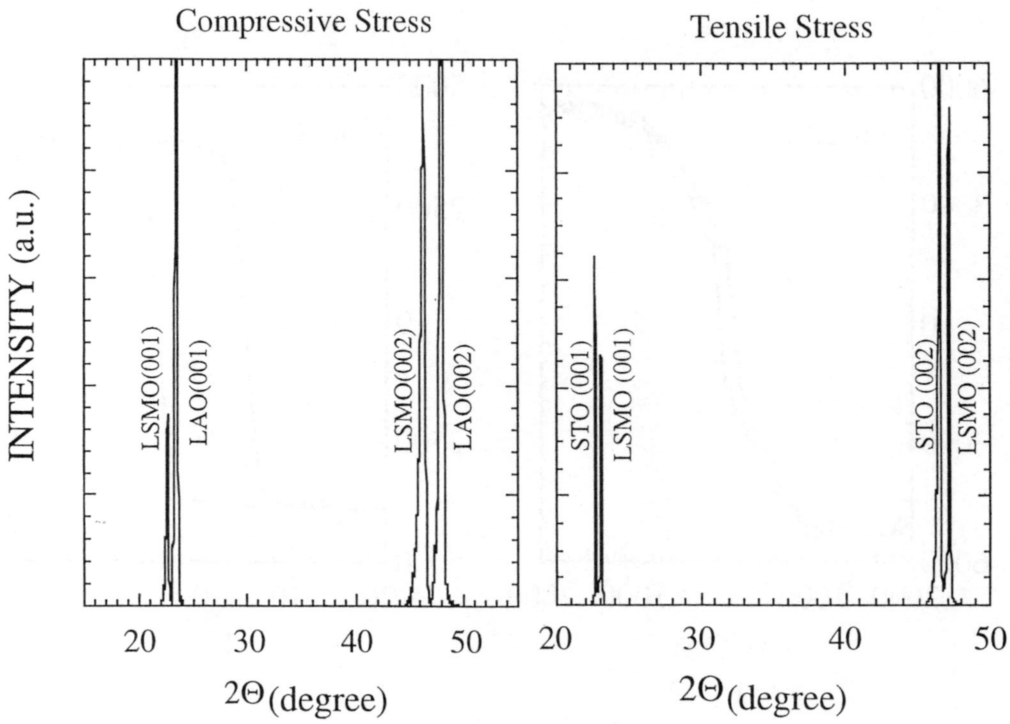

$C_{LSMO} = 3.926$ Å $\qquad C_{LSMO} = 3.856$ Å

Figure 3 : X-ray Bragg scans from epitaxial LSMO films grown on LAO and STO substrates illustrating the influence of substrate-film lattice mismatch on the film lattice parameters.

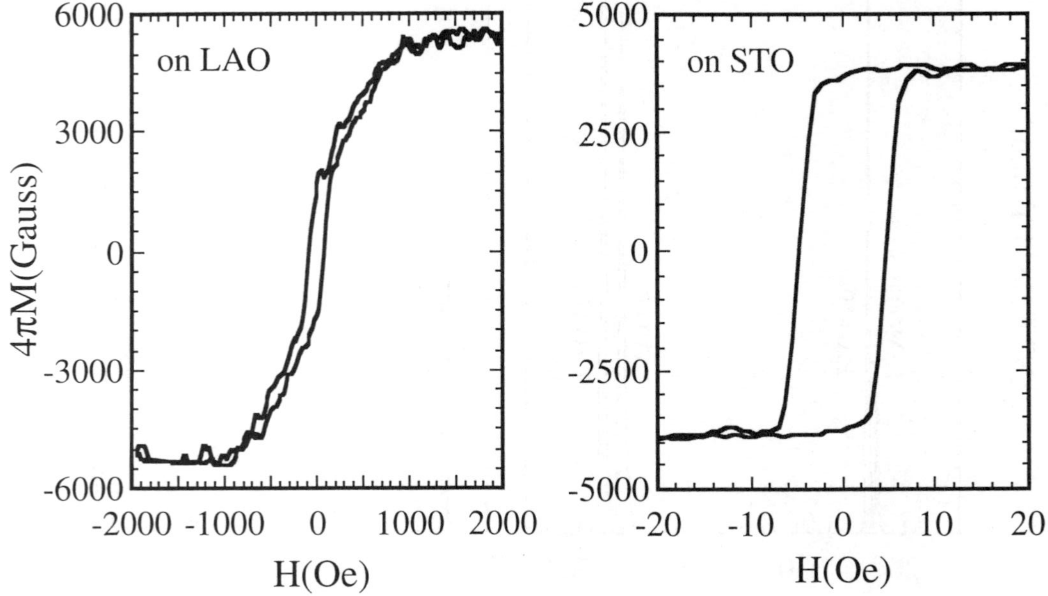

Figure 4 : Room temperature, in-plane hysteresis loops from epitaxial LSMO films grown on LAO and STO substrates showing the influence of strain on the preferred magnetization direction.

heterostructures are typically deposited by pulsed laser deposition. The reader should refer to one of the many articles in current literature for a detailed description of the thin film processing procedures as well as the details of the various experimental techniques that are being employed to characterize the materials.

Perovskite materials, even in the bulk, are very susceptible to lattice distortions, which may occur either spontaneously (e.g., ferroelectrics) or as a consequence of process induced variations of, for example, the oxygen stoichiometry. When grown in thin film form, they are even more susceptible to lattice strains, especially when constrained by heteroepitaxy to the substrate lattice. For example, the bulk lattice parameter of $La_{.7}Sr_{.3}MnO_3$ (LBMO) is 3.88Å while that of one common single crystal oxide substrate, $LaAlO_3$ (LAO) is 3.79Å, leading to a 2.4% compressive strain in the plane of the film[6]. On the other hand, if the same material is grown on $SrTiO_3$ (STO), the lattice mismatch is smaller (1.0%) and is tensile. This leads to out-of-plane strains in the film, as illustrated in Fig.3 where we see that 1000Å thick films grown on these two types of surfaces show different out-of-plane lattice parameters. What is interesting to note is that in both cases, the films do not seem to have relaxed into their bulk lattice dimensions, even after growth of 1000Å. Since these materials do not have any other source of magnetic anisotropy, they strongly exhibit stress (or conversely strain) induced magnetic anisotropy. For example, in the absence of any magnetocrystalline anisotropy, one would expect the easy direction of magnetization to be controlled by the magnetostatic energy, thereby inducing an in-plane easy magnetization direction (or easy magnetization plane). This is indeed seen in magnetic hysteresis, measured in the plane of the film, of the films grown on STO, Fig.4. However, the films grown on LAO are actually preferentially magnetized out of the plane, see Fig.4, with the magnetization vector possibly lying at an angle to the substrate surface. This conclusion is furthermore authenticated by magnetic force microscopy (MFM) studies of the film surface, which is illustrated in Fig.5. The films

Tensile
(STO)

Compressive
(LAO)

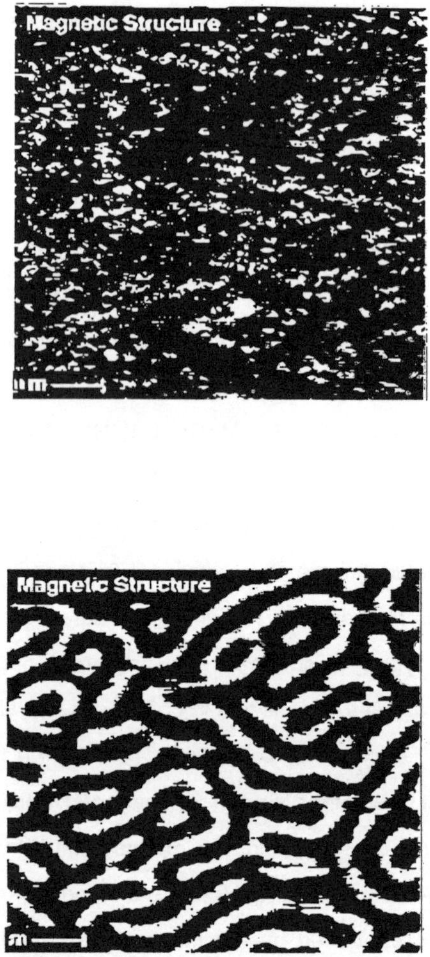

Figure 5 : Magnetic force microscopy images of the film surface in the case of STO and LAO substrates illustrating the effect of mismatch strain on the preferred magnetization direction and consequently the domain structure.

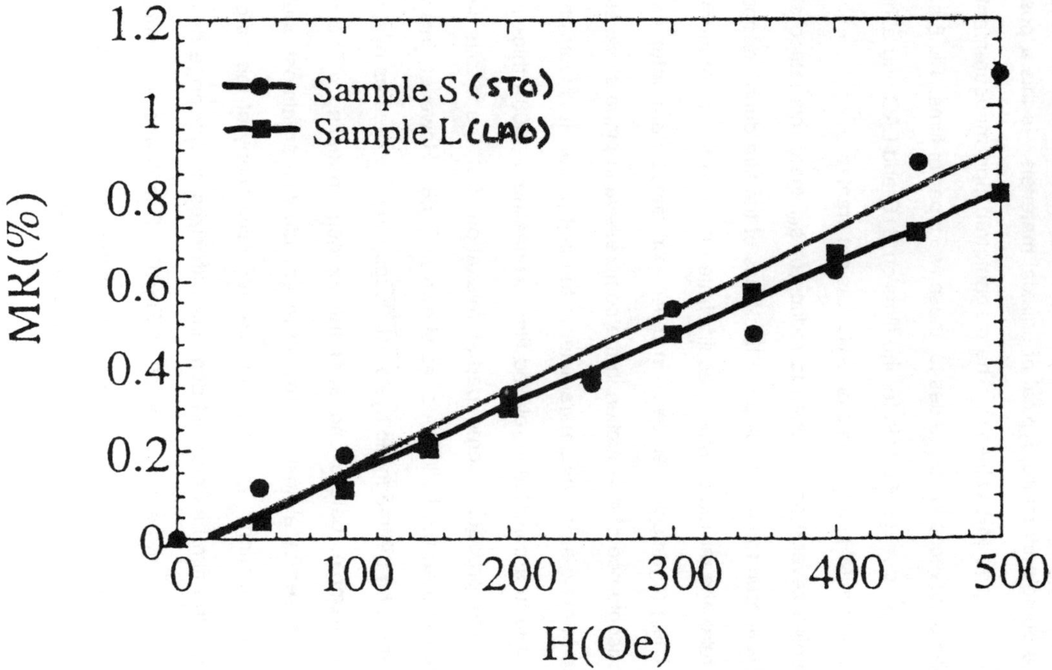

Figure 6 : Magnetoresistance as a function of applied field for the epitaxial films grown on STO and LAO showing that the mismatch strain has only a minimal influence on the MR properties.

grown on STO show a "feathery" surface structure (reflecting the fact that there is no magnetization vector normal to the film surface) while the films on LAO show the characteristic maze domain structure, typical of uniaxial materials. Is this a possible ingredient in obtaining large MR values ? The magnetotransport measurements do not indicate dramatic differences in MR values for these two types of films. Fig.6 shows MR as a function of applied magnetic field for the films on STO and LAO and although there is a small increase in MR for the films on STO, it is not dramatic.

We have studied several approaches to reduce or eliminate this lattice strain. Among them, we have found that changing the thickness of the film during deposition can progressively relax the mismatch strain, as illustrated through the x-ray diffraction spectra in Fig.7. The progressive decrease in the c-axis lattice paremeter can be interpreted as a consequence of a reduction in the compressive in-plane stresses that lead to an increase in the c-axis lattice parameter in the first place. It is worth noting that this does not lead to a completely relaxed film. Annealing a 1000Å thick film in oxygen at 900°C also produces some partial relaxation, but the c-axis lattice parameter is still larger than the bulk value, as shown in Fig.8. However, this does have a significant effect on the magnetic hysteresis loops. Fig.9 compares hysteresis loops for the same sample, before and after the oxygen annealing. The initial susceptibility has increased by almost an order of magnitude with an almost doubling of the saturation magnetization. Transmission electron microscopy of the as-grown samples indicate that the film has ordered domains with sizes in the range of 10-30 nm, as illustrated in the dark field micrograph in Fig.10. This image was obtained using the superlattice reflection which is circled in the diffraction pattern shown in the inset. The regions in bright contrast correspond to one set of ordered domains while the regions in dark contrast correspond to the other. We also observe that the domain size increases considerably with annealing in oxygen. The interfaces between these domains, i.e., the antiphase boundaries, can act as scattering sites during transport.

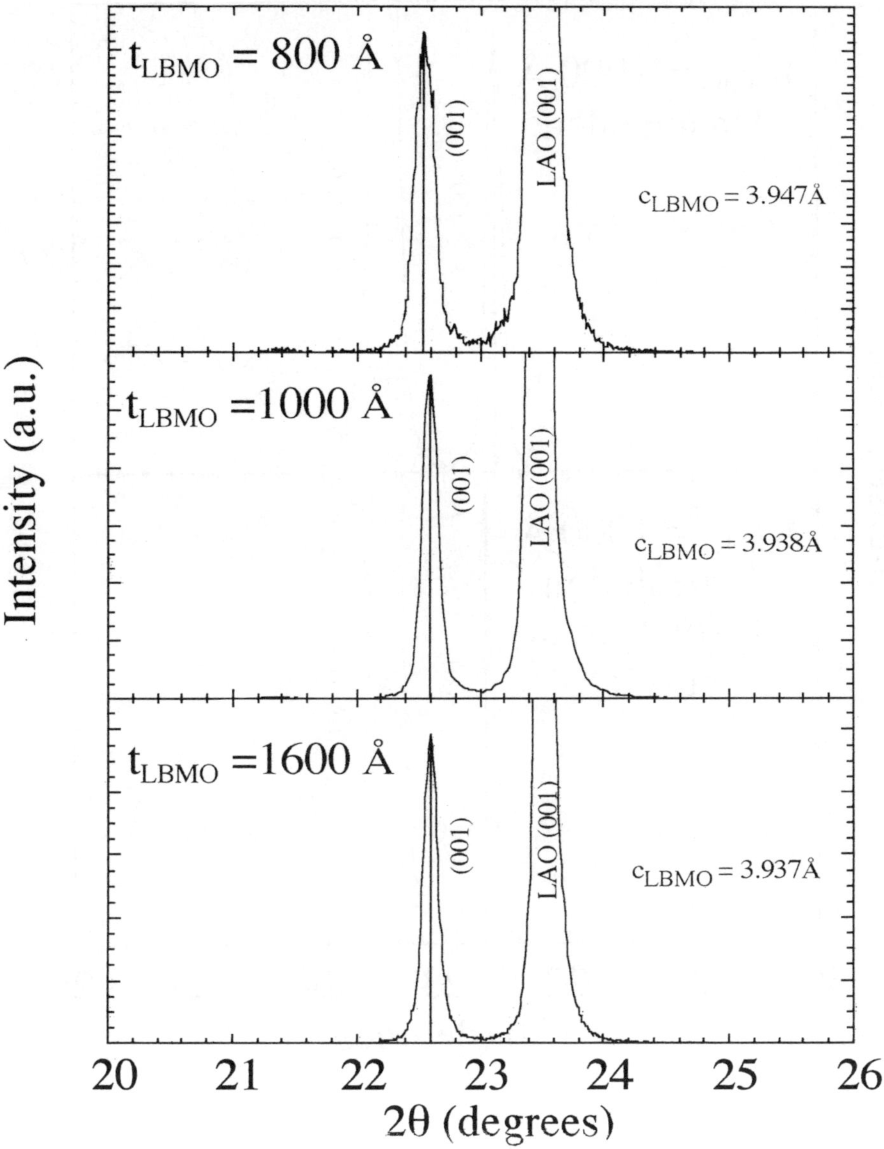

Figure 7 : X-ray diffraction spectra as a function of thickness of the LSMO layer, showing the partial relaxation of lattice mismatch strain with increasing thickness.

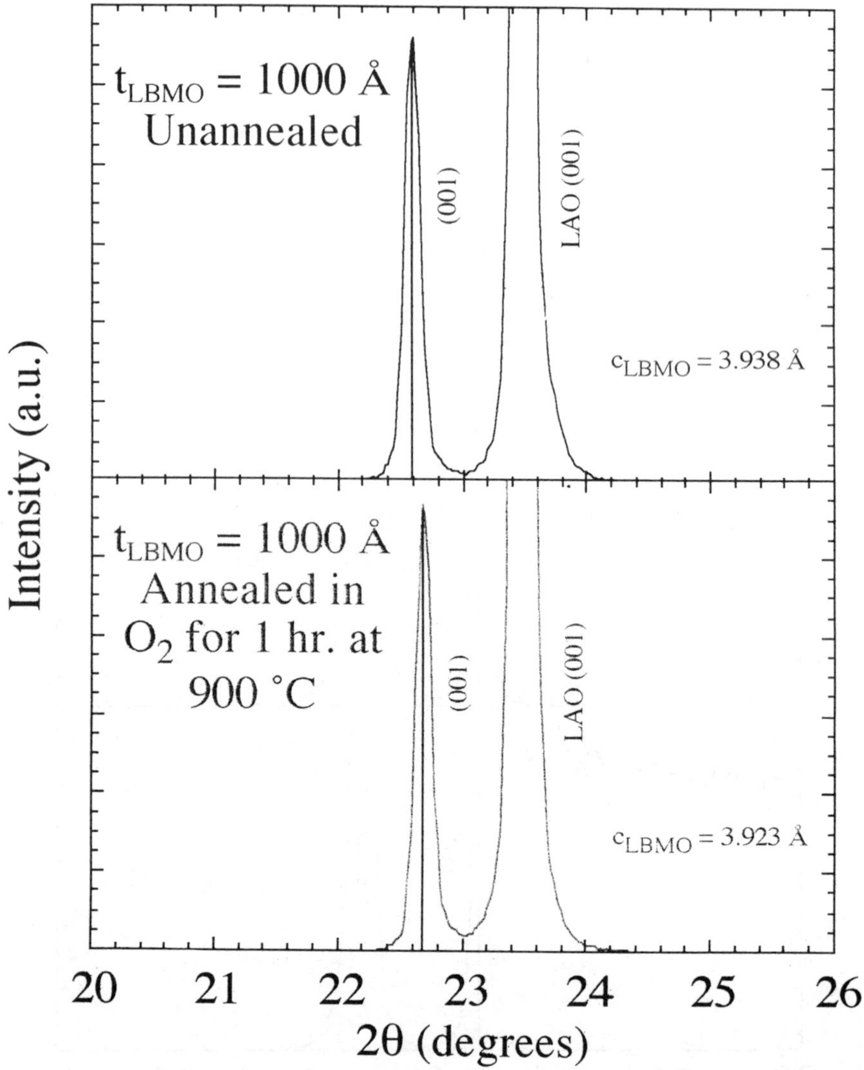

Figure 8 : X-ray diffraction spectra of an epitaxial LSMO film before and after annealing at 900°C showing the reduction in the out-of-plane lattice parameter due to relaxation of the lattice mismatch strain.

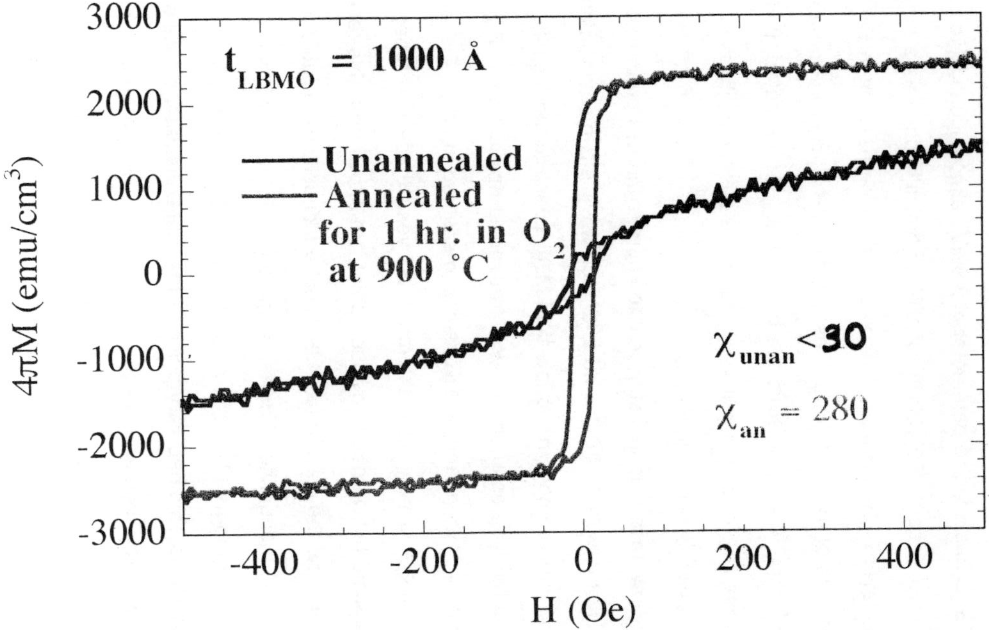

Figure 9 : In-plane hysteresis loops of the LSMO film before and after annealing at 900°C showing the significant improvement in the squareness as well as the low field susceptibility.

The effect of these types of defects on the magnetotransport properties need to be understood through further systematic experiments.

One of the most effective ways to alleviate the effects of lattice mismatch strain is to use buffer layers. A very suitable buffer layer is epitaxially grown $SrTiO_3$(STO) on a [001] $LaAlO_3$ substrate. STO has a lattice parameter of 3.905Å, almost identical to that of LBMO (3.9Å). A thin layer (about 700-1000Å) of STO is typically used as a buffer layer on which the LBMO layer is then grown. The x-ray diffraction patterns in Fig.11 clearly show the effectiveness of the STO layer. The LBMO layer now has an out-of-plane lattice parameter of 3.907Å, very close to the bulk value. As expected, this is also reflected in the magnetic properties, primarily the initial susceptibility and the saturation magnetization, which is shown in Fig.12.

The obvious question now arises : is this sufficient to improve the MR properties of the CMR layer, especially at low fields ? Our dc and microwave MR measurements indicate that there is some improvement in the low field MR values, but not significant enough to impact an actual implementation in low field sensing applications. Fig.13 shows a comparison of the microwave and dc MR values, measured at 1800 Oe, and clearly demonstrates that there is very little difference between these two types of measurements. An interesting, but perhaps obvious, point that arises from this result is that the magnetotransport and the consequent MR effect are directly associated with the electronic properties of the manganite. Further supporting evidence for this is also available from optical measurements [7]. It is also of interest to note that in these epitaxial films, the highest MR is obtained very close to the metal-insulator transition, although there is still considerable debate over the actual location of the peak in the MR value vis-a-vis the ferromagnetic Curie temperature[8]. This is to be contrasted with data that will be presented later on non-epitaxial films that show a very different MR behavior. Finally, it is to be noted that at this field level, the MR values are still only of the order of a few %, which is still not sufficient for low field magnetic sensing

Figure 10 : A dark field transmission electron micrograph of the as-grown LCMO film showing the existence of ordered anti-phase domain structure. The inset shows the electron diffraction pattern and the superlattice reflection that was used to form this dark field image.

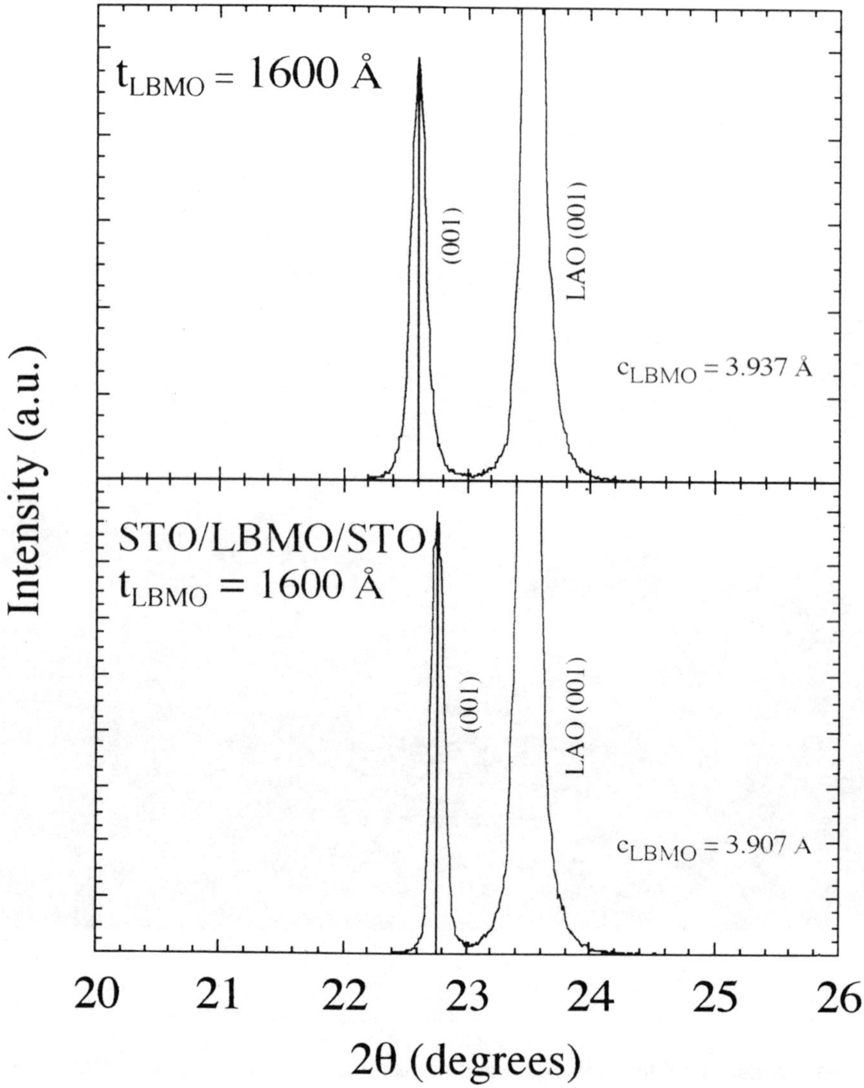

Figure 11 : X-ray diffraction patterns from an epitaxial LBMO thin film grown on a [001] LAO substrate with and without a STO buffer layer, showing the beneficial effect of the STO layer in relaxing the mismatch strain.

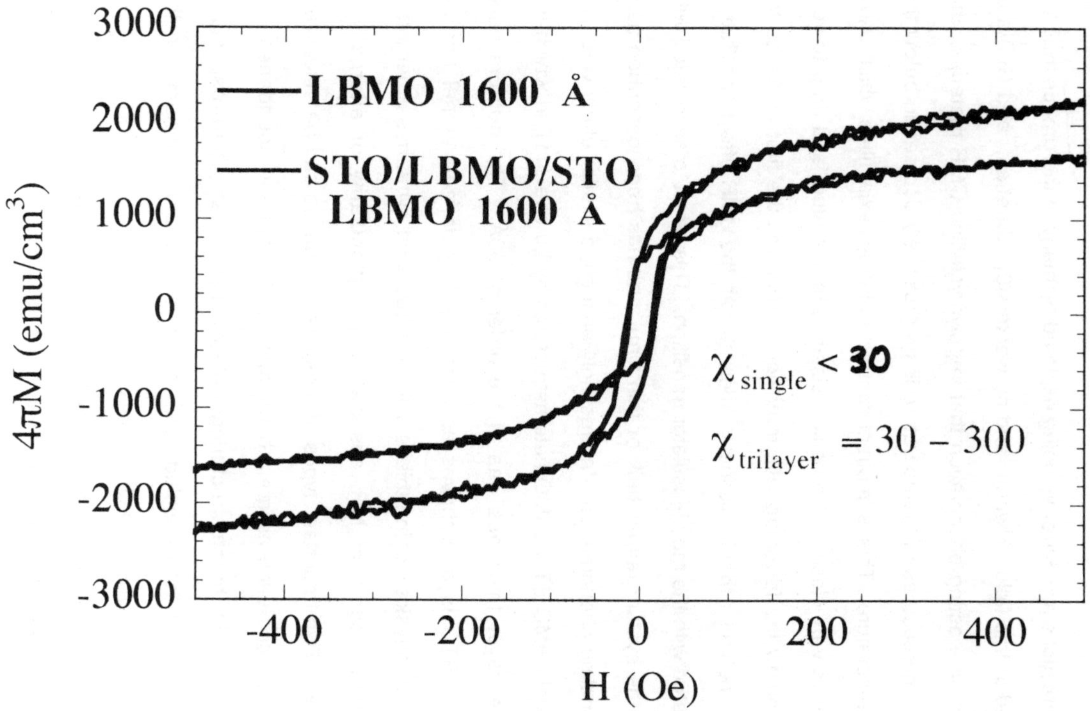

Figure 12 : In-plane hysteresis loops of the LBMO film grown with and without the STO buffer layer, showing the strong enhancement of low field susceptibility.

applications such as in read heads, etc. Obtaining large MR values at low fields clearly requires further ingenuity and engineering of the thin films and their microstructures.

Two different directions are currently being explored by many of the groups that are intensively involved in this field. Hwang and co-workers[9], Ju et.al., and Gupta et al.[10] made a very interesting observation that in polycrystalline CMR materials, the MR values actually increase as the temperature is progressively decreased below the peak resistance temperature. This is in direct contrast to the epitaxial films which show the largest MR values very close to the peak resistance temperature and drop to very small values well below the ferromagnetic transition. This was the first evidence for magnetoresistance behavior that is distinctly different from the CMR effect and that is not closely associated with the metal-insulator transition. This effect is associated with the granularity of the system (either bulk or thin film) and has been described as a spin polarized tunneling phenomenon. Interestingly enough, a similar effect was also observed in epitaxial LBMO/STO/LBMO heterostructures by Lu et.al.[11], in which the STO tunnel barrier layer thickness was of the order of 50Å. The authors have suggested that the STO layer behaves as such a spin polarized tunnel barrier. Although the detailed physics behind this is still being resolved, what has emerged is that there is another source of MR, associated with granularity or alternatively structural interfaces. Two important types of structural interfaces can be visualized, namely, natural interfaces such as grain boundaries and antiphase boundaries and those created artificially by heterostructuring. We discuss the structural aspects of these two types next and then conclude with some results and comments on the magnetotransport behavior across such interfaces.

Four distinct classes of films can be generated, based on the type of substrate structure, surface chemistry and deposition conditions. For example,m by systematically changing the nature of the substrate surface we can obtain these four

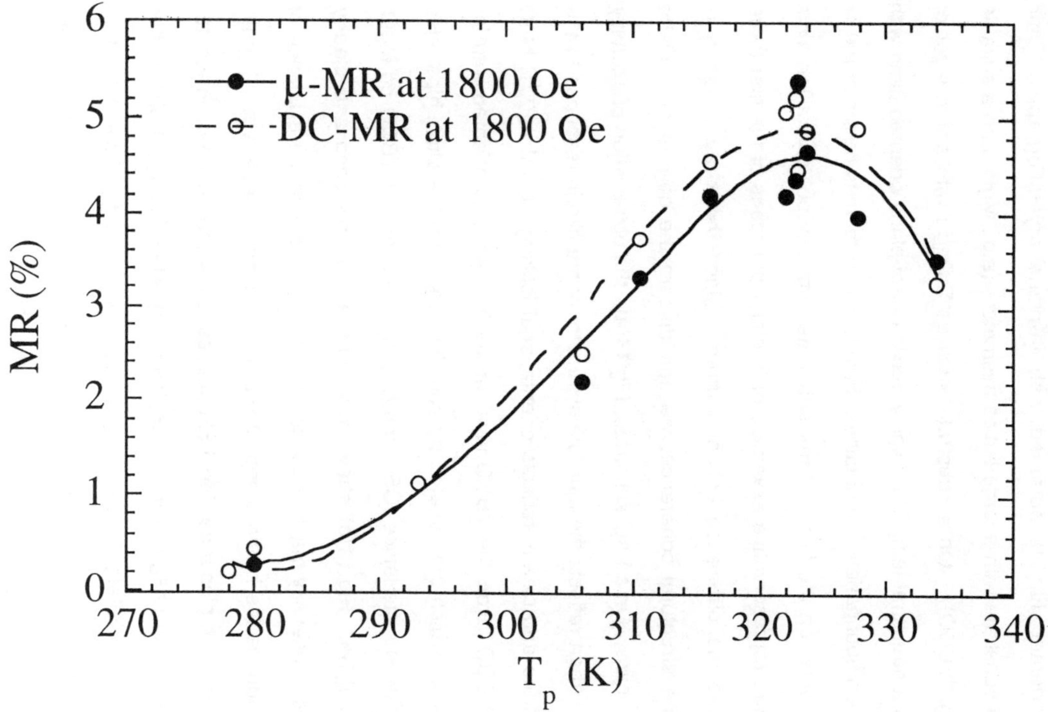

Figure 13 : Comparison of the dc and microwave MR of an epitaxial LBMO film at an applied field of 1800Oe, clearly demonstrating that the two types of measurements yield the same MR value.

distinct microstructures. For example, a fully epitaxial film of the CMR material (e.g., LSMO) on an almost lattice matched substrate such as LAO or STO can be obtained, in which case there is full in-plane orientational locking (i.e., $[100]_{LSMO} \parallel [100]_{LAO}$). A highly c-axis oriented film is obtained with in-plane crystallographic grain boundaries using a substrate with a large lattice mismatch (e.g., MgO) and a suitable template layer (e.g., YBCO). On a substrate such as SiO_2/Si with a thin layer of bismuth titanate as a structural template, highly c-axis out-of-plane oriented films with random in-plane grain boundaries are obtained. Finally, when there is no structural or chemical similarity between the film and the substrate, typically polycrystalline films result. Fig.14 shows representative examples of in-plane phi-scans of the first three types and illustrates the difference between them. Since the x-ray phi-scan is sensitive to in-plane structural coherence, we can discern the degree of in-plane registry from these scans. On a LAO substrate, Fig.14(a), the observation of intensity maxima spaced 90° apart reflects the four-fold symmetry of the film in response to the same for the substrate and clearly reflects orientational locking. On the [001] MgO substrate with the YBCO layer, the YBCO layer grows highly c-axis oriented with in-plane grain boundaries which are dictated by specific geometrical relationships, also known as the coincidence site lattice (CSL) model [or in the case of YBCO on MgO, the near-CSL model][12]. Fig.14(b) shows the phi-scan for this case and clearly shows the presence of intensity maxima spaced 45° apart, indicating the existence of 45° rotation grain boundaries in the plane of the film. Indeed, these type of grain boundaries are observed in planar section TEM images, shown for example in Fig.15. Fig.15(a) shows the actual image of this grain boundary while Fig.15(b) shows the same image after being digitally filtered. Figs.15(c,d) show light optical diffraction patterns obtained from the two grains showing that they are of a single orientation, while the composite optical diffractogram in Fig.15(e) shows clearly the 45° rotation between them. Returning to Fig.14, the third x-ray phi-scan shows the case of a LSMO

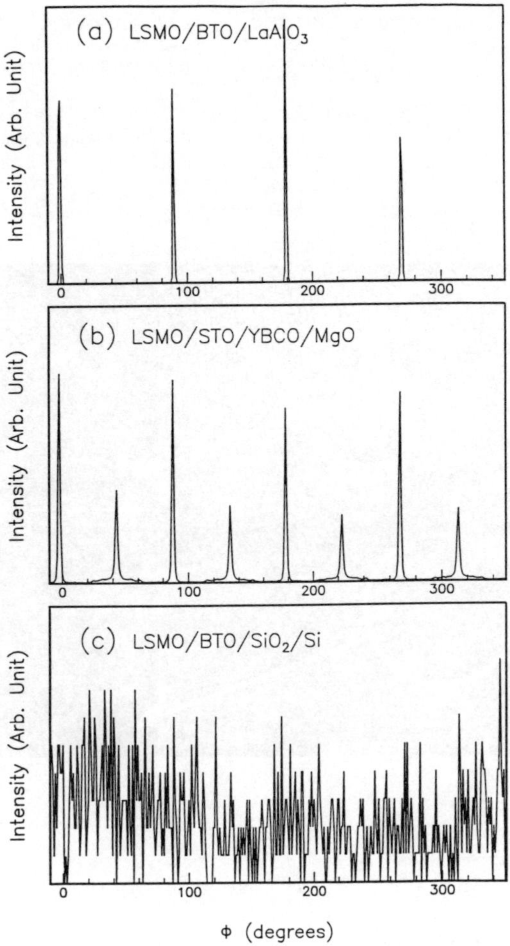

Figure 14 : X-ray phi-scans from three types of films : (a) epitaxial film on LAO substrate showing the full in-plane orientation locking ; (b) highly [001] oriented film on MgO with a predominance of 45° in-plane grain boundaries ; (c) highly [001] oriented film on SiO_2//SI with no in-plane orientation locking.

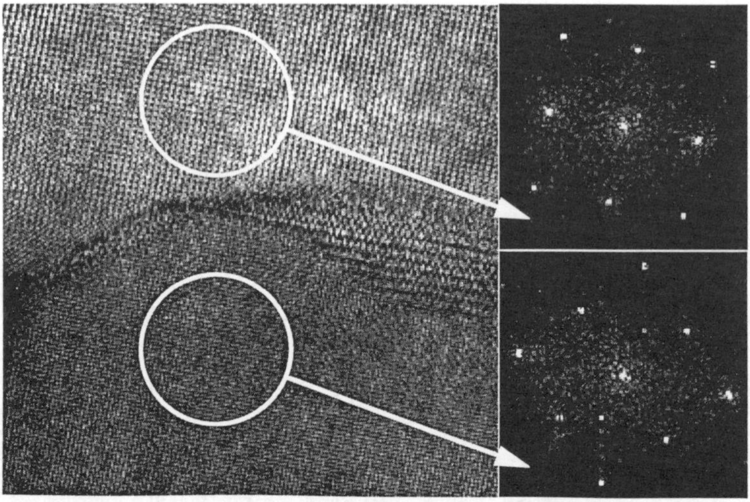

Figure 15 : Planar section transmission electron micrographs of the film grown on MgO showing the characteristic45° grain boundaries, identified from the optical diffraction patterns.

film grown on $BTO/SiO_2/Si$ where the film is c-axis oriented but there is no in-plane orientational locking.

Fig.16 shows the effect of these grain boundaries on the low field magnetotransport properties, at a temperature (77K) well below the Curie temperature or the maximum resistance temperature. The epitaxial film shows almost no MR; however, the films with grain boundaries show a much larger, negative magnetoresistance compared to the epitaxial films. This is one of the interesting aspects of transport in these materials that still needs detailed elucidation. Fig.17 shows schematically a possible model to explain the large field dependent MR in the films with grain boundaries. We suspect that the intrinsic structural disorder at the grain boundaries (the degree of which depends on the nature of the grain boundary, namely whether it is a CSL-type semi-coherent grain boundary or a fully random grain boundary) leads to spin disorder very close to the grain boundaries. Upon application of a small magnetic field, the ferromagnetic material inside the grain becomes fully magnetized while the grain boundary regions are still not fully aligned in the applied field direction[13]. The effective field at the grain boundaries is of the order of $H_{external} + 4\pi.M$ and eventhough the external field, $H_{external}$, is only a few hundred Oersteds the total field is of the order of a few thousand Oersteds. It is still unclear if this effective internal field is sufficient to cause the alignment of spins in the grain boundary regions, and further systematic studies are required.

A second approach to low field MR in these oxides is through heterostructuring. Two types of heterostructure devices, namely spin valves[14] and spin tunnel devices are being explored in our group. A typical example of a spin valve is shown schematically in Fig.18. Such a device typically consists of two ferromagnetic(FM) layers (significantly different coercive fields) separated by a metallic, non-ferromagnetic layer. Case 1 in this figure shows the condition where the two FM layers are polarized in the same direction. Under these conditions, if a spin polarized

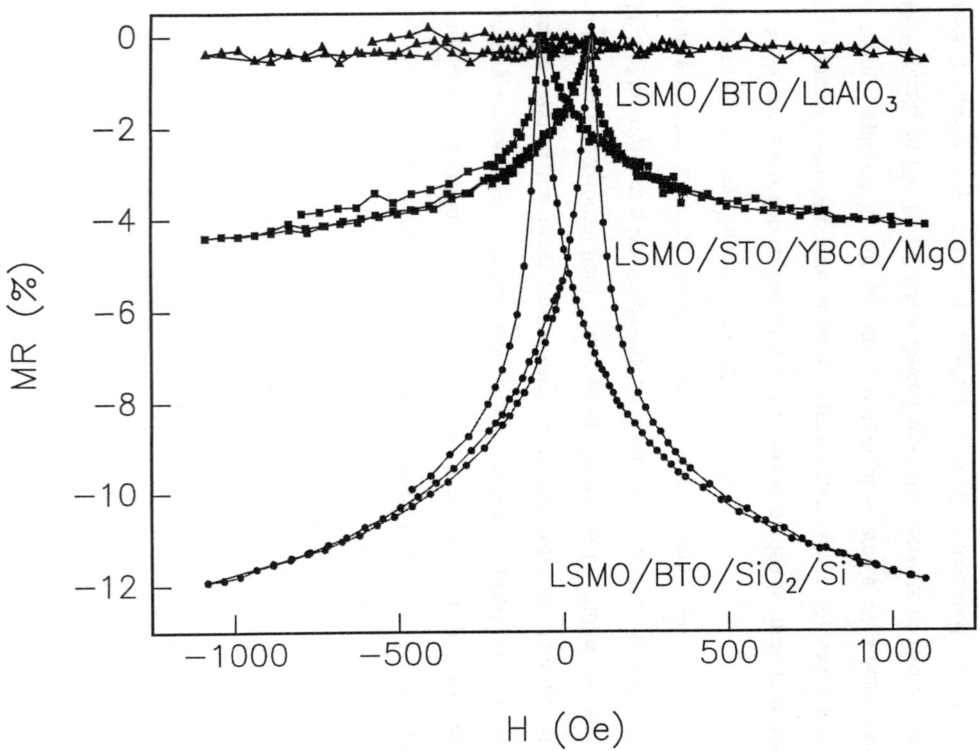

Figure 16 : Low field MR for the three types of films described in Fig.14, illustrating the beneficial effects of the in-plane grain boundaries.

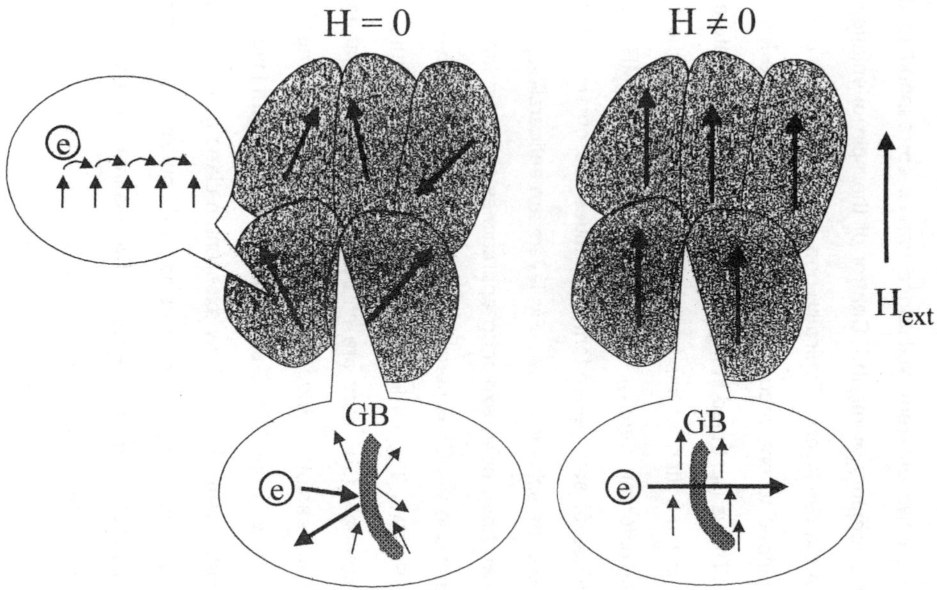

Figure 17 : A schematic illustration of the possible mechanism by which the large MR effects are observed in the granular films. At zero field, the magnetization vectors in each grain is randomly oriented in addition to the intrinsic spin disorder at the grain boundaries ; when a field is applied the spins within the grains reorient and thus reduce the scattering within the grain, leading to a lower resistance.

electron is launched from the bottom FM layer, then it can pass into the upper FM layer through the normal metal layer without any hinderance leading to a low resistance. However if Case 2 is true the spin polarized electron launched from the bottom FM layer will get scattered at the upper FM/metal interface and the resistance will be higher. In some sense, these two different resistance values correspond to two different logic states (as in a memory element). Control of the spin orientation is through the coercive fields of the two layers. Typically, one of the layers has a much higher coercive field or is pinned to one magnetization state by exchange biasing through a layer underneath. This is illustrated in Fig. 19 through schematic hysteresis loops. The hysteresis loop in the bottom of this figure shows the characteristic split indicative of the two distinctly different coercive fields. The corresponding MR plot is shown in Fig.20. Large values of spin dependent MR values are obtained in the intermediate field regime where the spins in the two FM layers are antiparallel.

Using these concepts, we have been exploring approaches to obtain large MR values in heterostructures that use the CMR perovskites as one of the FM layers, as schematically illustrated in Fig.21. The intrinsic similarity in crystal structure and chemistry between these materials (i.e., they are all based on oxygen octahedra) means that heteroepitaxy is feasible. Indeed, x-ray diffraction patterns from prototypical heterostructures grown using LSMO as the FM top and bottom layers and LSCO as the non-ferromagnetic metallic layer show excellent in-plane as well as out-of-plane structural coherence. Ferromagnetic hysteresis loops obtained from these heterostructures show clear proof of the existence of two distinct coercive fields, Fig.22, which is one prime requirement for a spin valve to operate. Transport measurements are currently underway and the details of magnetotransport in these spin valves should become clearer over the next year.

CONCLUDING REMARKS : Colossal magnetoresistance in perovskite manganates

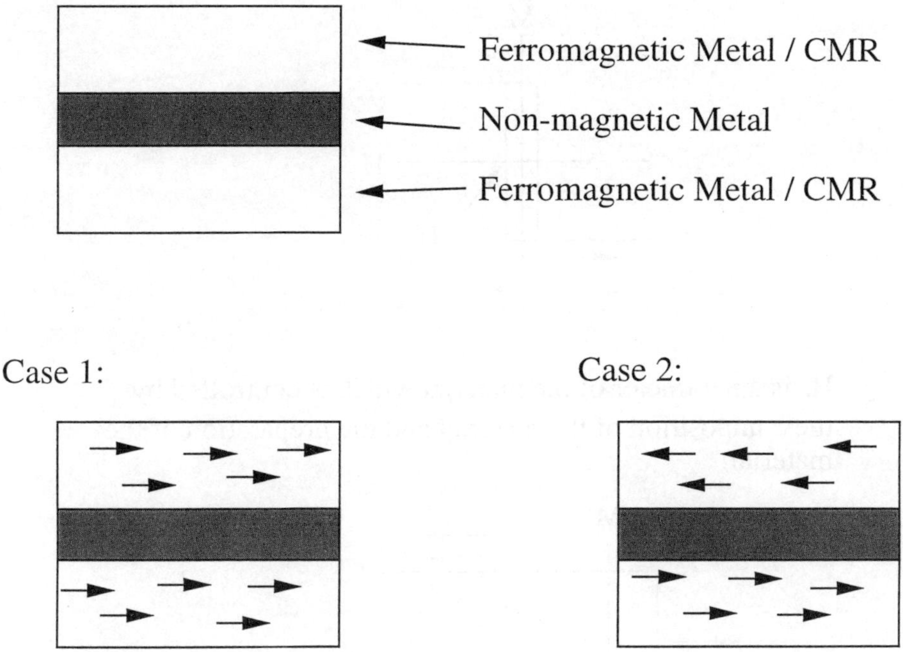

Resistance measured vertically through the stack depends on the relative spin orientation of the top and bottom ferromagnetic layers. If they are parallel (Case 1) there is no scattering and the resistance will be lower than when they are anti-parallel (Case 2) and there is scattering.

Figure 18 : A schematic illustration of the tri-layer structure of a spin valve illustrating the origin of scattering that is dependent on the relative orientations of the spin polarizations in the two FM layers.

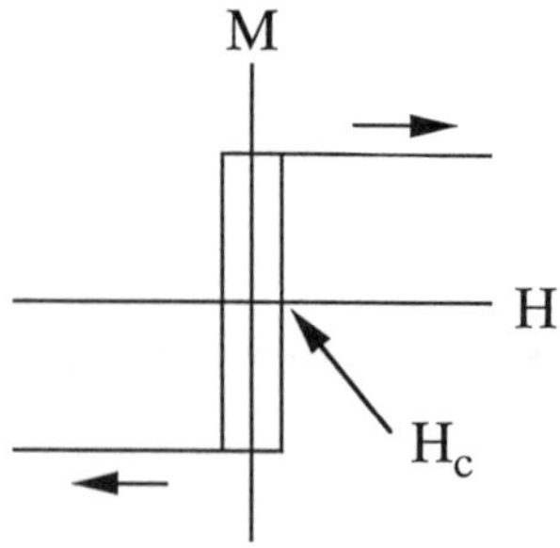

H_c is a parameter of the material which is controlled by the composition of the material and the preparation of the material.

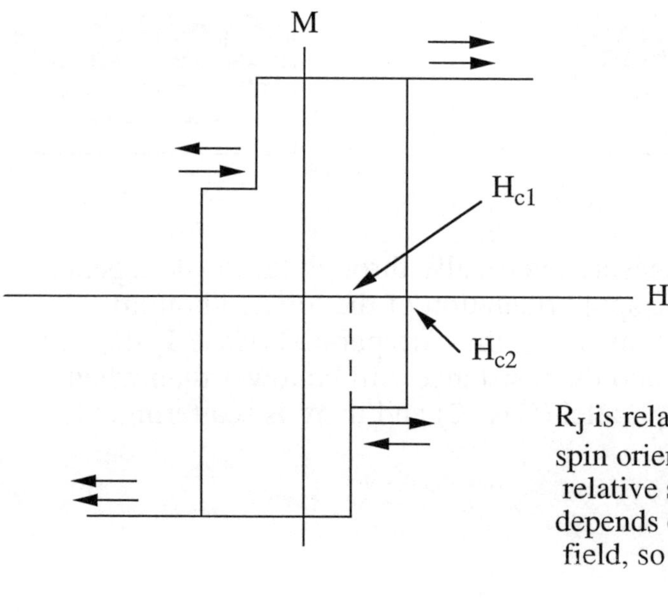

R_J is related to the relative spin orientation and the relative spin orientation depends on the magnetic field, so $R_J = R_J(H)$

Figure 19 : A schematic diagram of the method by which the spin polarizations in the two layers are controlled through the coercivities of the two FM layers.

$$MR(\%) = \frac{R_J(H) - R_J(0)}{R_J(0)} * 100$$

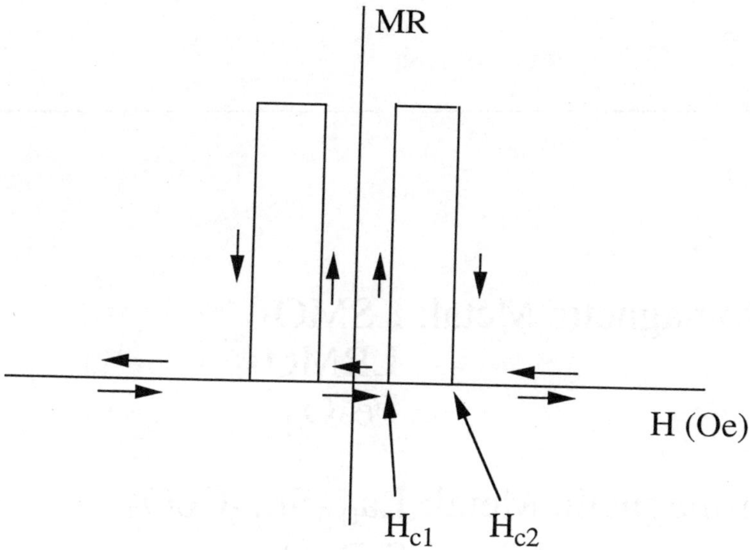

Figure 20 : The MR corresponding to the hysteresis loop shown in Figure 19 for the spin valve illustrated in Figure 18.

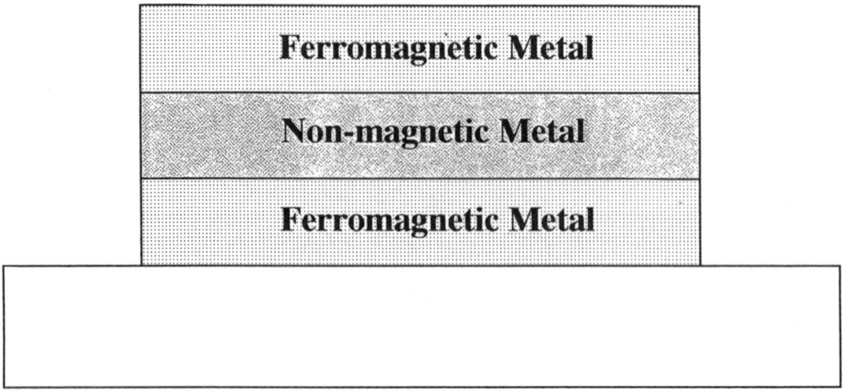

Ferromagnetic Metal: LSMO
LBMO
Fe_3O_4

Non-magnetic Metal: $La_{0.5}Sr_{0.5}CoO_3$
$SrRuO_3$

Figure 21 : A schematic diagram of the all-oxide spin valve that is being explored in our program.

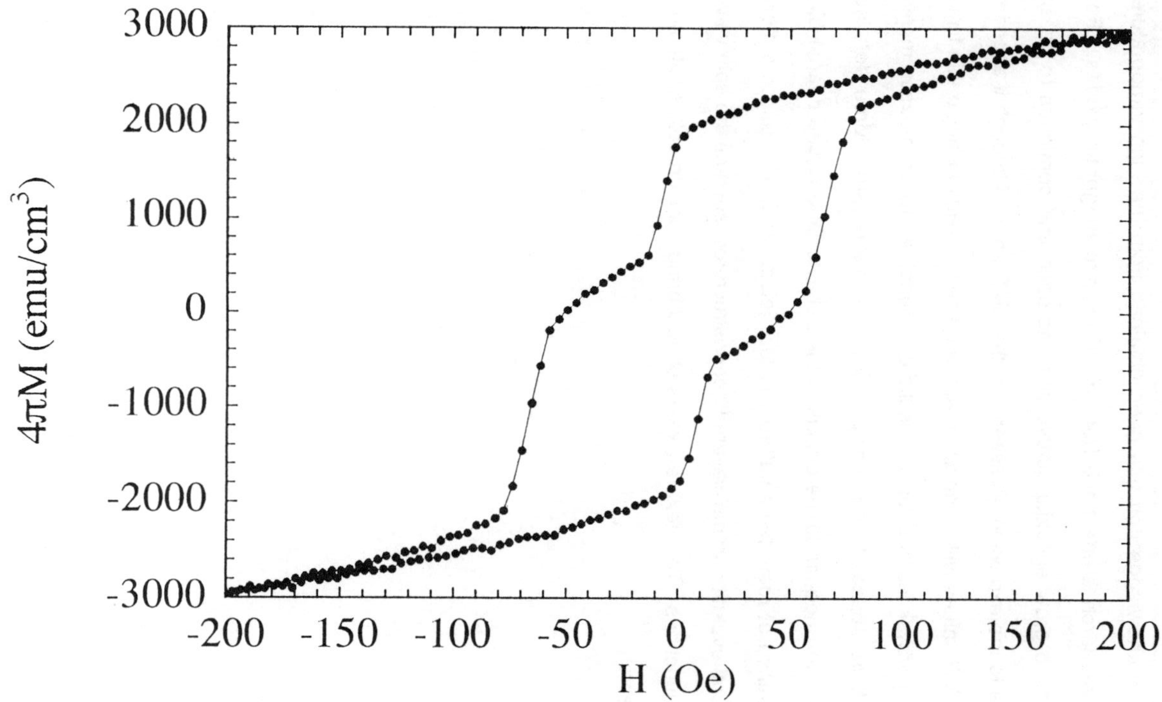

Figure 22 : Magnetic hysteresis loop of a LSMO/LSCO/LSMO trilayer spin valve showing the two distinct coercive fields that is a basic requirement for a spin valve operation.

is not a new phenomenon in the same way that the materials themselves aren't new. What has happened over the past 3-4 years is that new materials processing approaches and the knowledge that has been obtained from the high temperature superconductivity research arena have transferred over to magnetic oxides. For example, the ability to grow epitaxial perovskites has been and continues to be an important prequiste to the explosion of research in the CMR area. Although it appears doubtful that large MR values will be obtained at low magnetic fields in single thin films layers of the CMR perovskites, the recent results of magnetotransport measurements in polycrystalline films appear to be very promising in this regard. More detailed studies are required to understand the origins of the behavior and to see if large MR values can be obtained at room temperature and low fields. What has become more exciting is the natural evolution to multilayered heterostructures, such as the spin valve or the spin tunnel junctions. The next few years should bring considerable excitement to this area of research.

RERERENCES

1. J. Goodenough, Prog. in Solid State Chem. 5, 145(1972).

2. K. Chahara, et.al., Appl. Phys. Lett., 63, 1990 (1993).

3. R. Von Helmholt, et.al., Phys. Rev. Lett., 71, 2331(1993)

4. S.Jin, T.H.Tiefel, M.McCormack, R.Fastnacht, R.Ramesh and L.H.Chen, Science, 264, 413(1994).

5. G. C. Xiaong, Q. Li, H. Ju, R. L. Greene and T. Venkatesan, Applied Phys. Lett. 66, 1427 (1995).

6. C. Kwon, K.C.Kim, M.C.Robson, J.Y.Gu, S.E.Lofland, S.M.Bhagat, M. Rajeswari, T.Venkatesan and R.Ramesh, Jl. of Magn. and Mag. Matls., in press.

7. S.Kaplan, et al., Phys. Rev. Lett., 77, 2081(1996).

8. K.Ghosh, et al., Phys. Rev., B, in press.

9. H.Y.Hwang et al., Phys. Rev. Lett., 77, 2041(1996).

10. H. L. Ju, J. Gopalakrishnan, J. L. Peng, Qi Li, G. C. Xiong, T. Venkatesan and R. L. Greene, Phys. Rev. B51, 6143 (1995) ; A Gupta et al., Phys. Rev. B54, 15629(1996).

11. Y.Lu et al., Phys. Rev. B, 54, 8357(1996) ; J.Z.Sun, et al., Appl. Phys. Lett., 69, 3266(1996)

12. T.S.Ravi, et al., Phys. Rev., B42, 10141(1990).

13. N.D.Mathur, et al., Nature, in press.

14. M. Johnson and R. H. Silsbee, "Interfacial Charge-Spin Coupling; Injection and Detection of Spin Magnetization in Metals", Phys. Rev. Lett. 55, 1790 (1985) ; M. Johnson, " Bipolar Spin Switch", Science , 260, 320 (1993). ; M.Johnson, "The All-Metal Spin Transistor", I.E.E.E. Spectrum 31 , 47 (1994).

Low-Field Magnetoresistance Induced By Grain Boundaries In Doped Manganese Perovskites

Arunava Gupta
IBM T. J. Watson Research Center, Yorktown Heights, NY 10598, USA

Doped manganese perovskites exhibit orders of magnitude change in resistance near the ferromagnetic transition temperature when they are subjected to a strong magnetic field in the Tesla range. For practical applications it is necessary to significantly reduce the field scale and also broaden the temperature range of magnetoresistance. One approach has been to fabricate structures which exploit the high degree of spin polarization of the conduction electrons in these materials. It has been observed that the presence of grain boundaries disrupts the local magnetic order and alters the electrical conduction of polarized electrons, resulting in a large low-field magnetoresistance. Results obtained on samples consisting of naturally occurring or artificially prepared grain boundaries are reviewed and possible mechanisms for grain boundary transport are discussed.

1 Introduction

In the past few years a lot of attention has been focused on the colossal magnetoresistance (CMR) properties of doped manganese perovskites. While the magnetoresistance in these materials can be of unprecedented magnitude, in most cases the large resistance changes are achieved only in a strong field in the Tesla range, thus severely limiting their practical utility. Reducing the field scale and increasing the operating temperature has been the goal of a number of research groups worldwide. There has been some recent progress in reducing the field scale by taking advantage of the high degree of spin polarization of the conduction electrons in these materials.[1] One approach has been to fabricate heteroepitaxial magnetic tunnel junction structures containing magnetic electrodes of the manganites separated by a thin insulating barrier layer, such as $SrTiO_3$.[2-4] These tunneling structures have shown MR values of 50-90% at few tens of Oe at 4.2 K. Even higher values have been observed in some structures, but with poor reproducibility, presumably due to defects in the barrier layer or at the interfaces.[4] A large low-field MR component has also been observed in polycrystalline manganite samples, where a disruption in the crystalline order at the grain boundaries induces a local spin disorder.[14-18] Using grain boundaries (GB's) to manipulate magnetic behavior thus proves to be a simple method for enhancing the low-field sensitivity of these materials.

Significant differences in the magnetoresistance (MR) properties of polycrystalline and single crystal samples have been reported since the initial discovery of CMR in the manganites. In particular, early work on bulk polycrystalline samples have shown a substantial MR at temperatures much below the Curie temperature (T_c), while the MR magnitude is usually very small in single crystals or epitaxial films of the same composition.[5-13] A number of subsequent studies, both on bulk and thin film samples, have been undertaken to specifically clarify the role of GB's on the MR behavior of manganites.[14-21] Most of these investigations have focused on the doped manganites $La_{0.67}Ca_{0.33}MnO_3$ (LCMO) and $La_{0.67}Sr_{0.33}MnO_3$ (LSMO), or in samples with closely related compositions. We review the results of some of these studies in bulk and thin film samples. Recent experiments using manganite films grown on bicrystal substrates, which have further enabled to isolate the contribution of a single grain boundary, are also discussed.

2 Comparison of Magnetotransport and Magnetic Properties

2.1 Bulk Polycrystalline and Single Crystal Samples

One of the first reports regarding the effect of particle size on the transport and magnetic properties of polycrystalline LCMO and LSMO has come from Mahesh et al.[14] They prepared samples with different particle sizes (ranging from 0.025-3.5 μm) by a citrate-gel route followed by heat treatment at different temperatures in an appropriate atmosphere. For a direct comparison of the properties of the samples as a function of grain size, it is important to ensure that the Mn^{4+} concentration remains similar in the different samples, since it is a crucial factor in controlling the transport and magnetic properties. The Mn^{4+} content has been determined by redox titration. The resistivity (ρ) of the LCMO samples, with comparable Mn^{4+} concentration, has been found to increase substantially with particle size. Furthermore, the ratio of the ρ at 4.2 K and at the peak (corresponding closely to T_c), decreases with increasing particle size. The T_c is also observed to be broadened with decreasing particle size, with the 0.025 μm sample not even exhibiting a well defined transition temperature. The authors have noted that despite these changes, the MR near the peak did not show any significant changes with the size of the particle. However, the MR at 4.2 K is found to have a low-field and a high-field component, with the former increasing with decrease in particle size. These results, coupled with the knowledge that single crystal and epitaxial films exhibit very small MR at 4.2 K,[10-13] lead them to conclude that that a substantial part of the MR at low temperatures arises from the grain boundaries.

A more direct comparison of the MR and field dependent magnetization at low temperatures for single crystal and bulk polycrystalline samples has been provided by Hwang et al. in their study of LSMO single crystal and polycrystalline samples.[15] Single crystal LSMO samples have been grown using the floating zone

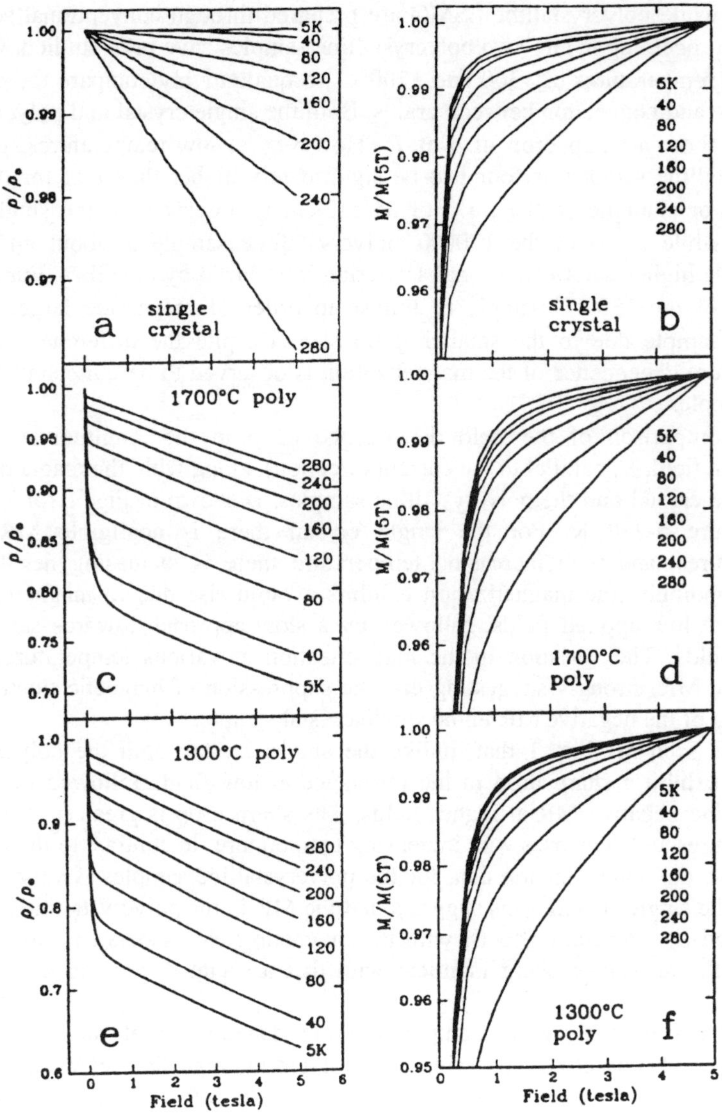

Fig. 1. Magnetic filed dependence of the normalized resistivity (a, c, e) and magnetization (b, d, f) at various temperatures for bulk single crystal and polycrystalline LSMO samples. From Ref. 15.

method, while polycrystalline LSMO are prepared through conventional solid-state reaction processing in air. Two polycrystalline samples have been studied, with final sintering temperatures of 1300 and 1700°C, to qualitatively compare the effects of grain size and connection between grains. Both the single crystal and polycrystalline samples show a sharp drop in ρ at T_c. However, at low temperatures, ρ of the polycrystalline samples is found to be significantly higher than that for the single crystal. For example, at 5 K, a ρ of 35 μΩ-cm is observed for the single crystal LSMO, while the ρ of the 1700°C polycrystalline sample is about an order of magnitude higher due to additional scattering introduced by the GB's. Furthermore, the ρ of the 1300°C sample is almost an order of magnitude larger than the 1700°C sample due to the smaller grain size. Despite the differences in ρ, the temperature dependence of the magnetization is observed to be quite similar for the three samples.

A comparison of the field dependence of ρ in the longitudinal geometry (magnetic field, *H*, parallel to the current direction), along with the magnetization of the single crystal and the polycrystalline samples, is shown in Fig. 1 for a range of temperature 5-280 K. For the single crystal there is negligible MR at low temperatures, and with increasing temperature there is increasing negative MR. Correspondingly, the magnetization exhibits a rapid rise due to magnetic domain rotation at low applied fields, followed by a slow approach towards saturation at higher fields. The variation in the magnetization at various temperatures closely tracks the MR, strongly suggesting that the suppression of magnetic fluctuations is the origin of the negative MR in the single crystal sample.

It is seen in Fig. 1 that, unlike the single crystal, both the polycrystalline samples exhibit a sharp drop in the resistance at low fields followed by a slower background negative MR at higher fields. The sharp drop is greatest at the lowest temperatures and decreases with increasing temperature. In contrast to the resistance variations, the magnetization data for the polycrystalline samples is very similar to that for the single crystal. This suggests that the MR in the polycrystalline samples is dominated by intergrain effects, with the magnetic field associated with the sharp drop in the resistance being identical with that associated with magnetic domain rotation.

In order to analyze the temperature (*T*) dependence of the low-field MR component, Hwang et al. have back-extrapolated the high field $\rho(H)/\rho_0$ to find the zero field intercept and calculate the magnitude of the MR associated with magnetic domain rotation. Their result shows quite clearly that the MR increases with decreasing temperature. A simple inverse temperature dependence does not accurately describe the temperature dependence of the MR. However, the data can be fit quite well to a functional form $a + b/(T + c)$, where a, b, and c are fitting constants; but the significance of this temperature dependence is not clear. Another significant observation noted by the authors, which is quite apparent from Fig. 1, is that above ~0.5T the MR in the polycrystalline samples appear to have the exact same field dependence for the entire temperature range 5 - 280 K. In addition to the visually obvious dominate H linear term, a weak H^2 dependence is also present at the high fields.

2.1 *Polycrystalline and Epitaxial Films*

At IBM we have systematically explored the properties of epitaxial and polycrystalline films of LCMO, LSMO, and $La_{0.75}MnO_3$ (LXMO) films.[16,17] To provide a direct comparison, both the polycrystalline and epitaxial films are grown under identical process conditions on the same substrate material ($SrTiO_3$). Consistent with the bulk results, we find that, unlike the epitaxial films, the polycrystalline films indeed show substantial MR over a wide temperature range in all three systems studied.

For the comparison, films, with thickness of about 1000Å, have been grown on (100)-oriented single crystal and polycrystalline $SrTiO_3$ substrates using pulsed laser deposition. The polycrystalline $SrTiO_3$ substrates are obtained by cutting and polishing sintered pellets (97-98% density) of the material. The time and temperature of sintering are varied to obtain samples with average grain size of 24 μm, 14 μm, and 3 μm. Transmission electron microscopy studies have confirmed the epitaxial and polycrystalline nature of the films grown on single crystal and polycrystalline $SrTiO_3$ substrates, respectively. In the case of the latter, a well-defined grain morphology reminiscent of the underlying substrate is observed.[16]

The temperature dependence of ρ at different magnetic fields are shown in Fig. 2 for polycrystalline LCMO films with three different grain sizes. The data for an epitaxial film is also included for comparison. As is commonly observed, the zero-field ρ for all the films first increases and then decreases as T is reduced. The peak resistivity (ρ_{max}) occurs at about the same T (~230 K) for all the samples and is very close to the T_c as determined from magnetization measurements. It is interesting to note that while the drop in ρ at lower temperatures is quite sharp for the epitaxial film, the decrease occurs much more slowly for the polycrystalline films. Moreover, the zero field ρ increases systematically with decreasing grain size over the whole range of T due to enhanced scattering at the GB's.

If the polycrystalline films are idealized to consist of low resistivity (ρ_1) grains of size L separated by thin GB layers of width L' and high resistivity ρ_2, then the effective resistivity is given by: $\rho \cong \rho_1 + (L'/L)\rho_2$. Indeed, an almost linear increase

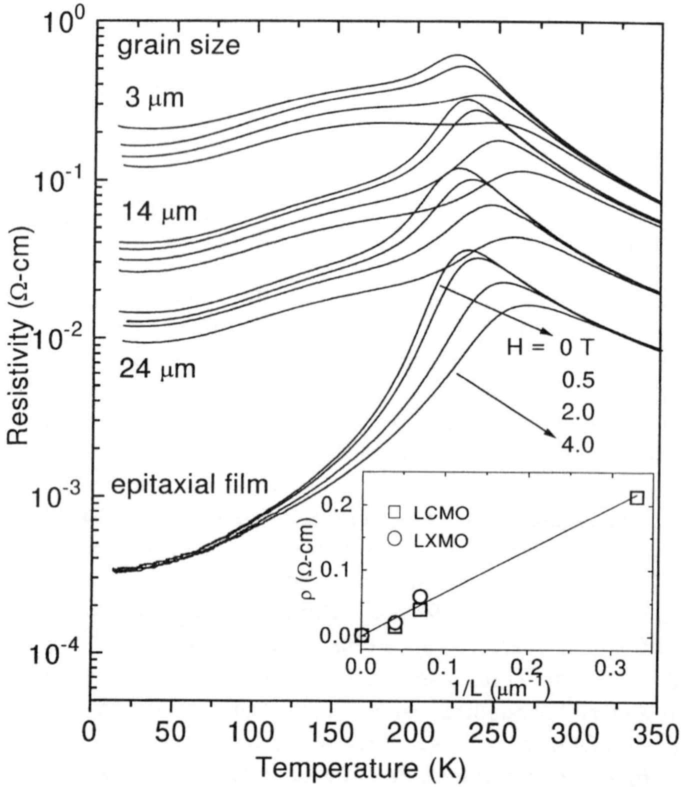

Fig. 2. Resistivity versus temperature for polycrystalline (3, 14, and 24 μm average grain size) and epitaxial LCMO films. The inset shows the zero field resistivity at 10 K as a function of the inverse average grain size. From Ref. 16.

in the residual ρ at 10 K is noted as a function of the reciprocal grain size for the LCMO (and also LXMO) films, as shown in the inset of Fig. 2. From the slope, the specific GB resistivity ($L'\rho_2$) has been estimated to be ~6×10^{-5} Ω-cm^2, with contributions from both magnetic and structural disorder. The structural disorder component has been separately estimated to be ~2×10^{-5} Ω-cm^2. By applying a magnetic field, the ρ drops significantly for all the films as seen in Fig. 2. However, unlike the epitaxial film where the drop in ρ occurs mostly near ρ_{max}, the decrease is broader, occurring over a much wider *T* range for the polycrystalline samples and persisting to low temperatures. For the epitaxial film, the MR ratio has a sharp peak at a *T* slightly below ρ_{max} and is very small at low temperatures. In contrast, the polycrystalline films at low temperatures show a large low-field component followed by a more gradual drop at higher fields, which is similar to the behavior observed for the bulk polycrystalline samples of LSMO.[15]

The temperature dependence of the low-field (H=1500 Oe) MR and the zero-field ρ for the polycrystalline LCMO and LSMO films with 14 μm average grain size are shown in Fig. 3.[17] The peak in the resistivity curves, at 230 K and 350 K for LCMO and LSMO, respectively, occur close to the T_c of these materials, as determined from magnetization measurements. The MR in both the systems show a steady decline with increasing T, interrupted by an increase near T_c, before finally dropping to zero above the respective transition temperatures. Note that the low temperature MR is somewhat higher in LSMO than LCMO. Nonetheless, the subsequent peak occurring close to the T_c is of significantly higher magnitude for the latter, which has the lower ferromagnetic transition temperature. The MR peak near T_c is identified as arising from the CMR component inside the grains. This is confirmed from the MR behavior of the epitaxial LSMO film as shown in Fig. 3. As expected, the MR for this film peaks near T_c and decreases to a negligible level at low T. The data in Fig. 3 supports the presence of two MR components in the

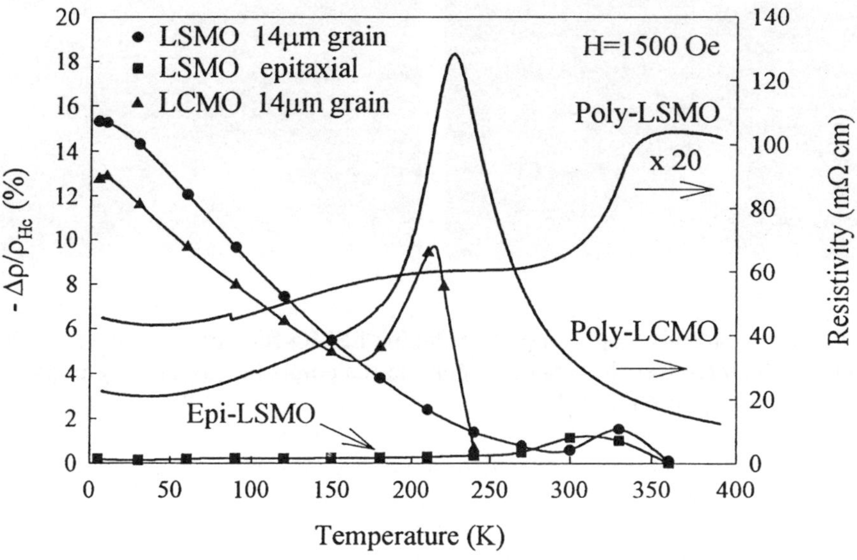

Fig. 3. Resistivity and MR as a function of temperature for polycrystalline LCMO and LSMO films (average grain size of 14 μm). The MR measured for an epitaxial LSMO film as a function of temperature is also shown for comparison. From Ref. 17.

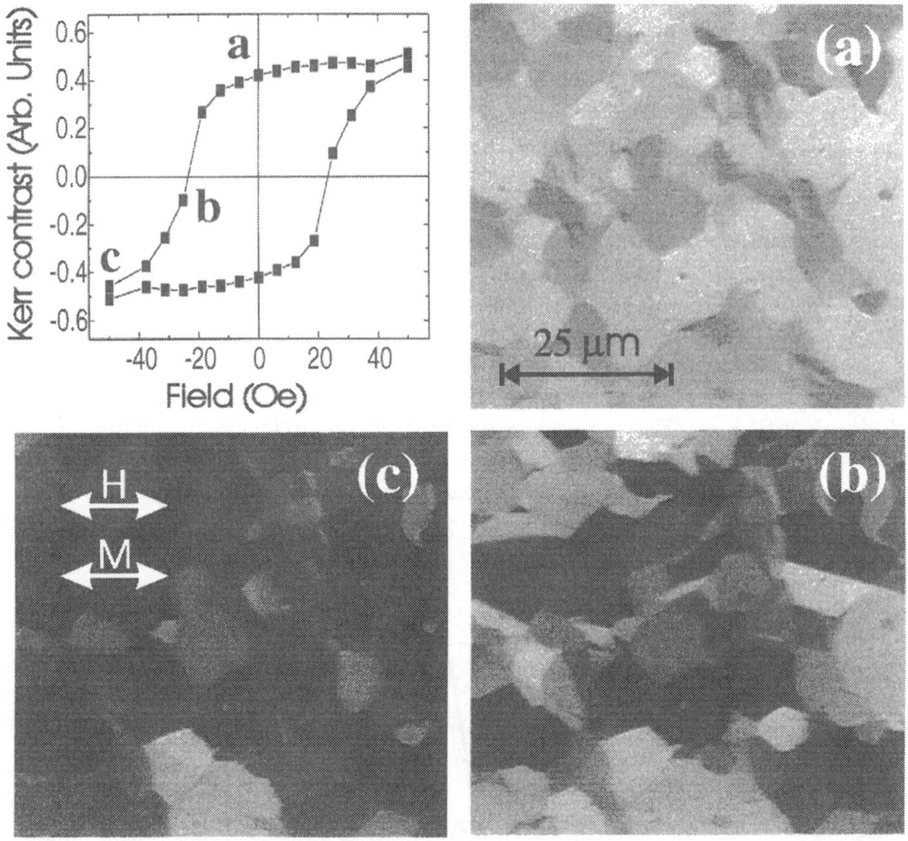

Fig. 4. Kerr hysteresis loop of a polycrystalline LSMO film averaged over number of grains. The three Kerr images correspond to the points (a) (b), and (c) marked in the hysteresis loop. From Ref. 16.

polycrystalline thin films - one near T_c from the regular CMR mechanism inside the grains, and the other most important at low T associated with the GB's.

Magnetic imaging using a wide angle Kerr microscope has shown that the domains in the polycrystalline manganite films are defined by the grains in the sample.[16,21] These domains, having different coercivities, are weakly coupled and they orient successively in an increasing field. Figure 4 shows a Kerr M-H loop of a 14 μm average grain-size polycrystalline LSMO film and three corresponding Kerr images of the film recorded at room temperature.[16] Image (a) displays the nearly uniform magnetic state of the sample at remanence. On moving close to the coercive point, half of the grains switch orientation as seen in image (b). Finally, in image (c) most of the grains have switched as the loop nears saturation. Individual grains can be observed to switch at different fields, and the loop data is an average over the many grains contained in the images. Some of the grains are also observed to switch by wall motion. For example, in the top right corner of (b), a wall can be seen to cross a GB. Wall motion in these samples has been observed to be impeded by surface defects such as scratches.

As pointed out earlier, in addition to the low-field MR there is also a high-field component associated with the GB's. The high-field (above 0.5T) MR effect in the polycrystalline samples is most likely related to alignment of spins in a magnetically disordered region near the GB's. For bulk polycrystalline samples, Hwang et al. have observed that the high-field MR is independent of temperature over a wide range from 5 to 280 K.[15] The spins are canted in the disordered regions with a distribution of angles and are aligned with increasing magnetic field. This is consistent with the fact that the magnetization continues to increase slowly with field in the polycrystalline samples.[16] On application of a large magnetic field, the increasing alignment of the neighboring spins in the disordered region leads to a gradual reduction in ρ.

Additional information regarding the behavior of polycrystalline films has come from Shreekala et al., who have studied the effect of crystallinity on the MR in $La_{0.67}Ba_{0.33}MnO_3$ (LBMO) and LCMO films.[18] They have grown epitaxial films on (001) $LaAlO_3$ and $Bi_4Ti_3O_{12}$ buffered (001) yttria-stabilized zirconia (YSZ) substrates. Polycrystalline films with a mixture of orientations have been obtained when deposited on polycrystalline YSZ substrate, whereas oriented polycrystalline films are formed on YSZ buffered Si substrates. While the MR in the epitaxial films peaks close to T_c, the polycrystalline films either show a monotonic increase or saturation with decreasing temperature. Moreover, as in the previous experiments, the polycrystalline films at low temperatures exhibit a sharp drop in resistance at low fields followed by a more gradual decrease at higher fields.

2.3 *Films on Bicrystal Substrates*

The above studies using polycrystalline films have probed the role of grain boundaries in samples consisting of multiple grains with different orientations over the measurement distance of the probe contacts. Bicrystal substrates consisting of a single grain boundary, with a well defined misorientation between the grains, are ideal for the growth of films in order to isolate the contribution of a single grain boundary. This has been successfully demonstrated in previous grain boundary studies of high T_c superconducting films.[22] However, unlike superconducting films where the grain boundary contribution can be quite readily measured because the grain resistivity is shorted below the transition temperature, the resistance of a non-superconducting oxide film across a single GB will in general be largely dominated by the ρ of the adjacent grains because of the geometry. One method to enhance the GB contribution to the total ρ is to increase the number of grain boundary intersections; for example, by patterning a meander track on the substrate in the GB region. Reduction of the grain region which is probed will lead to a further enhancement of the GB signal.

Mathur et al. have followed just such an approach to pattern thin film devices on a bicrystal substrate in order to study the contribution of a single grain boundary.[19] Epitaxial $La_{0.7}Ca_{0.3}MnO_3$ films, 2000Å thick, are grown by pulsed laser deposition at a growth temperature of about 800°C on bicrystal substrates consisting of two $SrTiO_3$ (001) crystals misaligned by 24°. To isolate and study the artificial grain boundary directly, they patterned a variety of highly symmetric Wheatstone-bridge structures into the bicrystal films using optical lithography and ion milling, as shown in Fig. 5. The symmetry of the bridge structures ensures that all resistance contributions balance to zero except those arising from the grain boundary. Bridge

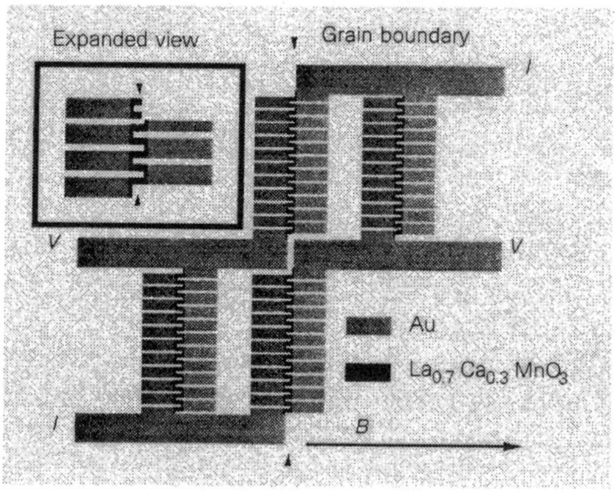

Fig. 5. The meander track Wheatstone-bridge geometry used for measurements on bicrystal substrates. From Ref. 19.

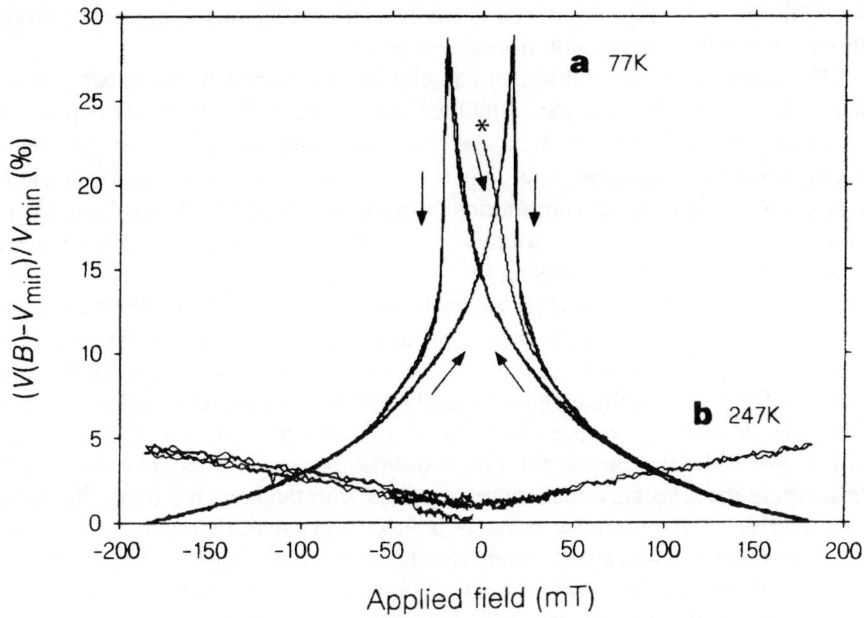

Fig. 6. Wheatsone-bridge measurement as a function of applied field at 77 K and 247 K. Voltages $V(B)$ and V_{min} are the field-dependent output voltage and the minimum output voltage measured, and represent a direct measure of the resistance introduced by the grain boundary. From Ref. 19.

structures are provided on each chip in which current is able to flow either across the grain boundary using a meander track or along it.

The devices have been characterized in a liquid nitrogen cryostat in a magnetic field up to 300 mT by passing a constant current through each bridge and measuring the voltage across the output terminals. This measured voltage represents a direct measure of the resistance introduced by the GB. A large bridge MR (27% at 77 K) is observed during magnetic field sweeps within ±200 mT over a range of temperatures down to 77 K, with a strong low-field hysteresis in bridge resistance at low temperatures (Fig. 6). At higher temperatures, the magnitude of the peaks in MR decreases and the peaks disappear altogether at temperatures around 230K. These results are qualitatively similar to those observed on polycrystalline films,[15] excepting for the magnitude of the low-temperature MR, which is about a factor of 2-3 higher for the single GB junctions. Interestingly, a small positive MR is observed above T_c (Fig. 6), which is masked in experiments using polycrystalline samples due to the contribution from the grains.

The use of bicrystal substrates has also been reported by Steenbeck et al. in their study of the MR in $La_{0.8}Sr_{0.2}MnO_3$ epitaxial films.[20] The films are deposited by sputtering on 36.8° $SrTiO_3$ bicrystal substrates and are patterned in meander structures with 11 crossings (total length l = 660 μm, width w = 8 μm) of the GB. To account for the grain contribution, measurements are performed on separate stripes patterned (l = 100 μm, width w = 20 μm) on the same substrate adjacent to the grain boundary but not crossing it.

The field-independent GB resistance has been estimated from the measurements to be 4.1×10^{-6} Ω cm^2, while the field-dependent resistance maximum has been found to have a value of 2.3×10^{-6} Ω cm^2. These are somewhat smaller than the values estimated for polycrystalline films, possibly because of different angles and widths of the GB. Another difference observed with respect to polycrystalline films is the field at which the maxima in the GB resistance occurs.[16,20] Unlike polycrystalline films, where the resistance maximum coincides with the coercive field, the maxima in the bicrystal GB resistance occur at a field larger than the coercive field, even after correcting for demagnetization effects. It has been suggested that these two values differ because the zero magnetization state does not coincide with the state where there is the maximum angle between the direction of magnetization of adjacent grains. The grain boundary MR, obtained after subtracting the grain contribution, is of a similar magnitude as observed by Mathur et al. for the 24° GB films.

3 Possible Mechanisms for Grain Boundary MR

All of the studies so far on naturally occurring or artificially prepared grain boundary samples have demonstrated that the local disruption of the magnetic order at the GB can alter the electrical conduction and induce a low-field magnetoresistive effect in the manganites. It is believed that spin-polarized transport effects play an important role in the conduction process in these materials due to the high degree of spin polarization of the electrons.[1] The almost completely polarized conduction electrons are easily transferred between pairs of Mn^{3+} and Mn^{4+} inside a magnetic domain. However, when these electrons travel across GB's, strong spin-dependent conduction at the boundaries will lead to a high zero-field ρ. Application of a moderately low field can readily align the domains associated with the GB's into a parallel configuration causing the ρ to drop substantially. While this picture can qualitatively explain the observed results, an understanding of the microscopic transport mechanism across a GB is necessary in order to quantify the magnitude and temperature dependence of the MR. A better understanding of the conduction process will also help in designing appropriate structures with optimized MR properties for practical applications.

Based on the results of polycrystalline films, we have suggested that the low-field MR is a consequence of spin-dependent scattering of polarized electrons at the grain boundaries which serve as pinning centers for the magnetic domain walls.[16,17] An external field reduces the scattering due to alignment of the domains associated with the grains in a parallel configuration. High-resolution electron microscopy of the grain boundary region in polycrystalline films has shown that the GB's are well structured with local disorder (~10Å) as expected for oxide grain boundaries.[16] Both magnetic and structural disorder scattering contribute to the ρ of the polycrystalline samples, with possibly different effective lengths existing for the two components. An alternative explanation for the low-field MR has been proposed by Hwang et al. based on their observations of bulk polycrystalline samples of LSMO.[15] They suggest that the conduction is dominated by intergrain spin-polarized tunneling rather than scattering.

In both the scattering and the tunneling models, ρ is expected to be a maximum at the coercive field and decrease as the relative orientation of the magnetization between grains changes with the application of a field. The MR hysteresis loops will therefore mirror the global magnetization, M, with the MR being proportional to $(M/M_s)^2$, where M_s is the saturation magnetization.[23-25] The MR(H) at 10 K for a polycrystalline LCMO film with 3 μm average grain size is shown in Fig. 7(a), together with the magnetic hysteresis loop of the sample measured at the same temperature. The MR data is fit quite well by a $(M/M_s)^2$ dependence as illustrated by the solid curve in Fig. 7(b). Nevertheless, since both the scattering and tunneling models predict the same dependence on M, it is not possible to distinguish between the two based on this fact alone.

To clarify the source of MR at low fields, it is instructive to measure the current-voltage (I-V) characteristics for the epitaxial and polycrystalline films. For tunneling to be the dominant conduction mechanism across the grain boundaries, one would expect strong non-Ohmic I-V characteristics at high bias voltages.[26] Polycrystalline films with grain sizes of 3 and 14 μm have been measured after being patterned to a small square (50 μm x 50 μm).[17] There are only 3-4 grains along the current path for the sample with 14 μm grain size, and about 17 grains for the 3 μm grain size. If tunneling is at play, one would expect the I-V curves to be very different for these samples, because the voltage drop across successive grains would be different. Nevertheless, a small decrease in the dynamic resistance (dV/dI) of about 5% at a bias voltage of 0.5 V has been observed for all the epitaxial and polycrystalline films. The lack of grain size dependence and the similarity in the behavior of epitaxial and polycrystalline films are strong evidence excluding tunneling between the grains as the dominant conduction mechanism. Recent

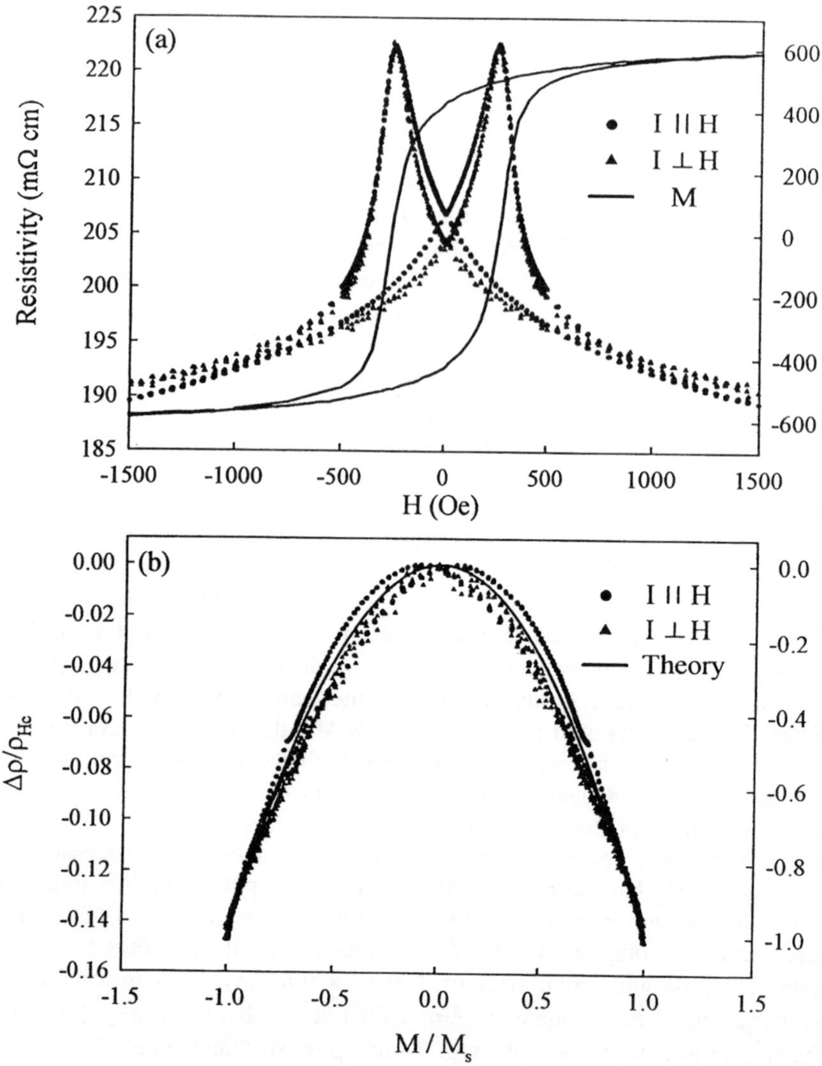

Fig. 7. (a) MR as a function of magnetic field for a polycrystalline LCMO film (3 μm average grain size) measured at 10 K. The magnetic hysteresis loop obtained using a SQUID at the same temperature is also plotted. (b) MR vs M/M_s from the above data. The solid curve is the $(M/M_s)^2$ fit to the data. From Ref. 17.

measurements using better contacts have shown that the polycrystalline films do show larger changes in the dynamic resistance as a function of bias voltage than the epitaxial films, but still quite modest as compared to conventional tunnel junctions. *I-V* characteristics across a single GB on bicrystal substrates are needed for further elucidating the conduction mechanism.

Additional support for a non-tunneling mechanism at the GB's comes from the temperature dependence of the resistivity for the polycrystalline samples. From measurement of polycrystalline films, the grain boundary ρ in LCMO has been estimated to be of the order of a few hundred Ω-cm, assuming a structural disorder thickness of ~10Å. This is significantly lower than the ρ of insulating barriers in traditional tunnel junctions.[26] Nonetheless, for tunneling-type behavior to be operative across the grains, the polycrystalline samples would be expected to display an increase in ρ with decreasing temperature, both because of the weak temperature dependence of pure tunneling and due to possible additional contribution from thermally activated processes. It is important to note that the increase in ρ with decreasing T will be apparent only when the grain contribution to the measured ρ is negligible. The MR observed in highly resistive Co-Al-O granular magnetic films has been suggested to result from tunneling of electrons.[24,27] A temperature dependence of the form $\ln\rho \propto T^{-1/2}$ is predicted based on a dominant contribution from the charging energy of the electrons, [23,24] and has actually been observed in this type of system.[27] Resistivity versus temperature data for all the manganite films have shown an increase in ρ with decreasing grain size, particularly at low T where the GB contribution dominates.[16] However, even for the sample with the smallest grain size (3 μm), ρ shows a decrease with T at low temperatures. Extrapolation to smaller grain sizes, obtained from plots of the peak and the lowest temperature ρ values, suggests that this trend continues. Measurements of the temperature dependence of the grain boundary ρ in bicrystal films will certainly be helpful in clarifying the conduction process.

Conclusions

Relatively large MR at low field have been observed in polycrystalline bulk and thin film samples of manganite perovskites. This low-field MR has a different temperature dependence than the CMR mechanism. It reaches a maximum of about 15% at 1500 Oe at low temperatures, but steadily declines at high T and vanishes at T_c. Even higher values have been observed in films consisting of a single GB with fixed misorientation, which have ben grown on bicrystal substrates, after subtracting the grain contribution. Initial results based on the magnetization dependence, *I-V* characteristics, and the temperature dependence coefficient of the resistivity suggest that the low-field MR in the polycrystalline samples results primarily from spin-dependent scattering of the polarized electron at the grain boundaries which serve as pinning centers for the magnetic domains. However, it is not possible to completely eliminate other conduction mechanisms, such as tunneling, occurring across GB's. It is conceivable that the GB's are inhomogeneous, with microscopic regions that are highly insulating separated by more conducting regions. Conduction in this parallel channel network can then occur by a combination of spin-scattering and tunneling processes, which displays not very non-linear *I-V* characteristics. Clearly, much more work is needed for a more complete understanding of the mechanism of conduction of polarized electrons across grain boundaries.

Acknowledgments

The work on the polycrystalline films has been done in collaboration with my colleagues at IBM, Brown University and Northwestern University, including G. Q. Gong, X. W. Li, G. Xiao, J. Z. Sun, P. R. Duncombe, P. Lecouer, P. Trouilloud, T. R. McGuire, Y. Lu, W. J. Gallagher, Y. Y. Wang, and V. P. Dravid.

References

1. W. E. Pickett and D. J. Singh, Phys. Rev. B **53**, 1146 (1996).
2. Y. Lu, X. W. Li, G. Q. Gong, G. Xiao, A. Gupta, P. Lecouer, J. Z. Sun, Y. Y. Yang, and V. P. Dravid, Phys. Rev. B **54**, R8357 (1996).
3. J. Z. Sun, W. J. Gallagher, P. R. Duncombe, L. Krusin-Elbaum, R. A. Altman, A. Gupta, Y. Lu, G. Q. Gong, and G. Xiao, Appl. Phys. Lett. **69**, 3266 (1996).
4. J. Z. Sun, L. Krusin-Elbaum, P. R. Duncombe, A. Gupta, and R. B. Laibowitz, Appl. Phys. Lett. **70**, 1769 (1997).
5. H. L. Ju, J. Gopalakrishnan, J. L. Peng, Q. Li, G. C. Xiong, T. Venkatesan, and R. L. Greene, Phys. Rev. B **51**, 6143 (1995).
6. R. Mahendiran, R. Ramesh, A. K. Raychaudhari, and C. N. R. Rao, Solid State Commun. **94**, 515 (1995).
7. G. Gong, C. Canedy, G. Xiao, J. Z. Sun, A. Gupta, and W. J. Gallagher, Appl. Phys. Lett. **67**, 291 (1995).
8. P. Schiffer, A. P. Ramirez, W. Bao, and S-W. Cheong, Phys. Rev. Lett. **75**, 3336 (1995).
9. H. L. Ju, C. Kwon, Q. Li, R. L. Greene, and T. Venkatesan, Appl. Phys. Lett **65**, 2108 (1994).
10. Y. Tokura, A. Urushibara, Y. Morimoto, T. Arima, A. Asamitsu, G. Kido, and N. Furukawa, J. Phys. Soc. Jpn. **63**, 3931 (1994).
11. K. Khazeni, Y. X. Jia, L. Lu, V. H. Crespi, M. L. Cohen, and A. Zettl, Phys. Rev. Lett. **76**, 295 (1996).

12. A. Gupta, T. R. McGuire, P. R. Duncombe, M. Rupp, J. Z. Sun, W. J. Gallagher, and G. Xiao, Appl. Phys. Lett. **67**, 3494 (1995).
13. M. F. Hundley, M. Hawley, R. H. Heffner, Q. X Jia, J. J. Neumeier, J. Tesmer, J. D. Thompson, and X. D. Wu, Appl. Phys. Lett. **67**, 860 (1995).
14. R. Mahesh, R. Mahendiran, A. K. Raychaudhuri, and C. N. R. Rao, Appl. Phys. Lett. **68**, 2291 (1996).
15. H. Y. Hwang, S-W. Cheong, N. P. Ong, and B. Batlogg, Phys. Rev. Lett. **77**, 2041 (1996).
16. A. Gupta, G. Q. Gong, Gang Xiao, P. R. Duncombe, P. Lecouer, P. Trouilloud, Y. Y. Wang, V. P. Dravid, and J. Z. Sun, Phys. Rev. B **54**, R15629 (1996).
17. X. W. Li, A. Gupta, G. Xiao, and G. Q. Gong, Appl. Phys. Lett. **71**, 1124 (1997).
18. R. Shreekala, M. Rajeswari, K. Ghosh, A. Goyal, J. Y. Gu, C. Kwon, Z. Trajanovic, T. Boettcher, R. L. Greene, R. Ramesh, and T. Venkatesan, Appl. Phys. Lett. **71**, 282 (1997)
19. N. D. Mathur, G. Burnell, S. P. Isaac, T. J. Jackson, B.-S. Teo, J. L. MacManus-Driscoll, L. F. Cohen, J. E. Evetts, and M. G. Blamire, Nature **387**, 266 (1997).
20. S. Steenbeck, T. Eick, K. Kirsch, K. O'Donnell, and E. Steinbeiβ, Appl. Phys. Lett. **71**, 968 (1997).
21. P. Lecouer, P. Trouilloud, G. Xiao, A. Gupta, X. W. Li, and G. Q. Gong, J. Appl. Phys. **82**, xxx (1997).
22. For example, see S. E. Babcock and J. L. Vargas, Annl. Rev. Mater. Sci. **25**, 193 (1995).
23. J. S. Helman and B. Abeles, Phys. Rev. Lett. **37**, 1429 (1976)
24. J. Inoue and S. Maekawa, Phys. Rev. B **53**, R 11 927 (1996).
25. J. Q. Xiao, J. S. Jiang, and C. L. Chien, Phys. Rev. Lett. **68**, 3749 (1992).
26. E. L. Wolf, *Principles of Electron Tunneling Spectroscopy* (Oxford University Press, Oxford, 1985).
27. H. Fujimori, S. Mitani, and H. Ohnuma, J. Magn. Magn. Mater. **156**, 311 (1996).

COLOSSAL MAGNETORESISTANCE WITHOUT DOUBLE-EXCHANGE: PYROCHLORES

M. A. SUBRAMANIAN

Dupont Central Research and Development, Experimental Station, Wilmington, DE 19880-0328, USA

A. P. RAMIREZ

Bell Laboratories, Lucent Technologies, 600 Mountain Avenue, Murray Hill, Nj 07974, USA

G. H. KWEI

Los Alamos National Laboratory, Los Alamos, NM 87545, USA

At present, we have two families of manganese oxides displaying colossal magnetoresistance (CMR): the substituted rare-earth manganite perovskites ($La_{1-x}D_xMnO_3$) and the thallium manganite pyrochlores ($Tl_2Mn_2O_7$). Structural determination using powder neutron and single-crystal x-ray diffraction data as well as x-ray absorption spectroscopic analysis of $Tl_2Mn_2O_7$ indicate that the compound is stoichiometric and do not show any evidence for the presence of Mn^{3+}/Mn^{4+} mixed valence. The observed Mn-O distance (1.90 Å) and the refinement of cation and oxygen occupancy factors indicate oxidation states very close to $Tl^{3+}{}_2Mn^{4+}{}_2O_7$. Thus, $Tl_2Mn_2O_7$ has neither mixed valence for a double-exchange magnetic interaction nor a Jahn-Teller cation such as Mn^{3+}, which are both known to play an essential role in CMR perovskite materials. An alternate mechanism for CMR in $Tl_2Mn_2O_7$ is proposed based on magnetic ordering driven by superexchange-type interaction originating from Mn-O sublattice while the electrical conduction resulting from the mixing of Tl-6*s* states at the Fermi level as implied by the band structure calculations. The observed CMR results from unusually large incoherent scattering from spin fluctuations accompanying ferromagnetic ordering. Thus, the pyrochlores represent a distinct new class of CMR materials.

1. Introduction

The strong correlation between magnetic moment and electrical resistance, and the resulting large magnetoresistance near room temperature in lanthanum manganese oxides ($La_{1-x}D_xMnO_3$ where D = Ca, Sr and Ba) is well known [1]. Recently, nearly 100% suppression of the resistance was observed in thin film manganite samples in a field of 6 Tesla [2] and the possible use of this effect for reading magnetic storage media, has increased the interest in these materials. The fundamental interaction in these manganese oxide materials is double-exchange, where electronic carrier hopping between heterovalent Mn pairs (Mn^{3+}/Mn^{4+}) is enhanced by the mutual alignment of the two magnetic moments. A ferromagnetic transition is stabilized by the kinetic energy gain of the carriers when a transition occurs from incoherent hopping to metallic conduction [3]. The high resistivity at temperatures above the ferromagnetic transition, and thus ultimately CMR in $La_{1-x}D_xMnO_3$, is explained by the strong Jahn-Teller effect in these compounds [4]. Although both heterovalency and strong Jahn-Teller effects are necessary to produce CMR in the $La_{1-x}D_xMnO_3$ compounds, CMR-type magnetoresistance is found in other systems but at much lower temperatures. Perhaps the best known example of this is $Eu_{0.95}Gd_{0.05}Se$ with $T_c \approx 20$ K [5]. Here, CMR is not driven by the same interatomic double-exchange mechanism as in the manganese oxides but rather by an intraatomic version of the process. This chapter presents evidence for a different mechanism that leads to CMR in $Tl_2Mn_2O_7$ and it's related phases adopting a pyrochlore structure. A brief summary covering the crystal chemistry and properties of $R_2Mn_2O_7$ (R = rare earth, Y, Sc, In or Tl) pyrochlores is also included.

2. $R_2Mn_2O_7$ Pyrochlores (R = rare earth, Y, Sc, In or Tl)

2.1 Structure

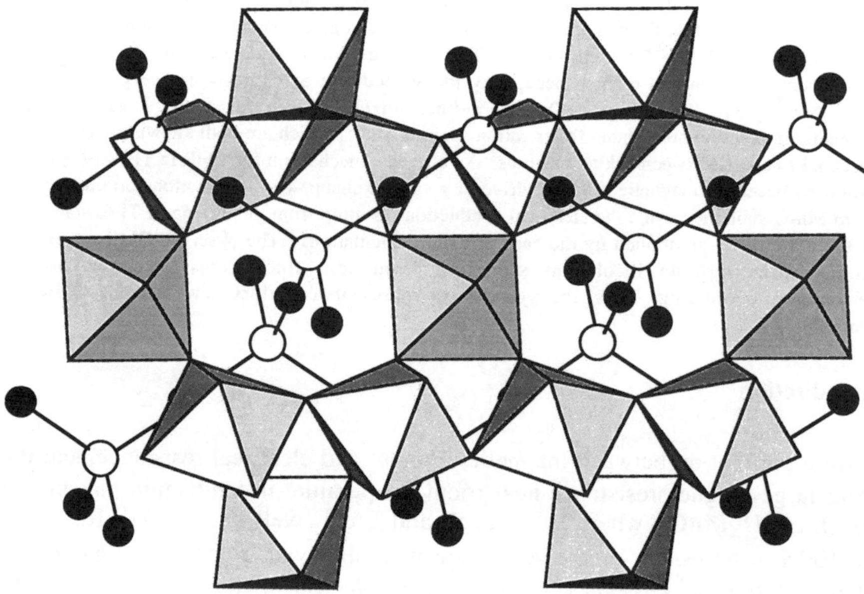

Figure 1: Schematic view of the structure of $Tl_2Mn_2O_7$ emphasizing the structural arrangement of the $MnO(II)_6$ octahedra. Corner-shared oxygens connect the octahedra, forming an Mn-O-Mn bond angle of about 133°. Small filled circles represent Tl ions, while the large open circles represent the O(I) oxygens.

The pyrochlore structure of $R_2Mn_2O_7$ (or more appropriately $R_2Mn_2O_6O'$) is based on a network of corner-sharing MnO_6 octahedra as in CMR perovskites (Figure 1). However, the manner of linking octahedra leads to Mn-O-Mn angles in the range close to 180° in the perovskite structure (for $La_{1-x}Ca_xMnO_3$, this angle is ~160°) but near 134° in the pyrochlore structure. The site symmetry at the Mn site in the pyrochlore structure is $\overline{3}$. Thus, while all of the Mn-O distances are equal, the O-Mn-O angles are not constrained to be 90°. The MnO_6 octahedron is distorted and is actually a trigonal antiprism as it is in all compounds with the pyrochlore structure [6]. This distortion is, however, not a Jahn-Teller type distortion, where opposite bonds lengthen or shorten relative to others. The R atoms are located in the center of the hexagonal rings formed by oxygens (O, located at the 48f positions) with two more oxygens (O', at the 8b positions) located above and below the ring. Hence, the A atom environment can be considered as distorted cube. The pyrochlore structure can also be viewed as two interpenetrating networks : one with the formula MnO_3 and the other with the formula R_2O'. Because the MnO_3 network forms the backbone of the structure, vacancies can occur only on the R and O' sites [6]. However, unlike in perovskite structure, the Mn atoms (as well as A atoms) in these compounds are located in the corners of a tetrahedral units which share corners to form infinite, intersecting chains. Such an arrangement leads to a very high degree of magnetic frustration if the nearest-neighbor interactions are antiferromagnetic.

2.1 Properties

Electrical measurements indicate $R_2Mn_2O_7$ where R = rare earth, Y or Sc are insulators with room temperature resistivities in the range 10^5 to 10^8 Ωcm and activation energy 0.4 to 0.5 eV [7]. All the compounds in this series show similar susceptibility data and obey a Curie-Weiss law at higher temperatures. The θ_c value for all the phases are large and positive indicating strong ferromagnetic exchange interactions between Mn moments. However, low field dc susceptibility data for these phases (R = Sc, Y, Ho, Yb and Lu) exhibit clear field-cooled (FC) , zero-field-cooled (ZFC) irreversibilities below "T_c" and each with a broad maximum. [8]. Ac susceptibility results also show a similar behavior and the positions of the maxima are frequency dependent [8]. Heat capacity studies down to 2K for $Y_2Mn_2O_7$ show no sharp anomaly and hence no evidence for long-range magnetic order. In addition, low temperature neutron diffraction as well as small angle neutron scattering data of $Y_2Mn_2O_7$ seems to indicate the presence of both near neighbor antiferromagnetic and ferromagnetic (second and third neighbor) short range correlations. These materials show many complex magnetic properties in common with spin glasses in spite of the fact that there is no evidence for chemical site disorder [8]. As mentioned earlier, the

Mn atoms in $R_2Mn_2O_7$ occupy the corners of a tetrahedra. Spins at two points of the tetrahedron may couple in an antiparallel fashion, but it is then impossible for the remaining two atoms of the tetrahedron to align their spins antiparallel to the first two. This high degree of frustration seems to be responsible for the above observed unusual as well as complex magnetic behavior. It appears that there exists a strong competition between ferromagnetism and frustrated antiferromagnetism, with superexchange dominating the magnetic interaction.

Magnetic data for $Tl_2Mn_2O_7$ indicate the compound is ferromagnetic [9, 10] and, unlike $R_2Mn_2O_7$ (A = Dy-Lu, Y or Sc) phases, there is no appreciable difference between FC and ZFC susceptibilities [10]. In addition, SANS clearly show evidence for the long-range magnetic order which is indicative of strong ferromagnetic interactions between Mn moments [10]. Electrical measurements show the compound is a good conductor with a room temperature resistivity in the range 0.1 to 0.01 Ωcm. This is again in contrast to the insulating behavior exhibited by $R_2Mn_2O_7$ phases (A = rare earth, Y or Sc). The resistivity as a function of temperature for $Tl_2Mn_2O_7$ shows a very sharp drop in resistivity at the magnetic ordering temperature (T_c~150 K) as observed in CMR $LaMnO_3$-type perovskites. It is interesting to note that the properties of $In_2Mn_2O_7$ falls in-between the insulating $R_2Mn_2O_7$ (R = rare earth, Y, Sc) and conducting $Tl_2Mn_2O_7$ pyrochlores. Although the $In_2Mn_2O_7$ shows a ferromagnetic ordering around 150 K, the compound is insulating above and below the T_c and there is no coupling between conduction and magnetism [10, 11].

3. CMR behavior in $Tl_2Mn_2O_7$

3.1 Absence of Mn^{3+}/Mn^{4+} double-exchange

Recently, CMR behavior comparable to that of $La_{1-x}D_xMnO_3$ perovskites has been found for $Tl_2Mn_2O_7$ pyrochlore [12-14] (Figure 2). Because $Tl_2Mn_2O_7$ also contains Mn as the magnetic constituent, it was thought that this compound would also contain significant amounts of the Jahn-Teller cation Mn^{3+} and that it was again Mn^{3+}/Mn^{4+} double-exchange that gave rise to ferromagnetism. Mixed valence could result from oxygen deficiency; in fact, the lattice constant of $Tl_2Mn_2O_7$ is greater than expected when compared to the isostructural compounds where Tl is completely replaced by a rare-earth ion. The larger lattice constant could result from the formulation $Tl_2Mn^{4+}{}_{2-2x}Mn^{3+}{}_{2x}O_{7-x}$ because Mn^{3+} is significantly larger than Mn^{4+} and oxygen deficiency is well established in many compounds with the pyrochlore structure [6].

Structural analysis [14] with powder neutron and single-crystal x-ray diffraction data for a sample of the $Tl_2Mn_2O_7$ pyrochlore, exhibiting CMR, shows no

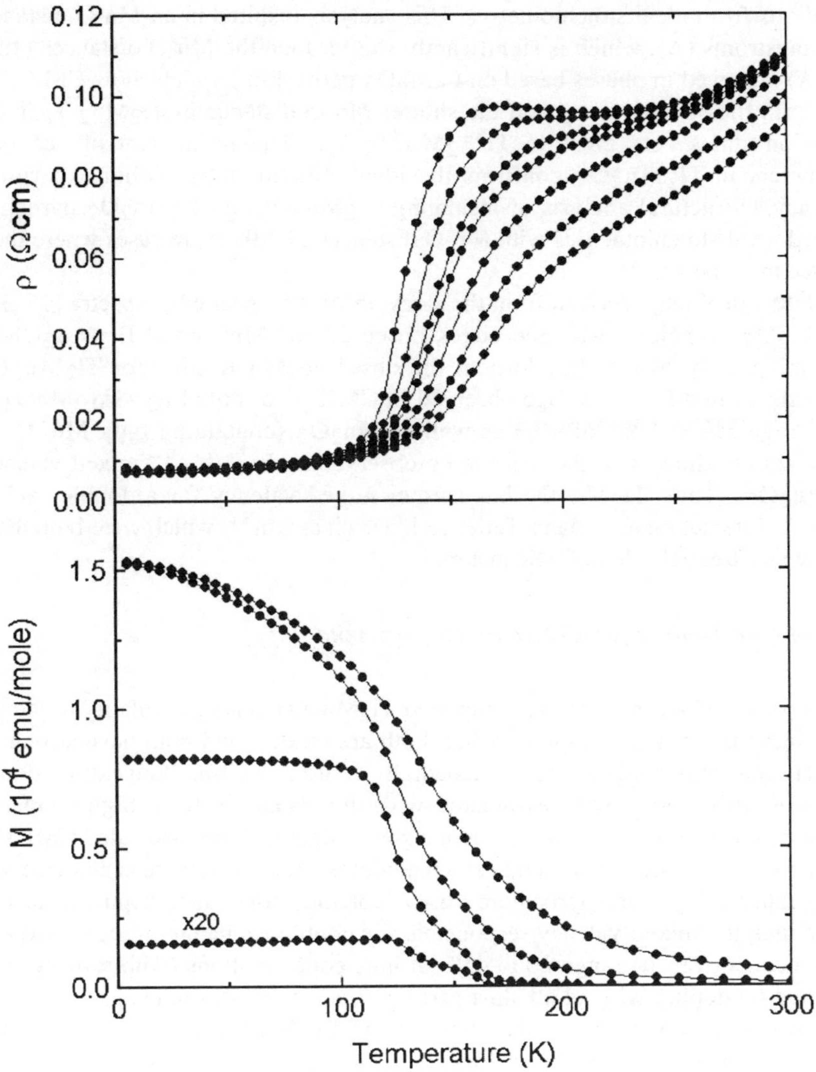

Figure 2: Temperature dependence of the resistivity (upper panel) and the magnetization (lower panel) for $Tl_2Mn_2O_7$. The resistivity data are for 0 (top), 0.1, 1, 2, 4, 6 and 8 (bottom) T. The magnetization data are for 0.001 (bottom), 0.1, 1 and 4 (top) T.

deviations from ideal stoichiometry. This analysis resulted in an Mn-O distance of 1.90 angstroms (Å), which is significantly shorter than the Mn-O distances (1.94 to 2.00 Å) observed in phases based on $LaMnO_3$ perovskites which show CMR (Table 1). Both the stoichiometry and the shorter Mn-O distance in $Tl_2Mn_2O_7$ indicate oxidation states very close to $Tl^{3+}{}_2Mn^{4+}{}_2O_7$. The refinement of occupation parameters in $Tl_2Mn_2O_7$ confirms the ideal stoichiometry within experimental accuracy. Structural analysis of insulating $Er_2Mn_2O_7$ and $Y_2Mn_2O_7$ pyrochlores showed ideal stoichiometries with Mn-O distances of 1.91 Å, in cases where the Mn valence must be 4+ [7].

The Mn K edge derived from the x-ray absorption near edge spectra [15, 16] of $Tl_2Mn_2O_7$ coincided with the edges observed for MnO_2 and $Er_2Mn_2O_7$ both containing only Mn^{4+}. In addition, the observed Mn K edge for $Tl_2Mn_2O_7$ is markedly shifted from the edge observed in CMR perovskite $La_{0.7}Sr_{0.3}MnO_3$ with an average Mn valence of +3.3 as well as Mn_2O_3 (containing only Mn^{3+}). The above results show that the there is no observable Mn^{3+}/Mn^{4+} mixed valency in $Tl_2Mn_2O_7$. Thus, $Tl_2Mn_2O_7$ has neither mixed valency for a double-exchange magnetic interaction nor a Jahn-Teller cation such as Mn^{3+}, which were both thought to play an essential role in CMR materials.

3.2 CMR mechanism: pyrochlore versus perovskite

Although, the CMR pyrochlore compound $Tl_2Mn_2O_7$ bears a similarity to the CMR manganese perovskite compounds (i.e. both are oxides, and both have strong local moment magnetism arising from octahedrally coordinated Mn, both exhibit dramatic decreases in the resistivity associated with the transition from high-temperature paramagnetic to low-temperature ferromagnetic state and the associated CMR) there are differences between two families: on a microscopic level, there is no evidence for significant doping in the pyrochlore Mn-O sublattice [14]. Such doping is necessary to produce the mixed valency responsible for double-exchange in the perovskites - CMR occurs over the range 20 to 45% of hole concentrations (with respect to Mn), obtained by doping with an alkaline earth on the rare-earth site [17]. Second, there is no evidence for Jahn-Teller distortions among the Mn-O octahedra, consistent with the stoichiometry of the compound and the approximately homovalent Mn population thus implied. In addition, neutron diffraction [11, 16] and x-ray absorption fine structure analysis of $Tl_2Mn_2O_7$ above and below T_c showed no structural anomaly (lattice parameter, Mn-O bond length, Tl-O bond length, and Mn-O-Mn bond angle) near T_c and/or metal-insulator transition temperature indicating that the spin-lattice/charge-lattice correlations are very weak or absent in this compound. This is in contrast to CMR perovskites, where one clearly sees a large anomaly in similar structural parameters near the metal-insulator transition and/or magnetic ordering

Table 1: Observed Mn-O bond distances (Å) and Mn-O-Mn (°) bond angles in CMR manganese oxides: $Tl_2Mn_2O_7$ (pyrochlore) and $La_{0.7}Ca_{0.3}MnO_3$ (perovskite).

$Tl_2Mn_2O_7$	$La_{0.7}Ca_{0.3}MnO_3$
Mn-O: 1.901(2)	Mn-O1 : 1.962(1); Mn-O2 : 1.951(2) Mn-O2 : 1.982(2)
Mn-O-Mn: 133.8(3)	Mn-O1-Mn : 159.8(1); Mn-O2-Mn : 160.2(1)

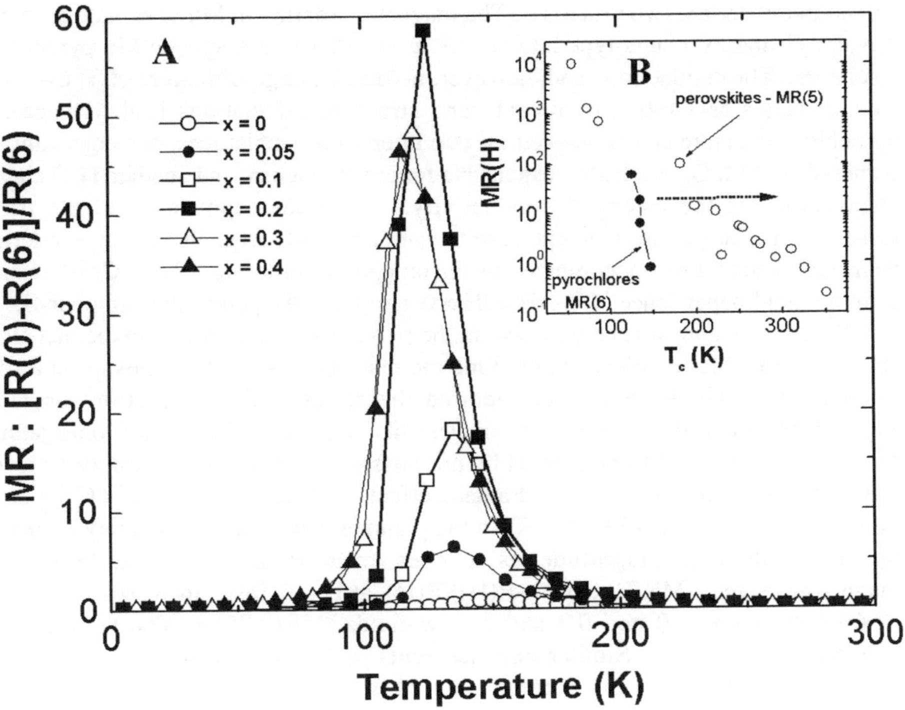

Figure 3: (A) Magnetoresistance versus temperature of $Tl_{2-x}Sc_xMn_2O_7$ pyrochlores for various values of x. (B) The maximum magnetoresistance versus ordering temperature for the pyrochlores (open symbols) and the data for manganite perovskites (closed symbols). The arrow symbolizes the expected shift of the pyrochlore data if the exchange interaction among Mn-site ions could be increased.

temperatures indicating a strong spin/charge-lattice coupling [18]. Third, the above mentioned tendency for Tl to form 6*s* conduction bands is unlike the perovskite case where the rare-earth levels are electronically inactive. From a macroscopic perspective, the saturation moment of 2.74 μB is below the value expected for Mn^{4+}, consistent with an absence of Mn^{3+} and weakly covalent character among the Mn valence electrons. Finally, the resistivity of $Tl_2Mn_2O_7$ in the paramagnetic state is metal like ($d\rho/dT > 0$), unlike the perovskites where polycrystalline samples typically exhibit hopping-type conductivity ($d\rho/dT < 0$).

The origin of CMR in $Tl_2Mn_2O_7$ is fundamentally different from that of the perovskites. Instead of a single mechanism, double-exchange, driving both the conduction and magnetic ordering processes as in the perovskites, in the pyrochlore compound, there are two processes. The magnetic ordering in Mn-O sublattice seems driven by superexchange-type interaction, as in other ferromagnetic Mn pyrochlore insulators. The conduction band, however, involves a large admixture of Tl 6*s*-based valence states as implied by recent band structure calculations [19]. Hence the pyrochlore structure can be viewed as two intertwined sublattices, one consisting of corner-shared MnO_6 octahedra responsible for the magnetism and another Tl-O chains responsible for the electrical conductivity. The interdependence of electrical resistivity and magnetic moment results from unusually large incoherent scattering from spin fluctuations accompanying ferromagnetic ordering. Then CMR results from the field-dependence of T_c, $dT_c/dH > 0$, similar to the perovskite compounds.

Because magnetism and transport in the pyrochlores originate from distinct atom sites, it is possible to independently tune the materials parameters relevant in a spin fluctuation model. Recently, we reported the effect of Sc-substitution for Tl in $Tl_{2-x}Sc_xMn_2O_7$ [20]. Scandium is trivalent like Tl, but with a smaller ionic radius. Most importantly, it differs from Tl by not possessing an accessible empty 6*s* level. We find that Sc-substitution has a dramatic effect on electronic transport. Over a large range of temperatures (150 - 200 K) in the paramagnetic phase, resistivity increases by five orders of magnitude as x goes from 0 to 0.4. In addition, the magnetoresistance, MR(H) : (R(0)-R(H))/R(H), at H = 6 Tesla, increases by a factor of 60 between x = 0 and 0.2 and at low fields, MR(0.05) is also significantly enhanced (Figure 3A). Similar enhancements in MR(H) are found in the Mn-O perovskites but are always accompanied by strong suppression (~100%) of the ferromagnetic Curie temperature, T_c [2]. In the Sc-substituted $Tl_2Mn_2O_7$ pyrochlores, however, the suppression of T_c accompanying the rise in MR(H) is much smaller (~15%) (Figure 3B). This demonstrates the possibility, in principle, to induce large increases in MR while leaving T_c effectively unchanged. This property, a desirable feature for applications, is characteristic of the two electronic-sublattice pyrochlore compound and motivates further materials searches.

References

1. G.H. Jonker and J.H. Van Santen, *Physica* **16**, 337 (1950) and G.H. Jonker, *Physica* **22**, 707 (1956); J. Volger, *Physica* **20**, 49 (1954); C.W. Searle and S.T. Wang, *Can. J. Phys.* **47**, 2703 (1969); *Can. J. Phys.* **48**, 2023 (1970).
2. S. Jin. T. H. Tieffel, M. McCormack, R. A. Fastnacht, R. Ramesh and L. H. Chen, *Science* **264**, 413 (1994).
3. C. Zener, *Phys. Rev.* **82**, 403 (1951); P.W. Anderson and H. Hasegawa, *Phys. Rev.* **100**, 675 (1955); P.-G. de Gennes, *Phys. Rev.* **118**, 141 (1960).
4. N-L.H. Liu and D. Emin, *Phys. Rev. Lett.* **42**, 71 (1979); A.J. Millis, P.B. Littlewood and B.I. Shraiman, *Phys. Rev. Lett.* **74**, 5144 (1995); H. Röder, J. Zhang and A.R. Bishop, *Phys. Rev. Lett.* **76**, 1356 (1996).
5. S. von Molnar and S. Methfessel, *J. Appl.Phys.* **38**, 959, (1967).
6. M.A. Subramanian, G. Aravamudan and G.V. Subba Rao, *Prog. Solid State Chem.* **15**, 55 (1983)
7. M.A. Subramanian.,C. C. Torardi, D. C. Johnson, J. Pannetier and A. W. Sleight, *J. Solid State Chem.* **72**, 24 (1988).
8. J. N. Reimers, J. E. Greedan, C. R. Cremer, E. Gmelin and M. A. Subramanian, *Phys. Rev.* **B43** (1991) 3387; N. P. Raju, J. E. Greedan, A. Maignan, Ch. Simon, J. S. Pederson, A. M. Niraimathi, A. Gmelin and M. A. Subramanian, *Phys. Rev.* **B54**, 7189 (1996); J. E. Greedan, J. Avelar and M. A. Subramanian, *Solid State Commun.* **82** 797 (1992); J. E. Greedan, N. P. Raju and M. A. Subramanian, *Solid State Commun.* **99**, 399 (1996).
9. H. Fujinaka, N. Ninomura, M. Koizumi, Y.Miymoto and S. Kume, *Mat. Res. Bull.*, **14** 1133 (1979).
10. N.P. Raju, J.E. Greedan and M.A. Subramanian, *Phys. Rev.* **B49**, 1086 (1994).
11. Y. Shimakawa, Y. Kubo, T. Manako, Y. V. Suchko, D. N. Argyriou and J. D. Jorgensen, *Phys. Rev.* **B55**. 6399 (1997).
12. Y. Shimakawa, Y. Kubo, T. Manako, *Nature* **379**, 538 (1996).
13. S-W. Cheong, H. Y. Hwang, B. Batlogg, and L. W. Rupp, Jr., *Solid State Commun.*, **98**, 163 (1996).
14. M. A. Subramanian, B. H. Toby, A. P. Ramirez, W. J. Marshall, A. W. Sleight and G. H. Kwei, *Science*, **273**, 81 (1996).
15. H. D. Rosenfeld and M. A. Subramanian, *J. Solid State Chem.* **125**, 278 (1996)
16. G. H. Kwei, C. H. Booth, F. Bridges and M. A. Subramanian, *Phys. Rev.* **B55**, R688 (1997).
17. P. Schiffer, A. P. Ramirez, W. Bao, and S-W. Cheong, *Phys. Rev. Lett,* **75**, 3336 (1995).
18. P. G. Radaelli, D. E. Cox, M. Marezio, S. -W. Cheong, P. E. Sciffer and A. P. Ramirez, *Phys. Rev. Lett.* **75**, 4488 (1995). D. N. Argyriou, J. F. Mitchell, C. D.

Pottor, D. G. Hinks, J. D. Jorgenson and S. D. Badder, *Phys. Rev. Lett.* **76**, 3826 (1996). S. J. L. Billinge, R. DiFrancesco, G. H. Kwei, J. J. Neumeier and J. D. Thomson, *Phys. Rev. Lett.* **77**, 175 (1996).
19. D.-K. Seo, M.-H. Whangbo and M. A. Subramanian, *Solid State Commun.*, **101**, 417 (1997); D. J. Singh, *Phys. Rev.* **B55**, 313 (1997); S. F. Matar, M. A. Subramanian and J. Etourneau, *J. Mater. Chem.* in press.
20. A. P. Ramirez and M. A. Subramanian, *Science*, **277**, 546 (1997)

FIRST-ORDER INSULATOR-METAL TRANSITIONS IN PEROVSKITE MANGANITES WITH CHARGE-ORDERING INSTABILITY

H. KUWAHARA
Joint Research Center for Atom Technology (JRCAT), Tsukuba, Ibaraki 305, Japan

Y. TOKURA
Department of Applied Physics, University of Tokyo, Hongo, Tokyo 113, Japan
and
Joint Research Center for Atom Technology (JRCAT), Tsukuba, Ibaraki 305, Japan

The metal-insulator (M-I) phase transitions relevant to charge/orbital-ordering have been investigated for single crystals of perovskite-type manganites with controlled doping level (x) and one-electron bandwidth (W). The magnetic-field-induced I-M transition accompanied by several orders of magnitude change in resistivity is observed concomitantly with metamagnetic transitions and lattice structural change for relatively small-W system, $Nd_{1/2}Sr_{1/2}MnO_3$ crystal. This is due to the field-induced transition from antiferromagnetic charge-ordered insulator to ferromagnetic metal. In the case of further reduced-W system, e.g., $Pr_{1-x}Ca_xMnO_3$, charge-ordered ground states are dominant for a wider range of x around x=0.50 and the robustness of the charge-ordered state can be controlled by degree of discommensuration (i.e., deviation of x from 0.50). With decrease of W, the antiferromagnetic spin fluctuation subsists even above the ferromagnetic transition temperature and enhances the electron-lattice coupling. The antiferromagnetic fluctuation that may be associated with the orbital correlations and collective Jahn-Teller distortions can be suppressed by application of a relatively low magnetic field, producing the low-field colossal magnetoresistance effect.

1 INTRODUCTION

Since the re-discovery of the so-called colossal magnetoresistance (CMR) phenomena in perovskite-type manganites, extensive studies have been performed in manganese oxides with structures of pseudo-cubic[1,2] and layered perovskite,[3,4] and pyrochlore[5] for a wide variety of composition from the viewpoints of exotic spin-charge coupled phenomena. For those manganese oxide compounds, there exit three crucial control parameters which modify the electron correlation strength and resultant properties of transport, magnetism, and lattice structure. One is a carrier density (hole-doping level) x or band filling $(1-x)$ of the conduction band. In the case of the perovskites, $RE_{1-x}AE_x MnO_3$, where

RE is a trivalent rare earth and AE a divalent alkaline earth element, chemical substitution for RE^{3+} by AE^{2+} introduces holes into the e_g orbitals, which are mobile and mediate ferromagnetic interaction between the Mn atoms (so-called double-exchange (DE) interaction).[6] The electronic and magnetic properties can be changed by hole-doping level x. One such example of doping effects is the controlled variation of robustness of charge-ordering with x (see Sec. 4).

The second control parameter is an effective one-electron bandwidth (W) or equivalently e_g-electron transfer interaction (t). The magnitude of W is changed by varying the averaged ionic radius of the (RE,AE) site,[7,8,9] or equivalently the tolerance factor (f), $f = (<r_A>+r_O)/[\sqrt{2}(r_{Mn}+r_O)]$. Here, $<r_A>$, r_{Mn}, and r_O are (averaged) ionic radii for the respective elements from tabulated values.[10] With decrease of f (or $<r_A>$) the averaged bond angle α of Mn-O-Mn shows a deviation from 180° in the orthorhombic lattice, as empirically expressed as $\alpha = 172.6 - 347.5(1 - f)$ in the case of x=0.30. [9] The W-value for the manganite is governed by the degree of the hybridization between the Mn $3d$ e_g state and O $2p$ σ state which approximately scales with $\cos^2\alpha$ due to the d-electron supertransfer process. Therefore, the smaller the averaged ionic radius of the A site or f is, the smaller W is. Another way to control W is application of an external pressure (P): the W-value can be more finely increased with P through the contraction of Mn-O bond length or straightening of a Mn-O-Mn bond angle.[11,12,13]

For perovskite manganites with a fixed x, transport properties including CMR phenomena can be drastically modified by changing W. For examples, we show in Fig. 1 an electronic phase diagram as a function of tolerance factor f (or $<r_A>$). The doping level of the present $RE_{1/2}AE_{1/2}MnO_3$ systems is fixed at a commensurate value of x=1/2, i.e. one e_g-hole (or electron) per two Mn-sites, and hence the charge-ordered state tends to be most stabilized. The DE interaction arises from the itinerancy of the doped holes, and therefore the reduction in W favors the charge-ordered state, while destabilizing the ferromagnetic state. As one can immediately notice in Fig. 1, the ferromagnetic metallic (FM) state due to the DE interaction is dominant in the maximal-W system, $La_{1-x}Sr_xMnO_3$.[14] On the other hand, when W is decreased (f is decreased), the FM state becomes to be apparently absent, as observed for $Pr_{1-x}Ca_xMnO_3$.[15] Around a middle-W region ($f\sim0.975$ in Fig. 1), i.e., $(Nd,Sm)_{1/2}Sr_{1/2}MnO_3$ crystals, the subtle balance or competition between the ferromagnetic DE and the antiferromagnetic charge-ordering interactions result in versatile phenomena including striction-coupled CMR as will be discussed in Sec. 5.

The third control parameter is a dimensionality: the lowering of the electronic dimensionality causes a variety of essential changes in the electronic

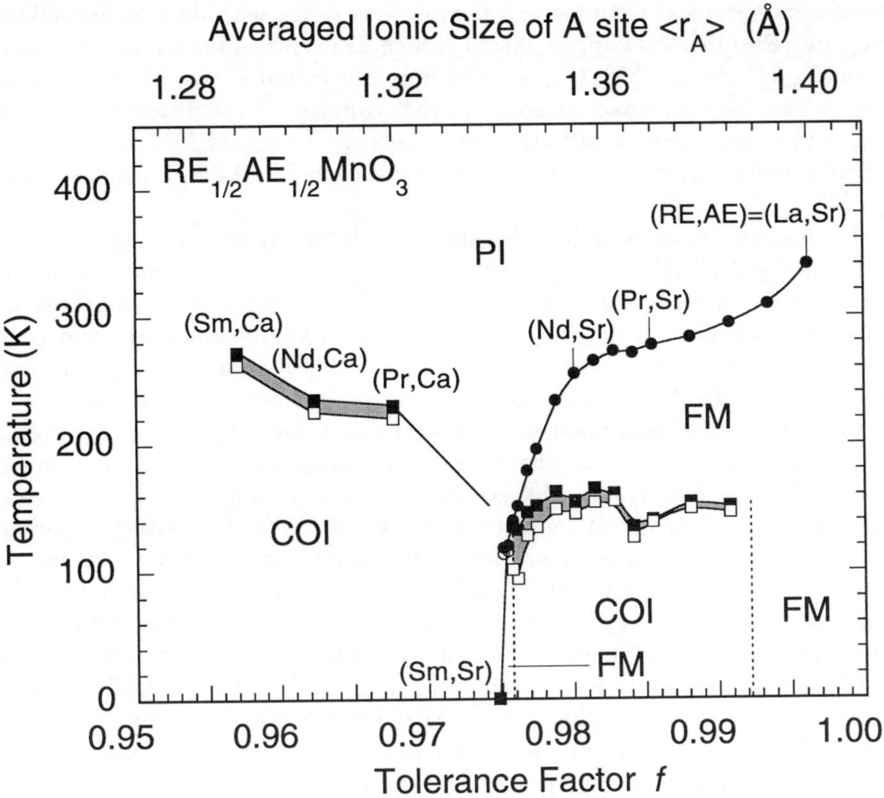

Figure 1: Phase diagram for perovskite-type manganites with a fixed doping level of x=1/2, $RE_{1/2}AE_{1/2}MnO_3$, where RE is a solid solution of rare earth elements and AE an alkaline earth element (Sr or Ca) (Ref. 34). The systems represented here are $Sm_{1/2}Ca_{1/2}MnO_3$ (Ref. 42), $Nd_{1/2}Ca_{1/2}MnO_3$, $Pr_{1/2}Ca_{1/2}MnO_3$, $(Sm,Nd)_{1/2}Sr_{1/2}MnO_3$, $(Nd,Pr)_{1/2}Sr_{1/2}MnO_3$, and $(Pr,La)_{1/2}Sr_{1/2}MnO_3$. The abbreviations mean paramagnetic insulator (PI), ferromagnetic metal (FM), and (antiferromagnetic) charge-ordered insulator (COI). The temperatures for phase transitions to respective states were determined from the resistivity measurements. The hatched area represents a thermal hysteresis, open and solid symbols are transition points obtained in cooling and warming, respectively. There is a crystallographic phase boundary from orthorhombic to rhombohedral at $f\sim$0.992. A neutron scattering study (Ref. 24) revealed that there are two types of magnetic structures for the antiferromagnetic state, CE- and A-type structures for $Nd_{1/2}Sr_{1/2}MnO_3$ and $Pr_{1/2}Sr_{1/2}MnO_3$, respectively. The disappearance of the COI states at temperatures below T_c around f=0.975 is explained in terms of the supercooling of (metastable) state, although the COI state is thermodynamically stable at low temperatures (see Ref. 30 for details of this novel phenomena).

properties. A typical example is a homologous series with layered perovskite structure (Ruddlesden-Popper phase), which are represented by the chemical formula, $(RE, AE)_{n+1}Mn_nO_{3n+1}$. The unit cell includes two sets of n-MnO_2 sheets which are connected via apical oxygens and separated by an intergrowth (RE, AE)O layer of rock salt structure. Recently, low field MR due to inter-layer tunneling of spin-polarized electron is discovered in $(La,Sr)_3Mn_2O_7$ (n=2) crystal.[4]

In this article, we overview the metal-insulator (M-I) transitions relevant to the charge/orbital-ordering (CO/OO) phenomena under variations of W and doping x as control parameters. In Sec. 2, we briefly review a prototype of the CO transition in the $Nd_{1/2}Sr_{1/2}MnO_3$ crystal. As one of major origins of the CMR effect, we present magnetic field destruction of charge-ordered state in Sec. 3. In this section, we also discuss the electronic phase diagram in the temperature (T) and magnetic field (H) plane, focusing on the metastable state (hysteretic region). In the case of $Pr_{1-x}Ca_xMnO_3$ system with much reduced W (see Fig. 1), the charge-ordered states expand over a wider concentration range. In Sec. 4 we discuss the robustness of CO with the doping level x deviated from 0.50 (i.e., increasing discommensurability). In Sec. 5, we describe the striction-coupled CMR in $(Nd_{1-y}Sm_y)_{1/2}Sr_{1/2}MnO_3$ crystals, which locates on the verge of the ferromagnetic and antiferromagnetic phase boundary: the ferromagnetic transition temperature T_c steeply changes as a function of f as shown in Fig. 1. The role of the antiferromagnetic fluctuation as well as its field suppression is emphasized in the CMR process. Summary is given in Sec. 6. To avoid the carrier scattering at the grain boundary of the conventional ceramic samples and unravel the intrinsic transport properties, the present systematic investigation was performed by using high-quality single crystals prepared by the floating zone method.

2 CHARGE/ORBITAL-ORDERING TRANSITIONS

It has recently become known that a charge/orbital-ordering phase transition is widely seen in the perovskite (or layered perovskite) type lattice of the transition metal oxides, such as $La_{2-x}Sr_xNiO_4$,[16] $La_{1-x}Sr_{1+x}MnO_4$,[17] and $La_{1-x}Sr_xFeO_3$,[18] when the doped hole concentration (x) takes a commensurate value, e.g., x=1/3, 1/2 and 2/3. With decreasing W, the interatomic Coulomb interaction becomes relatively important and the charge-ordered state is stabilized.[15,17,19,20,21] Figure 2 shows a typical example of the charge-ordering (CO) transition in manganites, $Nd_{1-x}Sr_xMnO_3$ (x=1/2) crystal.[22] This system is derived by decreasing W from the maximal-W system, $La_{1/2}Sr_{1/2}MnO_3$ [14] (see Fig. 1). As a generic feature of hole-doped perovskites of manganese oxide, the

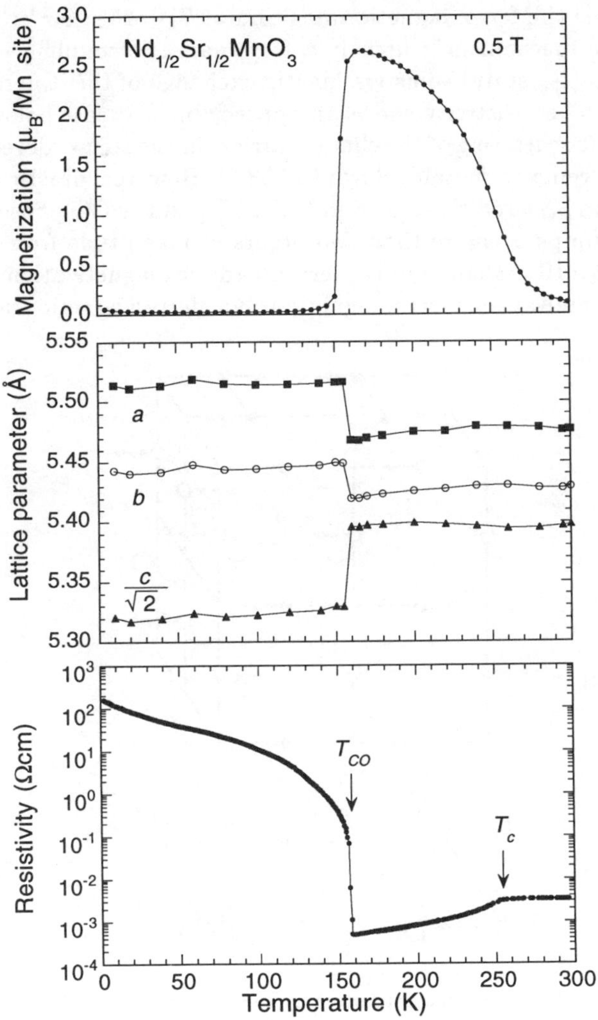

Figure 2: Temperature dependence of the magnetization (top panel; measured at a magnetic field of 0.5 T), lattice parameters (middle panel), and resistivity (bottom panel) in a crystal of $Nd_{1/2}Sr_{1/2}MnO_3$ (Ref. 22). T_c and T_{CO} represent the Curie temperature and the critical temperature for the charge-ordering transition, respectively. Upon the charge-ordering transition, resistivity shows a jump by several orders of magnitude. Simultaneously the ferromagnetic magnetization disappears due to the antiferromagnetic spin order and the lattice parameters in the orthorhombic structure (*Pbnm*; the so-called $GdFeO_3$ type) show an abrupt change characteristic of the first-order phase transition.

crystal undergoes the ferromagnetic transition at T_c(=255 K)[14,23] due to the double-exchange mechanism[6]; namely a ferromagnetic coupling between the local $3d$ electron (t_{2g} state) spins via kinetic exchange of the itinerant $3d$ electron (e_g state). The resistivity shows an appreciable decrease below T_c due to reduced magnetic scattering of the charge carriers in the spin-ordered ferromagnetic phase and remains metallic down to 158 K. However, drastic change due to CO occurs at T_{CO}=158 K, as seen in Fig. 2. Upon the CO phase transition, the resistivity jumps by more than two orders of magnitude from a typically metallic value (5×10^{-4} Ωcm) and the ferromagnetic magnetization disappears, indicating the simultaneous ferromagnetic-to-antiferromagnetic transition.

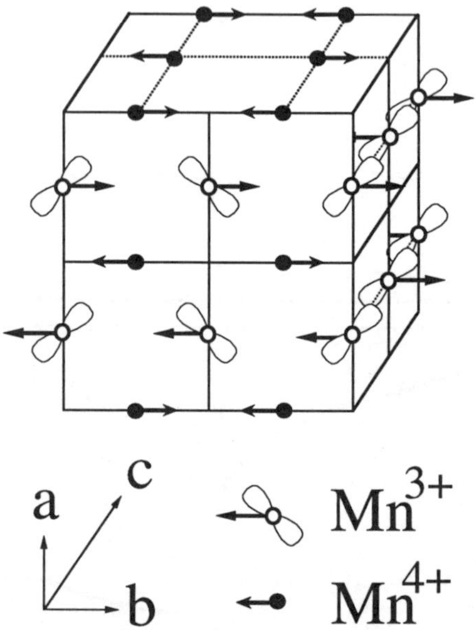

Figure 3: Schematic of the *CE*-type antiferromagnetic spin ordering and e_g-orbital ordering (Refs. 24 and 27).

In accord with these changes in electric and magnetic properties, the lattice parameters of the orthorhombically distorted perovskite show a distinct change as shown in the middle panel of Fig. 2. The CO transition is of the first order in nature and accompanies hystereses (not shown in Fig. 2 but see in Fig. 4). According to the neutron scattering study,[24] the ferromagnetic metal to antiferromagnetic insulator transition can be assigned to the phase transition to the charge/orbital-ordered state, in which the nominally Mn^{3+}

and Mn^{4+} species show a real space ordering (checkerboard-like) on the (001) plane of orthorhombic lattice (*Pbnm* notation) as illustrated in Fig. 3. The $3x^2-r^2/3y^2-r^2$ type orbitals show superlattice along *b*-axis on the same plane, as evidenced by neutron,[24] X-ray,[25] and electron diffraction measurements.[26] Reflecting such an orbital ordering, the spin ordering shows a complicated sublattice structure as shown in Fig. 3: The spin structure in charge-ordered state below T_{CO} is the $4 \times 4 \times 2$ unit cell in the pseudo-cubic perovskite setting and the stripes of the homovalent Mn ions are along the *c*-axis with collinear antiferromagnetic coupling of spins, what is the same as previously reported for $La_{1/2}Ca_{1/2}MnO_3$ [21,27] and $Pr_{1/2}Ca_{1/2}MnO_3$ [15,19,28,29] (called the *CE*-type).

3 FIELD-INDUCED MELTING OF CHARGE-ORDERED STATE

One of notable features of the perovskite manganites is that the charge-ordering (CO) transition is associated with the ferromagnetic-to-antiferromagnetic phase transition. Therefore, the charge-ordered state collapses under an external magnetic field, accompanying the metamagnetic, namely field-induced antiferromagnetic-to-ferromagnetic, transition. In other words, the "crystal state" of the doped holes can be melted and changed into a "liquid state" (i.e., a metallic state) by applying an external field. Figure 4 (a) displays the temperature dependence of resistivity under various magnetic field intensities for $Nd_{1-x}Sr_xMnO_3$ (x=0.50).[22,30] The resistivity near T_c (=255 K) is much reduced by applying an external field, since the field forcedly aligns the t_{2g} local spins and hence the double-exchange carriers gain a mobility due to a reduced spin scattering. This is the conventional magnetoresistance (MR) effect which is generally seen around T_c in the manganites showing the ferromagnetic-metallic ground state. In addition to such a canonical MR, a more remarkable field effect, i.e., six orders of magnitude change in resistivity, is seen in the charge-ordered state. The critical temperature of the CO transition, that accompanies the thermal hysteresis characteristic of the first-order phase transition, is decreased by increasing a magnetic field. The charge-ordered antiferromagnetic insulating state is totally extinguished over the whole temperature range by applying the field larger than 7 T. The isothermal MR at fixed temperatures below T_{CO} is shown in Fig. 4 (b), which clearly indicates the collapse of the charge-ordered state into the metallic state under a field. The measurements of the magnetization as well as neutron [24] and X-ray [25] diffraction studies under a field confirmed that the field-induced insulator-metal transition is accompanied by the metamagnetic transition as well as the collapse of the charge/orbital-ordered state.

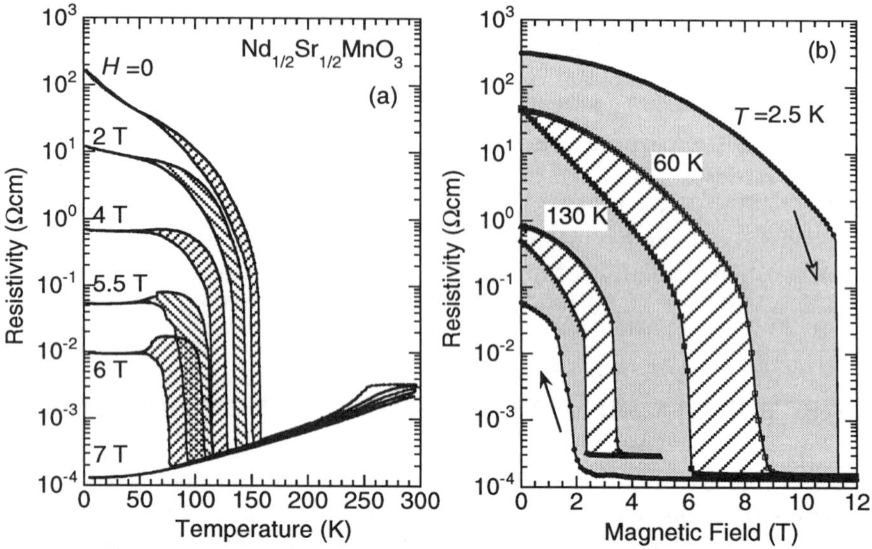

Figure 4: (a) Temperature dependence of resistivity for $Nd_{1/2}Sr_{1/2}MnO_3$ crystal under several magnetic field intensities (Ref. 30). (b) Isothermal magnetoresistance at fixed temperatures below T_{CO}=160 K (Ref. 22). The hatched areas are the hystereses against temperature and field.

Such a magnetic-field-induced destruction of charge-ordered state can be viewed as one of major origins of so-called colossal magnetoresistance effect. Reflecting the first-order nature of the CO transition, the isothermal MR shows a temperature dependent hysteresis against the field (indicated by hatches in Fig. 4 (b)). On the basis of these isothermal MR data, the insulator-metal phase diagram for $Nd_{1/2}Sr_{1/2}MnO_3$ crystal can be drawn in the $T-H$ plane (Fig. 5).[22] A hatched area in Fig. 5 shows a hysteretic region (between the lower and upper critical fields), that is a bistable region where the charge-ordered state and metallic state can coexist. One may immediately notice that the hysteretic region becomes anomalously broader when temperature is lowered, say below 20 K. In particular, the lower critical field for the metal-to-insulator transition is conspicuously decreased with decrease of temperature as shown in a magnified scale on the inset of Fig. 5.

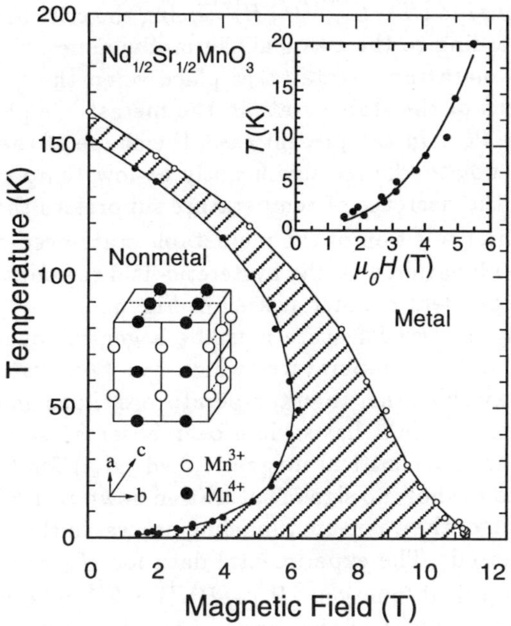

Figure 5: The electronic phase diagram of $Nd_{1/2}Sr_{1/2}MnO_3$ crystal which was determined by taking the critical fields for the insulator-to-metal and metal-to-insulator transitions in scanning an external field at a fixed temperature (Ref. 22). "Nonmetal" and "Metal" represent the antiferromagnetic state with the charge-ordering (as schematically depicted in the figure) and the ferromagnetic metallic state realized by the double-exchange interaction, respectively. The inset shows a magnification of the temperature dependence of the lower critical field for the metal-to-insulator transition.

The conspicuous change of the critical-field value with change of temperature can be interpreted as follows: In terms of the free energy as a function of the thermodynamic variable, i.e., the magnetization (M) in the present case, the charge "liquid" and "crystal" states correspond to the two local minima with large and small M-values, respectively, which are separated by a potential barrier (U). In this case, an external magnetic field (H) tends to stabilize the high-M state via Zeeman term, $-MH$, and hence U shape is subject to change with the field. It is easily shown that the barrier height U which is measured from the potential local minimum corresponding to the metastable state is proportional to $|H - H_c^{(0)}|^{3/2}$, $H_c^{(0)}$ being the critical value where U becomes zero. According to the classical thermodynamics, the first-order phase transition of the metastable state takes place when the formation energy of the critical nucleus of the stable state in the metastable phase is comparable with temperature (T). In the present case, the shape of the free energy function is supposed not to change significantly at low temperatures well below $T_{\rm CO}$. Therefore, the decrease of temperature suppresses the thermal fluctuation and hence the transition of the metastable state even near $H_c^{(0)}$. This is the qualitative explanation why the hysteretic-field region is expanded in such a temperature-dependent manner as seen in Fig. 5.

The presently observed feature must be a generic one of the first-order phase transition at low temperatures which can be caused by a thermodynamic intensive variable other than temperature, such as an external magnetic field and pressure, though it has seldom been observed so far. As seen in the inset of Fig. 5, the lower critical magnetic field (H_{cl}) for the metal-insulator transition appears to show no sign of saturation down to 1.5 K, indicating negligible contribution of the quantum tunneling process in the temperature region presently investigated. The experimental data for H_{cl} can be approximately scaled with an empirical function, $(H_{cl} - H_{cl}^{(0)}) \propto T^{\beta}$, where the critical exponent β was estimated to be in the range of 1/4–1/3 depending on the choice of $H_{cl}^{(0)}$. A solid line in the inset represents the case that β=4/15.

4 ROBUSTNESS OF CHARGE/ORBITAL ORDERING AGAINST CARRIER DISCOMMENSURABILITY

Let us come back to the electronic phase diagram shown in Fig. 1. When W is further reduced from $Nd_{1-x}Sr_xMnO_3$, the charge-ordered insulating state appears to be extended over higher-temperature and wider-composition regions, as seen for $Pr_{1-x}Ca_xMnO_3$. For $Pr_{1/2}Ca_{1/2}MnO_3$ (x=1/2), the charge- and orbital-ordering takes place at first in the same pattern as shown in Fig. 3 below $T_{\rm CO}$=240 K and no ferromagnetic metallic phase shows up at zero field.[15] The $Pr_{1/2}Ca_{1/2}MnO_3$ crystal undergoes the spin-collinear antiferromagnetic transition of the CE-type at 180 K: Onset of the spin ordering (T_N) emerges at a lower temperature than $T_{\rm CO}$. The charge-ordered state in $Pr_{1-x}Ca_xMnO_3$ dominates over the double-exchange ferromagnetic state in the fairly wide-x region ($0.30 \le x \le 0.70$) in contrast to the case of $Nd_{1-x}Sr_xMnO_3$ with a larger W-value where the charge-ordered state emerges in the limited region in the immediate vicinity of the commensurate x value (x=0.50±0.02).[31] With deviation of x from 0.50, the spin-canted phase arises in the lower temperature region below T_N. The spin-canting transition temperature ($T_{\rm CA}$) and the spin-

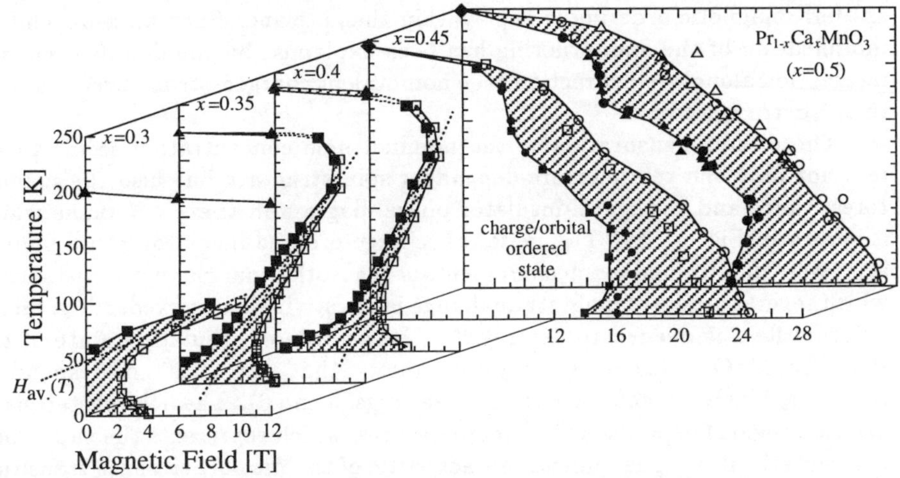

Figure 6: The charge-ordered state of $Pr_{1-x}Ca_xMnO_3$ (x=0.30, 0.35, 0.40, 0.45, and 0.50), which is plotted on the magnetic field and temperature plane (Refs. 15 and 32). The phase boundaries for x=0.45 and 0.50 have been determined by means of pulsed high magnetic fields up to 40 T (Ref. 32). The hatched area indicates the hysteresis region. In the cases of x=0.30, 0.35, and 0.40, the lines of $H_{av}(T)$ at which the free energy of both the charge-ordered and the ferromagnetic states are supposed to be equal are drawn by dashed lines.

canting angle tend to increase with departure of x from 0.50.[15,28] An observed variation of the transition temperatures T_N and T_{CA} with x can be interpreted in terms of the partial revival of the double-exchange carriers in the charge-ordered phase: The charge-ordered pattern in $Pr_{1-x}Ca_xMnO_3$ is observed to be always $(\pi, \pi, 0)$[28] as shown in Fig. 3 and hence the deviation of x from 0.50 may produce extra carriers which play a role of modifying the spin structure as observed. As revealed by a neutron scattering study,[28] the superlattice reflection due to the e_g-orbital ordering as shown in the Fig. 3 has been observed even for x=0.30 below its T_{CO} (~200 K) as well as for x=0.50. For $0.30 \leq x < 0.50$, however, the CE-type spin structure is modified by the deviation of x from 0.50: The spins are inclined out of the ab-plane and the coupling along the c-direction is not genuinely antiferromagnetic but canted. At x=0.30, the coupling along the c-direction becomes ferromagnetic in spite that the original antiferromagnetic ordering is kept within the *ab* plane. Such an x-dependent modification of the spin-ordering has been explained by the double-exchange interaction along the *c*-direction (the homovalent charged-stripe direction) due to the extra electrons.[28,29]

The discommensuration of the nominal hole concentration seems to affect not only the temperature-dependent spin structure but also the magnetoresistance and the metal-insulator phase diagram in the $T-H$ plane which is shown in Fig. 6. [15,32] The metal vs. charge-ordered-insulator phase boundaries were determined by measurements of the isothermal magnetoresistance in which several orders of magnitude change in resistivity is observed.[15,19] A larger external field is needed to destroy the charge-ordered insulating state of the $Pr_{1-x}Ca_xMnO_3$ (x=0.50) crystal than that of the corresponding $Nd_{1-x}Sr_xMnO_3$ (x=0.50) compound (see Figs. 5 and 6). The field-hysteresis (a hatched region) expands at low temperatures, which represents the supercooling (superheating) phenomena characteristic of the first-order phase transition as discussed in Sec. 3. The deviation of x from 0.50 (discommensuration), however, brings about a drastic modification in the phase boundary. The phase boundary is shifted to the lower-magnetic field side with deviation of x from 0.50, in particular at low temperatures: For example, the critical field for destruction of the charge-ordered state is ~30 T for x=0.50 and 6 T for x=0.30 at 4.2 K. This seems to be closely correlated with the discommensuration of the charge-ordered state and the resultant change of the spin structure to the canted antiferromagnetic state which is more amenable to an external field than the spin-collinear antiferromagnetic state at x=0.50. The reentrant behavior of the CO transition, e.g., in the field-cooled run at $x<0.50$, may arise from the excess entropy in the discommensurate charge-ordered state. In particular, for x=0.30 the field-induced transition from the charge-ordered insulator to the ferromagnetic metal in the isothermal process becomes irreversible below 40 K, as exemplified by the $T-H$ phase diagram of Fig. 6, accompanying the persistent change from an insulating value above 10^7 Ωcm to a typical metallic value (3×10^{-4} Ωcm) exceeding ten orders of magnitude.[19] Similarly to the case of the $Nd_{1/2}Sr_{1/2}MnO_3$, the measurement of the magnetization as well as the neutron scattering study [28] under a field has proved that the insulator-metal transition is accompanied by the metamagnetic ((spin-canted) antiferromagnetic to ferromagnetic) transition and lattice-structural change.

5 STRICTION-COUPLED MAGNETORESISTANCE NEAR CHARGE-ORDERING INSATIABILITY

The colossal magnetoresistance (CMR) thus typically appears with the collapse of the charge/orbital-ordering and antiferromagnetic spin ordering as described above. Even if such a long-range ordering is not established, however, the magnetic field suppression of the antiferromagnetic fluctuation (perhaps as well as of the orbital correlations) may play an important role in the CMR effect. In this section, we discuss such a phenomenon as typically observed for $(Nd_{1-y}Sm_y)_{1/2}Sr_{1/2}MnO_3$ crystals.[33,34] This system is located on the verge of the ferromagnetic-antiferromagnetic phase boundary as already shown in Fig. 1. The competition between the ferromagnetic double-exchange (DE) and antiferromagnetic CO interactions is sensitive to W. In a series of $(Nd_{1-y}Sm_y)_{1/2}Sr_{1/2}MnO_3$ crystals, the W can be finely controlled with change of y, $0 \leq y \leq 1$ ($0.980 \geq f \geq 0.975$).

Figure 7 shows overall features of $\rho(T)$ in zero magnetic field (upper panel) and inverse magnetization H/M versus T measured in a field of 0.5 T (lower panel) for $(Nd_{1-y}Sm_y)_{1/2}Sr_{1/2}MnO_3$. All the crystals undergo the ferromagnetic transition at different T_c's depending on y (or f). The resistivity shows an appreciable decrease below T_c owing to reduced magnetic scattering of the charge carriers in the ferromagnetic phase and the crystal shows metallic conductivity as far as the temperature is above the CO transition temperature T_{CO}. (The properties below T_{CO} are not shown in Fig. 7 to avoid complexity.) As the Sm concentration y increases (f decreases), T_c, at which the resistivity steeply drops, decreases. This is seemingly consistent with the simple expectation: the W-value (or e_g-electron transfer interaction t-value) decreases with y so that the ferromagnetic DE interaction should be weakened. As is clear from Fig. 1, however, the change of T_c with f is too steep to be ascribed to the change of W alone, indicating the presence of competing interaction(s), e.g., antiferromagnetic correlation. A similar correlation between the averaged ionic radius of A site (or f) and T_c has also been reported for a different doping level of x=0.30. [9] Another notable point for the systematics with varying y is a change in character of the resistive transition at T_c. The resistivity immediately above T_c remarkably increases with y (or with decrease of f). This may be partly due to the enhanced electron-lattice coupling, e.g. dynamic and collective Jahn-Teller coupling.[35] The resistivity drop at T_c is hence remarkably enhanced with increase of y. In the smaller-W case of the y=0.94 crystal, the temperature dependence of resistivity shows a thermal hysteresis around T_c (not shown in Fig. 7 but see the right panel of Fig. 9), indicating the occurrence of the first-order phase transition. It is noted that all the $\rho-T$ curves converge to the constant resistivity of 30 mΩcm around 400 K, a temperature higher enough than T_c. This is in contrast to the low temperature region in which only a small variation of W causes a crucial change in transport properties.

Corresponding to the change in character of the resistive transition with decrease of W, anomalies of magnetization are also observed. H/M for any y obeys the Curie-Weiss law for temperatures above ~300 K as shown in lower panel of Fig. 7. Their Weiss temperatures are ranging from 210 to 260 K perhaps reflecting the change of W. It should be noted that the

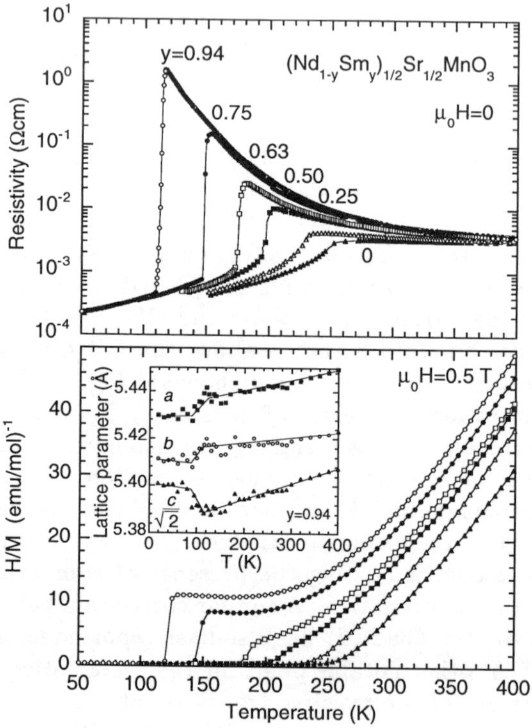

Figure 7: Temperature dependence of resistivity in zero magnetic field (upper panel) and inverse magnetization measured in a field of 0.5 T (lower panel) for $(Nd_{1-y}Sm_y)_{1/2}Sr_{1/2}MnO_3$ crystals; y=0, 0.25, 0.50, 0.63, 0.75, and 0.94 (Ref. 34). The resistivity and inverse magnetization jump upon antiferromagnetic charge-ordering transitions below T_c for crystals with y=0–0.75, which are omitted in the figure to avoid complexity. The inset shows the temperature dependence of lattice parameters for the y=0.94 crystal. All measurements were performed in warming run after zero-field cooling.

clear deviation from the Curie-Weiss law is observed for crystals with $y>0.50$ at temperatures below ~250 K. As y increases, the anomalous plateau becomes more pronounced, and above $y=0.50$ the transition to the ferromagnetic state becomes more like a first-order phase transition. The plateau or upturn of the H/M curve indicates that the antiferromagnetic interaction is competing with or dominating over the ferromagnetic DE interaction in this temperature range. This may imply not necessarily a static long-range order, but short-range order or antiferromagnetic fluctuation.[36,37] A similar phenomenon is observed in the much smaller-W system, $Pr_{1/2}Ca_{1/2}MnO_3$ ($f=0.968$), in which a charge/orbital-ordered state was confirmed to exist above spin-ordered temperatures by neutron scattering studies.[28] By analogy

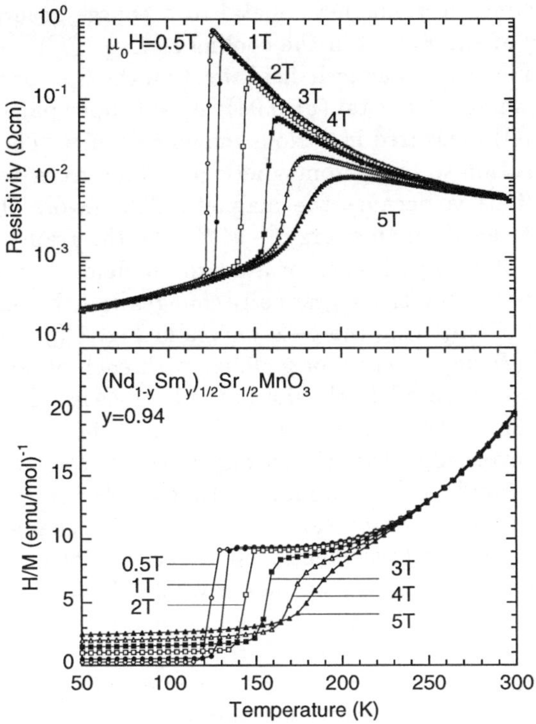

Figure 8: Temperature dependence of resistivity (upper panel) and inverse magnetization (lower panel) for the $(Nd_{1-y}Sm_y)_{1/2}Sr_{1/2}MnO_3$ ($y=0.94$) crystal measured in various magnetic fields (Ref. 34). Measurements were done in warming run.

to the case of $Pr_{1/2}Ca_{1/2}MnO_3$, the observed antiferromagnetic fluctuation for the higher-y crystals may be considered to originate from the charge/orbital ordering instability. The enhanced electron-lattice coupling, e.g., the dynamic and collective Jahn-Teller coupling,[35] appears to arise from the charge/orbital correlations and plays an important role in the CMR effect. The DE interaction appears to overcome the antiferromagnetic fluctuation at lower temperatures, giving rise to the abrupt phase switching to the ferromagnetic metallic state.

The simultaneous change of lattice structure at T_c has been confirmed by X-ray diffraction measurements for the y=0.94 crystal as shown in the inset of Fig. 7. It was found that the space group of the crystal structure is unchanged but the lattice parameters abruptly change at T_c: both the a and b shrink by ~0.14 % and the c expands by ~0.19 % at T_c in the cooling run. These changes seem to arise from a relaxation of distortion inherent to charge/orbital-ordering instability, since the similar but reverse changes are observed upon the transition from a ferromagnetic metal to a charge-ordered insulator (see the middle panel of the Fig. 2) in the cooling run.

We show in Fig. 8 the magnetic-field effect on the first-order ferromagnetic transition of the small-W crystal (y=0.94): $\rho(T)$ (upper panel) and H/M versus T (lower panel) measured in various magnetic fields. The critical temperature of the transition to the ferromagnetic metal increases with the magnetic field intensity. This is because the magnetic field favors the ferromagnetic state in terms of the Zeeman energy, $-MH$. As the transition temperature approaches ≈200 K by application of a magnetic field of more than 4 T, the character of the phase transition gradually changes and the resistivity decrease becomes gradual. Comparing the ρ–T curves in Fig. 7 (y-dependence in zero field) and Fig. 8 (H-dependence for y=0.94) with each other, one may immediately notice a nearly parallel behavior in the both cases. This is qualitatively interpreted as follows. According to the DE model,[6] the transfer interaction of e_g carriers (t) is governed by the relative angle (θ) of the local t_{2g} spins between neighboring Mn^{3+} and Mn^{4+} in a manner that $t = t_0 \cos(\theta/2)$. Here t_0 is the bare transfer interaction, which can be controlled by changing the averaged ionic size of the A site (or f) as mentioned in Sec. 1. The t-value (or W-value) can also be increased effectively by means of the H-induced reduction of θ. Thus, the *chemical* pressure and the magnetic field give rise to a similar effect in the vicinity of the ferromagnetic transition via the modification of t. This crossover from the extraordinary lattice-coupled first-order phase transition to the conventional lattice-irrelevant transition induced by a magnetic field can also be seen in the temperature dependence of magnetization as functions of y and H (compare the lower panels of Figs. 7 and 8).

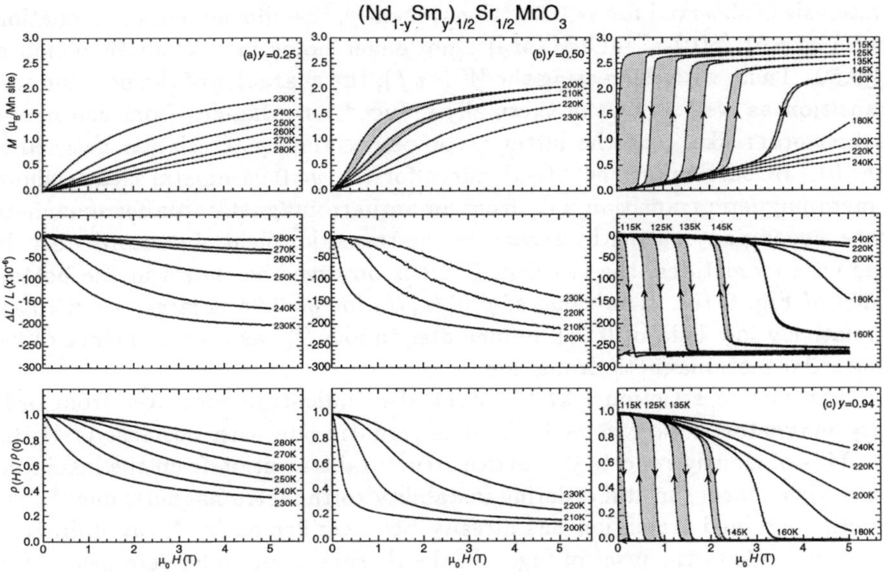

Figure 9: Magnetization M (top panel), striction $\Delta L/L$ (middle), and resistivity $\rho(H)$ (bottom; normalized by the resistivity in zero field $\rho(0)$) for $(Nd_{1-y}Sm_y)_{1/2}Sr_{1/2}MnO_3$ crystals as a function of an external magnetic field at fixed temperatures near above T_c; (a) y=0.25, (b) y=0.50, and (c) y=0.94 (Ref. 34). Hatched are hysteresis regions.

On these bases, we discuss the evolution from the conventional CMR to the extraordinary lattice-coupled CMR afforded by a relatively low magnetic field with decreasing of W. We show in Fig. 9 the magnetization, lattice striction, and resistivity as a function of magnetic field at fixed temperatures near above T_c for $(Nd_{1-y}Sm_y)_{1/2}Sr_{1/2}MnO_3$; (a) y=0.25, (b) 0.50, and (c) 0.94. The y=0.25 crystal exemplifies a case of the relatively weak electron-lattice coupling, like $La_{1-x}Sr_xMnO_3$ (the maximal-W system [14]). Above T_c (=230 K), the MR is well explained by an empirical relation,[38] $[\rho(H) - \rho(0)]/\rho(0) = -C(M/M_S)^2$, as observed for $La_{1-x}Sr_xMnO_3$, where $\rho(H)$ is the resistivity in a magnetic field, $\rho(0)$ in zero field, M_S the saturation magnetization ($\approx$3.5 μ_B/Mn site), and C the scaling constant (4.5 for the present case). The smaller the W is, the steeper the field-induced changes are for the respective quantities at temperatures near above T_c. In accord with the charge/orbital instability (or the enhanced electron-lattice coupling) subsisting above T_c, the crystal with $y \geq 0.63$ shows clear hystereses in magnetization, lattice striction, and resistivity as typically seen for case of the y=0.94 crystal in Fig. 9 (c), while no hysteresis is observed for $y<0.50$. Consequently, the above-mentioned relation, $[\rho(H) - \rho(0)]/\rho(0) = -C(M/M_S)^2$, no longer holds in a small W-region of $y \geq 0.50$. Thus, with decreasing the W (or f), the character of the field-induced transition as well as of the thermally-induced one changes from continuous to first-order like, and the latter accompanies the hysteresis (as hatched in Fig. 9). In particular, the $M-H$ curve for the y=0.94 crystal clearly shows a metamagnetic transition, i.e., from an antiferromagnetic to a ferromagnetic state, accompanying the hystereses in the transition field. In accord with the magnetic transitions, the low-field CMR is observed as shown in the bottom panel of Fig. 9 (c). The value of $\rho(0)/\rho(H)$ for y=0.94 is larger than 10^3 in a relatively low field of 0.5 T immediately above T_c, as a consequence of the first-order metamagnetic transition.

The lattice striction ($\Delta L/L$) data also indicate a crossover from ordinary magnetostriction to switching-like gigantic one with decrease of W (or f). This obviously reflects the lattice-structural transition from the insulating state with charge/orbital ordering instability to the ferromagnetic metal. The striction-coupled transition has already been confirmed by X-ray diffraction measurements in the inset of Fig. 7 in the thermal scan under zero field. Magnetostriction ($\Delta L/L$) for the y=0.94 crystal shows an abrupt shrink of the crystal length by ~0.025 % at the critical magnetic field for the resistive and metamagnetic transitions. (From the data on lattice structural changes in the thermal scan, the observed magnetic field dependence of $\Delta L/L$ is supposed to mainly reflect the change in the ab plane.) The magnetic field does not directly affect the crystal lattice but indirectly via localized t_{2g} spins. Therefore, such a magnetic-field-induced structural change indicates strong coupling between the spin and charge degrees of freedom via Hund's-rule coupling as well as the charge and lattice ones via general electron-lattice interaction, or Jahn-Teller coupling. As clearly seen in Fig. 9, the mutual coupling is strongly enhanced in the small W-region for $y \geq 0.50$ in $(Nd_{1-y}Sm_y)_{1/2}Sr_{1/2}MnO_3$. Thus, the colossal magnetoresistance effect appears to be closely related to the magnetic field suppression of the antiferromagnetic fluctuation as well as the electron-lattice (collective Jahn-Teller) coupling.

6 SUMMARY

We have reviewed our recent studies on the metal-insulator transitions relevant to the charge-ordering (CO) phenomena of perovskite-type manganites with use of single crystals in which the one-electron bandwidth (W) and the nominal hole-doping level (x) are finely controlled. It was proved by the systematic investigation on the magnetotransport and related magnetic and structural properties that collapse of the charge-ordered insulating state is one of the origins for the field-induced insulator-metal transition. The variations of W and x affect crucially the metal-insulator phase diagram in the temperature (T)−magnetic field (H) plane through the change in robustness of the charge-ordered state. In other words, a variety of magnetic-field induced resistive and structural transitions can be realized by control of the subtle balance between the antiferromagnetic CO and ferromagnetic double-exchange interactions in terms of fine tuning of W and x.

Beside the charge-ordered state with static long-range ordering, the charge/orbital-ordering instability with the antiferromagnetic spin fluctuation above T_c becomes prominent and enhances the electron-lattice coupling with decrease of W. As a result, the ordinary ferromagnetic transition is transformed to the striction-coupled first-order insulator-metal transition at T_c in the small-W region as observed in $(Nd_{1-y}Sm_y)_{1/2}Sr_{1/2}MnO_3$ ($y \geq 0.5$). We have demonstrated that a relatively low magnetic field can induce the transition from the insulator with antiferromagnetic spin fluctuation to the ferromagnetic metal, which accompanies a metamagnetic transition as well as a lattice-structural change. In the authors' view, the so-called CMR phenomena are closely related to the antiferromagnetic fluctuation competing with the double-exchange interaction, which should arise from the concomitant orbital and charge correlations, as typically seen in their long-range ordering in the CE-type charge-ordered state and its collapse under a magnetic field. Such a close interplay among the spin, charge, orbital, and lattice degrees of freedom in the composition-controlled manganites produces further intriguing metal-insulator phenomena which can be triggered by other external fields than magnetic field. In fact, it has recently been found that the transition of the metastable charge-ordered insulating state in $Pr_{1-x}Sr_xMnO_3$ (x=0.30) crystal can be induced not only by an external magnetic field and pressure [2,12,13] but also by irradiation of an X-ray [39] and a visible-IR pulsed laser,[40] and by application of an electric field or current.[41] These field- or photo-induced insulator-metal transitions may have potential for new magnetoelectronic functionalities.

Acknowledgments

We are grateful to Y. Tomioka, Y. Moritomo, A. Asamitsu, T. Kimura, R. Kumai, H. Yoshizawa, H. Kawano, N. Miura, M. Tokunaga, and N. Nagaosa for their collaborations and enlightening discussions at the respective stages of the research. The present work, partly supported by NEDO, was performed in the Joint Research Center for Atom Technology (JRCAT) under the joint research agreement between the National Institute for Advanced Interdisciplinary Research (NAIR) and the Angstrom Technology Partnership (ATP).

References

1. For examples at an early stage of the research; R. M. Kusters, J. Singleton, D. A. Keen, R. McGreevy, and W. Heyes, Physica B **155**, 362 (1989); S. Jin, T. H. Tiefel, M. McCormack, R. A. Fastnacht, R. Ramesh, and L. H. Chen, Science, **264**, 413 (1994); K. Chahara, T. Ohno, M. Kasai, and Y. Kozono, Appl. Phys. Lett. **63**, 1990 (1993); R. von Helmolt, J. Wecker, B. Holzapfel, L. Shultz, and K. Samwer, Phys. Rev. Lett. **71**, 2331 (1993); M. McCormack, S. Jin, T. H. Tiefel, R. L. Greene, and T. Venkatesan, Appl. Phys. Lett. **64**, 3045 (1994); G. C. Xiong, Qi Li, H. L. Ju, S. N. Mao, L. Senapati, X. X. Xi, R. L. Greene, and T. Venkatesan, Appl. Phys. Lett. **66**, 1427 (1995); G. C. Xiong, Qi Li, H. L. Ju, R. L. Greene, and T. Venkatesan, Appl. Phys. Lett. **66**, 1689 (1995); R. Mahendiran, S. K. Tiwary, A. K. Raychaudhuri, T. V. Ramakrishnan, R. Mahesh, N. Rangavittal, and C. N. R. Rao, Phys. Rev. B **53**, 3348 (1996).
2. For a review paper; Y. Tokura, Y. Tomioka, H. Kuwahara, A. Asamistu, Y. Moritomo, and M. Kasai, J. Appl. Phys. **79**, 5288 (1996).
3. Y. Moritomo, A. Asamitsu, H. Kuwahara, and Y. Tokura, Nature **380**, 141 (1996).
4. T. Kimura, Y. Tomioka, H. Kuwahara, A. Asamitsu, M. Tamura, and Y. Tokura, Science **274**, 1698 (1996); T. Kimura, Y. Tomioka, A. Asamitsu, M. Tamura, and Y. Tokura, Phys. Rev. Lett. (in press).
5. Y. Shimakawa, Y. Kubo, and T. Manako, Nature **379**, 53 (1996); S-W. Cheong, H. Y. Hwang, B. Batlogg, and L. W. Rupp. Jr., Solid State Commun. **98**, 163 (1996).
6. C. Zener, Phys. Rev. 82, 403 (1951); P. W. Anderson and H. Hasegawa, Phys. Rev. **100**, 675 (1955); P. -G. de Gennes, Phys. Rev. **118**, 141 (1960); K. Kubo and N. Ohata, J. Phys. Soc. Jpn. **33**, 21 (1972).

7. J. B. Torrance, P. Lacorre, A. I. Nazzal, E. J. Ansaldo, and Ch. Nidermayer, Phys. Rev. B **45**, 8209 (1992); Y. Tomioka, A. Asamitsu, H. Kuwahara, and Y. Tokura, J. Phys. Soc. Jpn. **66**, 302 (1997); Y. Tomioka, H. Kuwahara, A. Asamitsu, M. Kasai, and Y. Tokura, Appl. Phys. Lett. **70**, 3609 (1997).
8. Y. Okimoto, T. Katsufuji, Y. Okada, T. Arima, and Y. Tokura, Phys. Rev. B **51**, 9581 (1995).
9. H. Y. Hwang, S-W. Cheong, P. G. Radaelli, M. Marezio, and B. Batlogg, Phys. Rev. Lett. **75**, 914 (1995).
10. R. D. Shannon, Acta Crystallogr. Sect. A **32**, 751 (1976).
11. H. Y. Hwang, T. T. M. Palstra, S-W. Cheong, and B. Batlogg, Phys. Rev. B **52**, 15046 (1995).
12. Y. Moritomo, A. Asamitsu, and Y. Tokura, Phys. Rev. B **51**, 16491 (1995);Y. Moritomo, H. Kuwahara, and Y. Tokura, J. Phys. Soc. Jpn. **66**, 556 (1997);Y. Moritomo, H. Kuwahara, Y. Tomioka, and Y. Tokura, Phys. Rev. B **55**, 7549 (1997).
13. H. Yoshizawa, R. Kajimoto, H. Kawano, Y. Tomioka, and Y. Tokura, Phys. Rev. B **55**, 2729 (1997).
14. Y. Tokura, A. Urushibara, Y. Moritomo, T. Arima, A. Asamitsu, G. Kido, and N. Furukawa, J. Phys. Soc. Jpn. **63**, 3931 (1994); A. Urushibara, Y. Moritomo, T. Arima, A. Asamitsu, G. Kido, and Y. Tokura, Phys. Rev. B **51**, 14103 (1995).
15. Y. Tomioka, A. Asamitsu, H. Kuwahara, Y. Moritomo, and Y. Tokura, Phys. Rev. B **53**, R1689 (1996).
16. C. H. Chen, S. -W. Cheong, and A. S. Cooper, Phys. Rev. Lett. **71**, 2461 (1993).
17. Y. Moritomo, Y. Tomioka, A. Asamitsu, Y. Tokura, and Y. Matsui, Phys. Rev. B **51**, 3297 (1995); B. J. Sternlieb, J. P. Hill, U. C. Wildgruber, G. M. Luke, B. Nachumi, Y. Moritomo, and Y. Tokura, Phys. Rev. Lett. **76**, 2169 (1996).
18. P. D. Battle, T. C. Gibb, and P. Lightfoot, J. Solid State Chem. **84**, 271 (1990).
19. Y. Tomioka, A. Asamitsu, Y. Moritomo, and Y. Tokura, J. Phys. Soc. Jpn. **64**, 3626 (1995).
20. Y. Tomioka, A. Asamitsu, Y. Moritomo, H. Kuwahara, Y. Tokura, Phys. Rev. Lett. **74**, 5108 (1995).
21. P. Shiffer, A. P. Ramirez, W. Bao, and S-W. Cheong, Phys. Rev. Lett. **75**, 3336 (1995); A. P. Ramirez, P. Schiffer, S-W. Cheong, C. H. Chen, W. Bao, T. T. M. Palstra, P. L. Gammel, D. J. Bishop, and B. Zegarski, Phys. Rev. Lett. **76**, 3188 (1996); C. H. Chen and S-W. Cheong, Phys.

Rev. Lett. **76**, 4042 (1996); G. Xiao, E. J. McNiff. Jr., G. Q. Gong, A. Gupta, C. L. Canedy, and J. Z. Sun, Phys. Rev. B **54**, 6073 (1996); P. G. Radaelli, D. E. Cox, M. Marezio, and S-W. Cheong, Phys. Rev. B **55**, 3015 (1997).
22. H. Kuwahara, Y. Tomioka, A. Asamitsu, Y. Moritomo, and Y. Tokura, Science **270**, 961 (1995).
23. G. H. Jonker and J. H. Van Santen, Physica 16, 337 (1950).
24. H. Kawano, R. Kajimoto, H. Yoshizawa, Y. Tomioka, H. Kuwahara, and Y. Tokura, Phys. Rev. Lett. **78**, 4253 (1997).
25. S. Shimomura *et al.*, unpublished.
26. J.Q. Li and Y. Matsui, unpublished.
27. E. O. Wollan and W. C. Koehler, Phys. Rev. **100**, 545 (1955).
28. H. Yoshizawa, H. Kawano, Y. Tomioka, and Y. Tokura, Phys. Rev. B **52**, R13 145 (1995); J. Phys. Soc. Jpn. **65**, 1043 (1996).
29. Z. Jirak, S. Krupicka, V. Nekvasil, E. Pollert, G. Villeneuve, and F. Zounova, J. Mag. Mag. Mat. **15-18**, 519 (1980); Z. Jirak, S. Krupicka, Z. Simsa, M. Dlouha, and Z. Vratislav, J. Mag. Mag. Mat. **53**, 153 (1985); K. Knizek, Z. Jirak, E. Pollert, F. Zounova, and S. Vratislav, J. Solid State Chem. **100**, 292 (1992).
30. Y. Tokura, H. Kuwahara, Y. Moritomo, Y. Tomioka, and A. Asamitsu, Phys. Rev. Lett. **76**, 3184 (1996).
31. H. Kuwahara *et al.*, unpublished.
32. M. Tokunaga, N. Miura, Y. Tomioka, and Y. Tokura, submitted to Phys. Rev. B.
33. H. Kuwahara, Y. Tomioka, Y. Moritomo, A. Asamitsu, M. Kasai, R. Kumai, and Y. Tokura, Science **272**, 80 (1996).
34. H. Kuwahara, Y. Moritomo, Y. Tomioka, A. Asamitsu, M. Kasai, and Y. Tokura, J. Appl. Phys. **81**, 4954 (1997); H. Kuwahara, Y. Moritomo, Y. Tomioka, A. Asamitsu, M. Kasai, R. Kumai, and Y. Tokura, Phys. Rev. B **56**, 9386 (1997).
35. A. J. Millis, P. B. Littlewood, and B. I. Shraiman, Phys. Rev. Lett. **74**, 5144 (1995); A. J. Millis, B. I. Shraiman, and R. Mueller, Phys. Rev. Lett. **77**, 175 (1996).
36. Similar antiferromagnetic correlations (fluctuations) have been observed in the layered perovskite crystal $La_{2-2x}Sr_{1+2x}Mn_2O_7$ with x=0.4 above T_c by inelastic neutron scattering measurements; T. G. Perring, G. Aeppli, Y. Moritomo, and Y. Tokura, Phys. Rev. Lett. **78**, 3197 (1997).
37. M. Kataoka, Chechoslovak J. Phys. **46**, 1857 (1996); M. Kataoka and M. Tachiki, Physica B **237-238**, 24 (1997).
38. N. Furukawa J. Phys. Soc. Jpn. **63**, 3214 (1994).

39. V. Kiryukhin, D. Casa, J. P. Hill, B. Keimer, A. Vigliante, Y. Tomioka, and Y. Tokura, Nature (London) **813**, 813 (1997).
40. K. Miyano, T. Tanaka, Y. Tomioka, and Y. Tokura, Phys. Rev. Lett. **78**, 4257 (1997).
41. A. Asamitsu, Y. Tomioka, H. Kuwahara, and Y. Tokura, Nature (London) **388**, 50 (1997).
42. Y. Tomioka, A. Asamitsu, H. Kuwahara, Y. Moritomo, M. Kasai, R. Kumai, and Y. Tokura, Physica B **237-238**, 6 (1997).

STRIPED CHARGE & ORBITAL ORDERING IN PEROVSKITE MANGANITES

S-W. CHEONG [1,2] and C. H. CHEN [1]

1. Bell Laboratories, Lucent Technologies, Murray Hill, NJ 07974, USA
2. Department of Physics & Astronomy, Rutgers University, Piscataway, NJ 08855, USA

The rich physics of colossal magnetoresistance compound is dominated by the intriguing competition between ferromagnetic double exchange coupling and striped charge/orbital ordering. Charge localization, prerequisite of charge ordering, is mutually exclusive with ferromagnetism in the double exchange mechanism which requires real hopping of charge carriers from a Mn^{3+} ion across an intervening O^{2-} ion to an adjacent Mn^{4+}ion. The general ground state of mixed-valent perovskite manganites is, therefore, either ferromagnetic-metallic or charge-ordered-antiferromagnetic. However, we have surprisingly found the coexistence of ferromagnetism and incommensurate charge ordering in a narrow temperature window of $La_{0.5}Ca_{0.5}MnO_3$. Furthermore, it is against intuition that the ferromagnetic transition temperature is optimized at the commensurate Ca concentration of 3/8 in (La,Ca)MnO_3. It is noteworthy that the tendency of striped charge ordering is generic in doped Mott insulators.

[1] Introduction: striped charge ordering and stripe pattern formation in general

Colossal magnetoresistance (CMR) effects have been observed in mixed-valent perovskite manganites [1]. The mixed-valent manganites can be prepared by the partial substitution of La by various divalent ions such as Pb, Ca, Sr or Ba in perovskite $LaMnO_3$. Other rare earth ions with relatively small ionic size (e.g., Pr, Nd or Sm) also form in the perovskite structure, and the divalent substitution in the perovskites with these rare earths could also induce CMR. It turns out that (La,Ca)MnO_3 is the prototype of mixed-valent perovskite manganites in the following sense. First, the Ca ionic size is almost identical to the La ionic size, and thus true solid solution forms in the entire Ca concentration. Furthermore, the structure, unlike other perovskite manganites, remains to be orthorhombic below ~700 K in the entire doping concentration. Therefore, we have investigated primarily (La,Ca)MnO_3 system, and found (La,Ca)MnO_3 compound representative of doped perovskite manganites.

Both $LaMnO_3$ and $CaMnO_3$, forming in perovskite structure, are antiferromagnetic insulators. Mn valence in $LaMnO_3$ is 3+ ($t_{2g}^3e_g^1$), and $CaMnO_3$ is a Mn^{4+} (t_{2g}^3) compound. True solid solution forms in the entire concentration of $La_{1-x}Ca_xMnO_3$, and Mn ions in the solid solution system are mixed-valent. The ionic size of Mn^{3+} is significantly larger than Mn^{4+} ionic size, and, furthermore, Mn^{3+} is a good Jahn-Teller ion. The antiferromagnetic superexchange coupling exists between Mn^{4+} ions via an intervening oxygen, but the superexchange interaction between Mn^{3+} ions via an intervening oxygen can be antiferromagnetic or ferromagnetic depending on the relative orbital ordering. The real hopping of the e_g electron between Mn^{3+} and Mn^{4+} through an oxygen produces a ferromagnetic

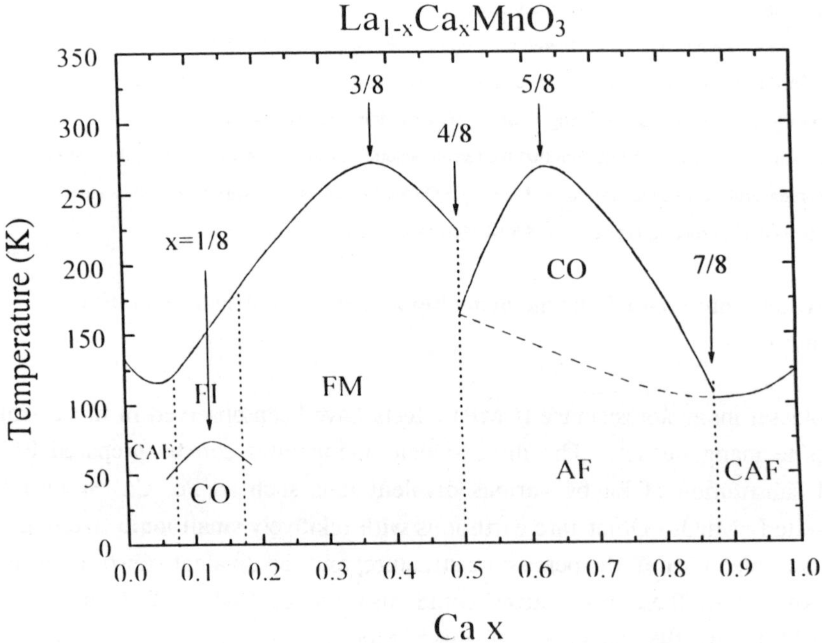

Fig. 1. The phase diagram of $La_{1-x}Ca_xMnO_3$. There exist well-defined anomalies at commensurate carrier concentrations of x=N/8 (N=1, 3, 5 and 7). FM: Ferromagnetic Metal, FI: Ferromagnetic Insulator, AF: Antiferromagnetism, CAF: Canted Antiferromagnetism and CO: Charge Ordering.

coupling in the double exchange mechanism [2]. Therefore, it is expected that there exists close interplay among charge carriers, magnetic couplings, and structural distortions in the mixed-valent (La,Ca)MnO_3.

Colossal magnetoresistance (CMR) has been observed in the Ca concentration of 0.2 - 0.5 (see the phase diagram of (La,Ca)MnO_3 in Fig. 1). These materials in this doping range exhibit a paramagnetic-to-ferromagnetic transition upon cooling which is accompanied by a sharp drop of resistivity (ρ). The colossal magnetoresistance occurs as a result of a rapid shift of ferromagnetic transition temperature (T_C) to higher temperatures in the presence of a magnetic field. For the high doping range with $x \geq 0.5$, the doped charge carriers localize, and, in fact, order with stripe modulations at low temperatures [3,4,5,6]. Thus, the ground state in this high doping range is an antiferromagnetic insulator with the ordering of doped

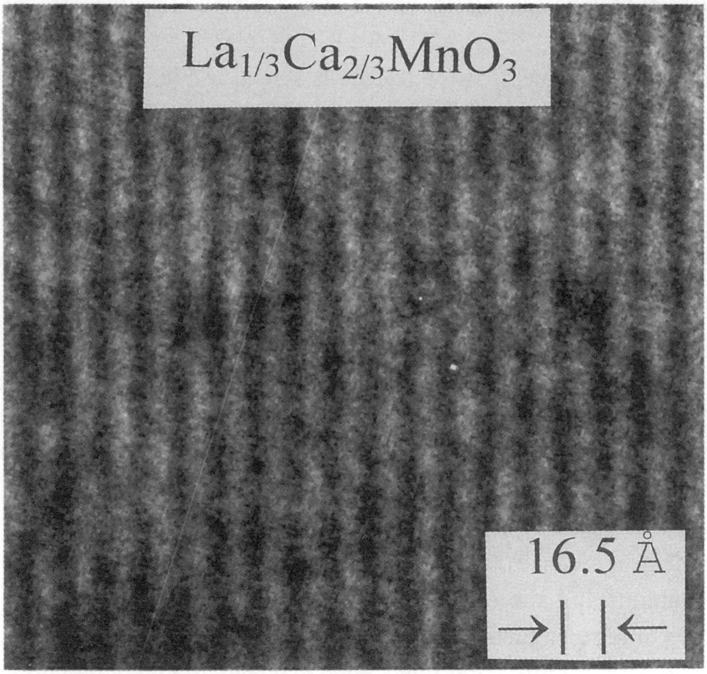

Fig. 2. High resolution lattice image of $La_{1/3}Ca_{2/3}MnO_3$ obtained at 95 K, showing the diagonal charge-ordered stripes with a periodic spacing of 16.5 .

charge carriers. The real space image of the charge ordered stripes, obtained from electron microscopy work for x=2/3 of $La_{1-x}Ca_xMnO_3$, is shown in Fig. 2.

Charge ordering is commonly observed in various mixed-valent transition metal oxides, and the charge ordering with stripe modulation is associated with a strong tendency toward electronic phase separation with stripe modulations when charge carriers are introduced into antiferromagnetic, insulating background. Interestingly, modulated pattern formations with diverse length-scale have been observed in a wide variety of physical and chemical systems. Particularly, phases with stripe modulations are common in two-dimensional systems, and sheet formations have been often observed in three dimensional systems. Note that the charge modulation can be stripe-like in two-dimensional systems, and sheet-like in tree-dimensional systems. However, in this chapter the term of stripe modulation is used even for sheet modulations in three-dimensional systems, especially because of simple stacking modulation along one of the crystallographic axes, as well as convenience.

One intriguing example of stripe texture is shown in Fig. 3(a). When CaO and ZrO_2 are mixed, melted at high temperatures, and then solidified with cooling, chemical composition modulates with sheet pattern, partially due to the existence of chemical miscibility gap [7]. In other words, two phases with one CaO-rich phase and the other $CaZrO_3$-rich phase alternates regularly with sheet texture. It is worthy to note that the modulation pattern in Fig. 3(a) resembles the charge/orbital ordering pattern in Fig. 2, but the modulation periodicity in Fig. 3(a) is much larger (~5 μ) than the one in Fig. 2 (~16.5 Å). Another well-known case of stripe texture formation is ferromagnetic domains [8]. As displayed in Fig. 3(b), magnetic domains in thin plate of ferromagnetic $BaFe_{12}O_{19}$ shows stripe patterns with periodicity of ~1 μ in zero field. This stripe pattern changes with increasing applied magnetic field.

These periodic special organizations in a wide set of physical and chemical systems can be analyzed in a unified framework of competing interactions favoring spatial inhomogeneities [9]. The competing interactions are often one short-range attractive coupling and the other long-range repulsive coupling. For example, stripe patterns of ferromagnetic domains result from the competition between short-range ferromagnetic coupling and long-range magnetic dipole interaction. The line tension at magnetic domain walls, originating from local ferromagnetic coupling, favors the reduction of magnetic domain walls. However, magnetic dipole

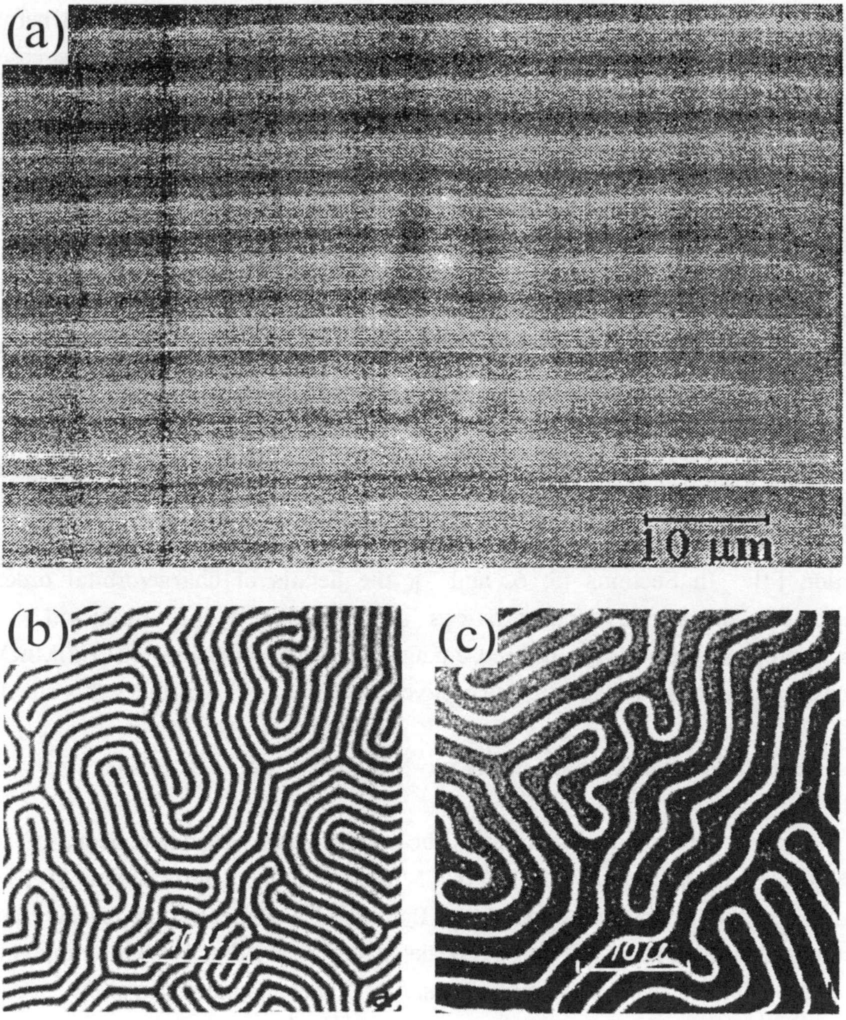

Fig. 3. (a) Optical photograph (using transmission illumination through a sample thickness of 1 mm) of a transverse section of a ZrO_2-$CaZrO_3$ eutectic fiber grown by the laser floating zone method. The light region is ZrO_2-rich, and the dark phase is mainly $CaZrO_2$ (Ref. 7). (b, c) Ferromagnetic domain structure observed for a thin plate of $BaFe_{12}O_{19}$ for different values of the applied field normal to the plane of the plate. (b) H = 0 Oe and (c) H = 2,700 Oe (Ref. 8).

interaction tends to increase magnetic domain walls. The competition between these two interactions leads to the striped domain formation. In the case of charge(/orbital) ordering in transition metal oxides, the long-range repulsive interaction is probably the repulsive Coulomb interaction between carriers, but the origin of the short-range attractive interaction is still an open question. This attractive interaction between doped carriers may originate from local interaction of charge carriers with lattice and spins.

The details of the competing interactions determine the instability, periodicity and also the dynamic nature of the modulated phases. The investigation of stability and dynamics of charge/orbital ordered states with striped (or sheeted) modulation will open new avenues toward the next decade condensed matter physics.

This chapter is organized as follows. The details of the phase diagram of (La,Ca)MnO_3 are discussed in Section [2], ionic ordering and charge ordering phenomena are compared in Section [3], and the schematics of charge/orbital ordering configurations for x=0, 1/2 and 2/3 in $La_{1-x}Ca_xMnO_3$ are presented in Section [4]. In Sections [5, 6, and 7], the details of charge/orbital ordering phenomena for various concentrations in (La,Ca)MnO_3 are discussed. The charge/orbital ordering for layered manganites and perovskite (Bi,Ca)MnO_3 is discussed in Sections [8] and [9], respectively. Section [10] concludes this chapter.

[2] Phase diagram of $La_{1-x}Ca_xMnO_3$: x=N/8 anomalies

There are well defined features at the commensurate carrier concentrations of x=N/8 (N=1, 3, 4, 5 and 7) in $La_{1-x}Ca_xMnO_3$. Careful determination of T_C and charge ordering transition temperature (T_{co}) with Ca concentration reveals that T_C becomes maximum at x=3/8, and T_{co} peaks at x=5/8 (see Fig. 1). The compound at the phase boundary with x=0.5 undergoes, first, a ferromagnetic transition, and then a simultaneous antiferromagnetic and charge ordering transition at a lower temperature. The system at x=1/8 concentration, also, undergoes two transitions: first, a ferromagnetic transition, and then an antiferromagnetic transition. This antiferromagnetic transition at x=1/8, probably, accompanies a charge ordering transition. There is another well-defined phase boundary at x=7/8; charge ordering occurs below x=7/8, and canted antiferromagnetism (or weak ferromagnetism) is observed for x>7/8.

It is very informative to view our results in terms of electron-hole symmetry. The active band of insulating $LaMnO_3$ is the narrow e_g band split by Jahn-Teller distortion and electron Coulomb correlation. In the case of insulating $CaMnO_3$, the e_g band is empty, and the Jahn-Teller splitting does not exist in this Mn^{4+} compound. Therefore, the initial Ca-substitution in $LaMnO_3$ produces hole carriers in the split e_g band, and the low doping of La in $CaMnO_3$ induces electrons in the broader e_g band. This view is consistent with the experimental results of Hall effect and thermoelectric power. For example, the hole nature of carriers is revealed in Hall effect measurements on $La_{0.7}Ca_{0.3}MnO_3$ at low temperatures [10]. In addition, above T_C, thermoelectric power for low Ca concentration region is positive, and for high Ca concentration range is electron-like [11]. In terms of the concentrations where physical anomalies occur, there exists a perfect electron-hole symmetry in hole-doped $LaMnO_3$ and electron-doped $CaMnO_3$. In other words, physical anomalies are observed at the carrier concentration of 1/8, 3/8 and 4/8 in hole-doped $LaMnO_3$ and electron-doped $CaMnO_3$. It should be also pointed out that in low doping region doped-carriers, both holes in $LaMnO_3$ and electrons in $CaMnO_3$, tend to induce ferromagnetic coupling, probably double exchange-type, in antiferromagnetic $LaMnO_3$ and $CaMnO_3$, leading to weak ferromagnetic ground states. Another interesting symmetry is that the transition temperatures of the ferromagnetic state in 3/8-hole-doped $LaMnO_3$ and the charge ordered state in 3/8-electron-doped $CaMnO_3$ are nearly identical (270±5 K).

However, there exists a pronounced electron-hole asymmetry in the ground state properties of $(La,Ca)MnO_3$. First of all, the ground state is a ferromagnetic metal for the hole concentration of $1/8<x<4/8$, but $(La,Ca)MnO_3$ with the electron concentration of $1/8<x<4/8$ shows charge ordering at low temperatures. Furthermore, in the carrier concentration of $0<x<1/8$, ferromagnetism is much more pronounced for the hole carrier case in comparison with the electron case, which is evident in M_s at 5K (e.g., $M_s \approx 4\mu_B$/Mn for $La_{0.9}Ca_{0.1}MnO_3$ and $M_s \approx 0.55\ \mu_B$/Mn for $La_{0.1}Ca_{0.9}MnO_3$). A general aspect is that holes in $(La,Ca)MnO_3$ tend to induce metallic and ferromagnetic behavior at low temperatures, and $(La,Ca)MnO_3$ system with electron carriers is susceptible to charge ordering. These behaviors can be ascribed to the presence of ferromagnetic superexchange coupling in $LaMnO_3$. Spins in the orthorhombic basal planes of $LaMnO_3$ are ferromagnetically coupled due to the antiferrodistortive arrangement of the Jahn-teller distortions of Mn^{3+} ions in the basal planes, but spins in $CaMnO_3$ are antiferromagnetically coupled in all three directions. It is, therefore, expected that the ferromagnetic ordering and metallic behavior are more readily produced in hole-doped $LaMnO_3$ than in electron-doped $CaMnO_3$. On the other hand, these results are contrary to the view that electrons exist in the broad e_g band without Jahn-Teller splitting. This puzzle remains to be understood.

When carrier-lattice coupling is taken into account, the charge ordering instability at the commensurate carrier concentrations of x=1/8, 4/8, 5/8 and 7/8 is somewhat natural in $(La,Ca)MnO_3$ with its three-dimensional quasi-cubic MnO_6 network. The charge ordering instability at x=N/8 can be ascribed to the tendency of the localization of N carriers per 2x2x2 cubic Mn lattice. However, this tendency of charge localization should persist even at x=3/8. Experimentally, it is clearly seen that charge ordering is detrimental to ferromagnetism at x=1/8, 4/8 and 7/8, and ferromagnetism is completely suppressed for x=5/8. In the simple double exchange mechanism which neglects the carrier-lattice coupling, the ferromagnetic coupling has to be reduced if carrier hopping is suppressed through localization or ordering of carriers. Why then is ferromagnetism maximized at x=3/8? The surprising implication is that carrier-lattice coupling at this 3/8 commensurate doping level enhances ferromagnetic exchange and electron itinerancy.

[3] Charge ordering vs. ionic ordering

One should not confuse the charge ordering with ionic ordering. Ionic ordering in alloys has been well investigated, especially in the field of metallurgy. For example a binary alloy of copper and gold forms a true solid solution in the entire composition range. When quenched from a high temperature, the Cu-Au alloy forms a disordered solid where Cu and Au ions are randomly arranged. However, when the specimen is cooled slowly from a high temperature, ordered structures are formed at Cu_3Au and CuAu [12]. This ionic ordering , reducing impurity scattering, results in lowering residual ρ. The dips in ρ at 25 % and 50 % Au concentrations in Fig 4 are due to the reduction of impurity scattering induced by Cu-Au ordering. This type of ionic order-disorder transitions tend to occur at high temperatures, much higher than room temperature, because the energy scale involved in the transitions is on the order of chemical bond energy. For example, the ionic order-disorder transition in Cu-Zn is about 460 °C.

One intriguing example of ionic ordering can be found in $La_2Cu_{1/2}Li_{1/2}O_4$ [13]. La_2CuO_4 crystallizing in the K_2NiF_4 structure, is the parent compound of layered high temperature superconductors. When one half of the copper ions are replaced by lithium ions, Cu-Li ions order in the alternated fashion in the Cu/Li-O_2 layers. This Cu-Li ordering occurs at the material-synthesis temperature near 1000 °C, and produces superlattice peaks of $2\pi/a(\delta,0,0)$ and $2\pi/a(0,\delta,0)$ with $\delta=1$ in electron diffraction pattern at room temperature. Through out this chapter, the main Bragg spots in the diffraction pattern are indexed with the orthorhombic crystal structure with $a \approx b \approx \sqrt{2}a_p$ (in the Pbnm perovskite case, $a \approx b \approx \sqrt{2}a_p$, and $c \approx 2a_p$) where $a_p \approx 3.9$ Å is the lattice parameter of the simple perovskite structure.

Charge ordering transition in Fe_3O_4, the so-called Verwey transition [14], has been studied half a century ago. Magnetite, Fe_3O_4, forms in the spinel structure of the formula AB_2O_4, and the formula unit of the Fe_3O_4 can be written as

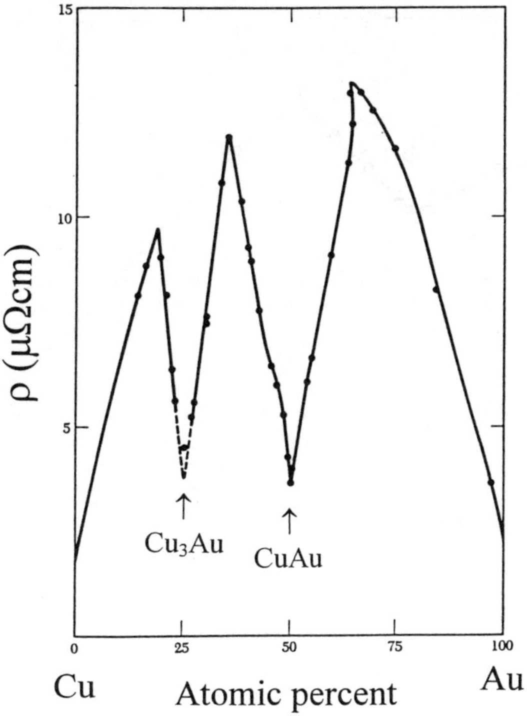

Fig. 4. Residual resistivity of a binary alloy of Cu_xAu_{1-x}. The alloy samples here have been annealed, and the compositions of low residual resistivity correspond to the ordered compounds Cu_3Au and CuAu.

$Fe^{3+}[Fe^{2+},Fe^{3+}]O_4$. In the spinel structure, A ions are tetrahedrally coordinated by four oxygens, and there exists octahedral oxygen coordination around B ions. In the Verwey model, electrons in B-site Fe ions hop around at high temperatures, but freeze and order at ~125 K. This charge ordering transition accompanies a sudden first-order increase of ρ, and the ordered structure proposed by Verwey is a simple valence alternation of 2+ and 3+ of B-site Fe ions. It turns out that the actual charge ordering, estimated from various diffraction data, is much more complicated than the Verwey model [15]. Major difficulty in understanding the Verwey transition lies in the structural complexity.

It had been mysterious that the nature of doped holes in $(La,Sr)_2NiO_4$ is remarkably different from that in isostructural $(La,Sr)_2CuO_4$. Both compounds belong to the class of charge-transfer insulators, characterized by an oxygen 2p-like

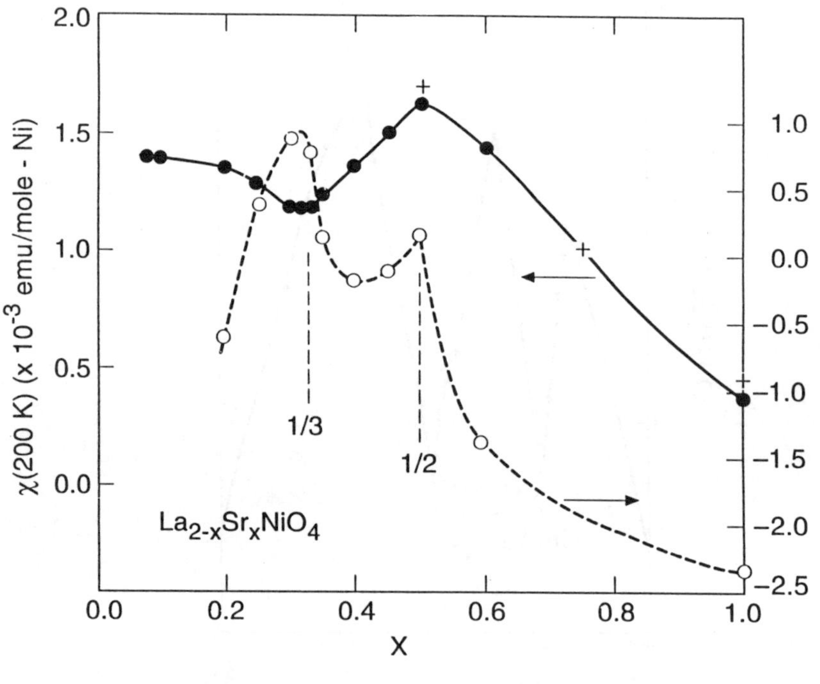

Fig. 5. Magnetic susceptibility and logarithmic resistivity at 200 K as a function of Sr doping x. Anomalies at x=1/3 and 1/2 are evident.

valence band and a 3d-like upper Hubbard conduction band which results from the strong correlation among the 3d electrons. However, much higher Sr concentration is required to induce metallic behavior in $(La,Sr)_2NiO_4$. Furthermore, superconductivity is never found in the layered nickelates [16]. It is now revealed that this different behavior is associated with the charge ordering with striped modulations in the layered nickelates. As shown in careful electron diffraction experiments as a function of Sr concentration, the charge ordering is most pronounced for x=1/3 and 1/2 in $La_{2-x}Sr_xNiO_4$ [17]. These trends corroborate the ρ variation as a function of concentration at a fixed temperature in the neckelate. As displayed in Fig 5, ρ peaks at x=1/3 and 1/2 [18]. This peaking of ρ at the concentrations where charge ordering clearly occur contrasts the ρ minimum at ionic ordering concentrations in alloys (see Figs. 4 and 5). This charge ordering transition for x=1/3 occurs at ~ 240 K, at which charge gap opens up, magnetic susceptibility (χ) drops rapidly, and superlattice peaks show up in electron diffraction patterns. It is worthy to note that there are no superlattice peaks at room temperature, indicating no ionic ordering of La and Sr at high temperatures. On the contrary to ionic ordering, charge ordering usually occurs at low temperatures, often below room temperature. We also note that recent neutron scattering results indicate that there exists π-shift of spin ordering at charge stripes, i.e., charge-ordered stripes become magnetic anti-phase boundaries [19,20].

[4] Schematics of charge/orbital ordering for x=0, 1/2, and 2/3 in $La_{1-x}Ca_xMnO_3$

Similar to $(La,Sr)_2NiO_4$, doped charge carriers in $(La,Ca)MnO_3$ tend to localize and order at low temperatures. In fact, the carrier-lattice coupling is even stronger in doped manganites than that in doped nickelates so that charge localization tendency is very strong in manganites. Furthermore, there exist orbital degrees of freedom in the case of the e_g electrons in Mn^{3+} ions, and the orbital ordering can lower electronic energy through Jahn-Teller mechanism. Therefore, on the contrary to doped nickelates there can be orbital ordering in mixed-valent manganites in addition to charge ordering. As discussed in Section [2], the tendency of charge/orbital ordering prevails in the physics of doped manganites. In this section, we discuss briefly the charge/orbital ordering configurations in the cases of x=0, 1/2 and 2/3. The structure of perovskite $(La,Ca)MnO_3$ is orthorhombic below ~700 K in the entire concentration range. It turns out that the charge/orbital ordering configuration is simple stacking along c-axis in the orthorhombic unit cell. Therefore, we will discuss what happens in the orthorhombic basal plane as far as charge/orbital ordering is concerned.

All of the Mn ions in $LaMnO_3$ are trivalent, and $LaMnO_3$ undergoes a cooperative Jahn-Teller distortion at ~750 K. The orbital ordering configuration is depicted in Fig. 6(a). It has been estimated that the energy associated with Jahn-Teller distortion of Mn^{3+} in $LaMnO_3$ is on the order of 0.25 eV, but the transition temperature of ~750 K is much lower than the energy scale of 0.25 eV, indicating the ~750 K transition is an order-disorder-type [21].

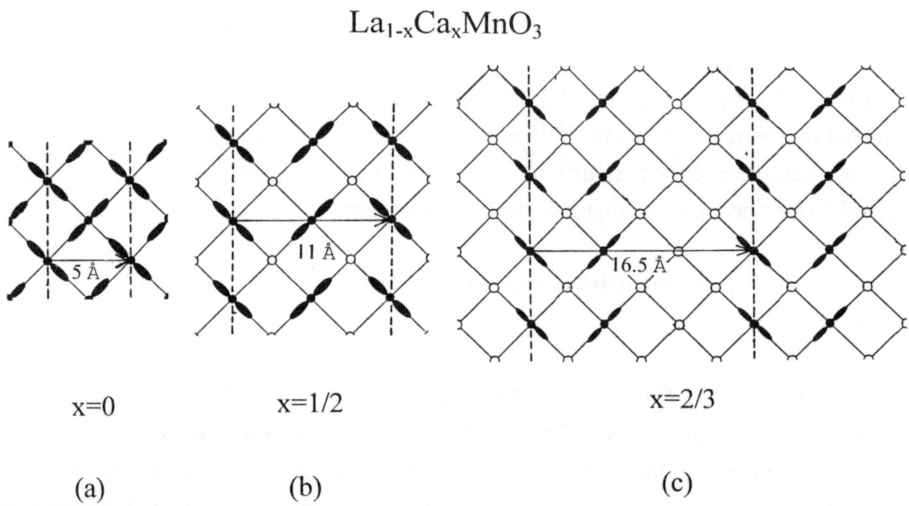

Fig. 6. The charge and orbital ordering configurations in the orthorhombic basal plane for x=0, 1/2 and 2/3 of $La_{1-x}Ca_xMnO_3$. Open circles are Mn^{4+} and the lobes show the orbital ordering of e_g electrons of Mn^{3+}. The new charge modulation wave lengths are ~5.5, ~11 and ~16.5 Å for x=0, 1/2 and 2/3, respectively.

In $La_{1/3}Ca_{2/3}MnO_3$, there are twice as many $Mn^{4+}(3d^3)$ ions as the $Mn^{3+}(3d^4)$ ions and the ordering of diagonal rows of Mn^{4+} and Mn^{3+} ions plus the orientational ordering of the d_z^2 orbitals in Mn^{3+} gives rise to the striped pattern shown in Fig. 6(c). In the schematics in Fig. 6(c), diagonal charge stripes are evident, and their periodicity is ~16.5 Å. These ~16.5 Å charge stripes form the pattern in Fig. 2. Shown in Fig. 6(b) is the similar charge/orbital ordering scheme for x=0.5, where there are just as many Mn^{4+} ions as Mn^{3+} ions. In this case, the diagonal charge stripes adopt a wave vector δ=0.5 with a spacing of ~11 Å.

In the following sections, we will discuss the detailed nature of charge/orbital ordering in $(La,Ca)MnO_3$.

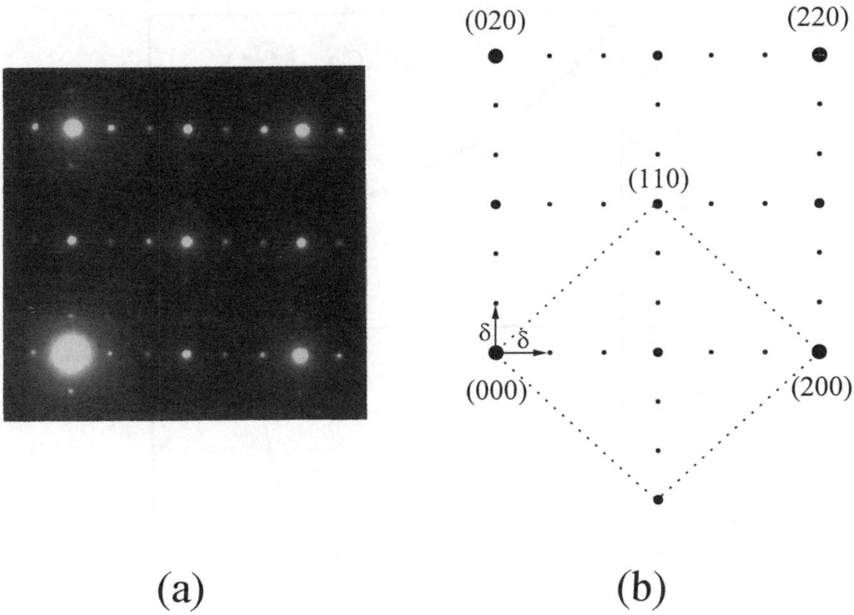

Fig. 7. (a) [001] zone axis selected area electron diffraction pattern for $x=La_{1/3}Ca_{2/3}MnO_3$ obtained at 95 K. The presence of superlattice spots with modulation wave vector (1/3,0,0) and (0,1/3,0) is evident. (b) A Schematic representation for the first quadrant of the diffraction pattern (a), showing the locations of the superlattice spots with fractional indices. The unit cell corresponding to the cubic perovskite structure with lattice parameter of $a_p \approx 3.9$ Å is also indicated by dash lines.

[5] Charge/orbital ordering for $La_{1/3}Ca_{2/3}MnO_3$

The electron diffraction evidence of the charge ordering in $La_{1/3}Ca_{2/3}MnO_3$ is shown in Fig. 7, which represents an [001] zone-axis electron diffraction pattern obtained from a selected area at a temperature of 95 K. The sharp weaker spots that can be indexed with fractional indices are superlattice spots due to charge/orbital ordering. For clarity we also show in Fig. 7(b) a schematic representation of the essential features of Fig. 10(a), in which the modulation wave vector and the cubic perovskite unit cell are also indicated. the modulation wave vector of the charge ordering as measured from the nearest fundamental Bragg peak can be written as q=(δ,0,0) or (0,δ,0) in reciprocal lattice unit with δ=1/3. It should be noted that δ

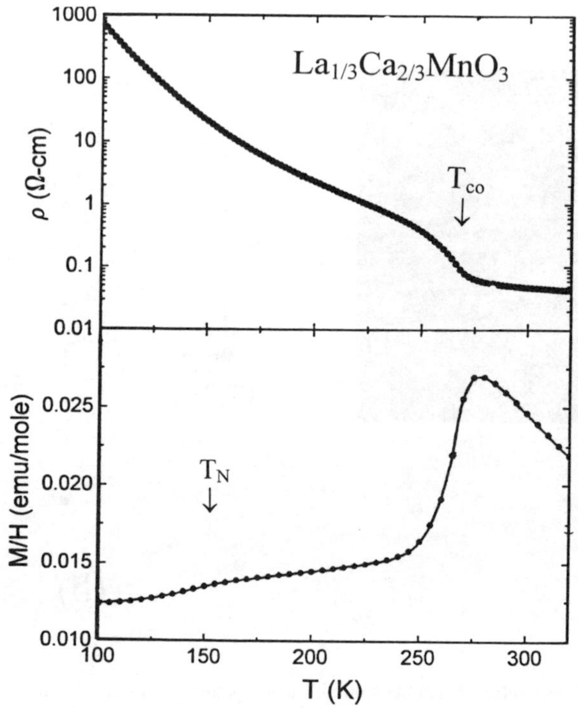

Fig. 8. Temperature dependence of logarithmic resistivity (a) and magnetic susceptibility (b) for $La_{1/3}Ca_{2/3}MnO_3$. The sharp changes at ~270 K are due to charge ordering and weak features at ~150 K are due to antiferromagnetic spin ordering.

tends vary slightly (10 %) from area to area within the sample and may not always adopt the commensurate value of 1/3. We have also determined that the two perpendicular modulation wave vectors actually originate from different areas of the sample. As shown in fig. 8, there exits a rapid increase of ρ and a drop of χ at ~270 K, suggestive of charge ordering (see Section [9] for the origin of χ drop at charge ordering transition temperature (T_{co})). Hopping doped carriers, probably the type of the Jahn-Teller polarons [22], produce relatively low ρ above T_{co}, and when these doped carriers freeze and order, ρ rises rapidly. The detection of superlattice spots at low temperatures provides direct structural evidence of charge ordering as suggested from indirect transport and χ measurements. The superlattice intensity disappears above the temperature ~270 K, consistent with the transport and magnetic measurements.

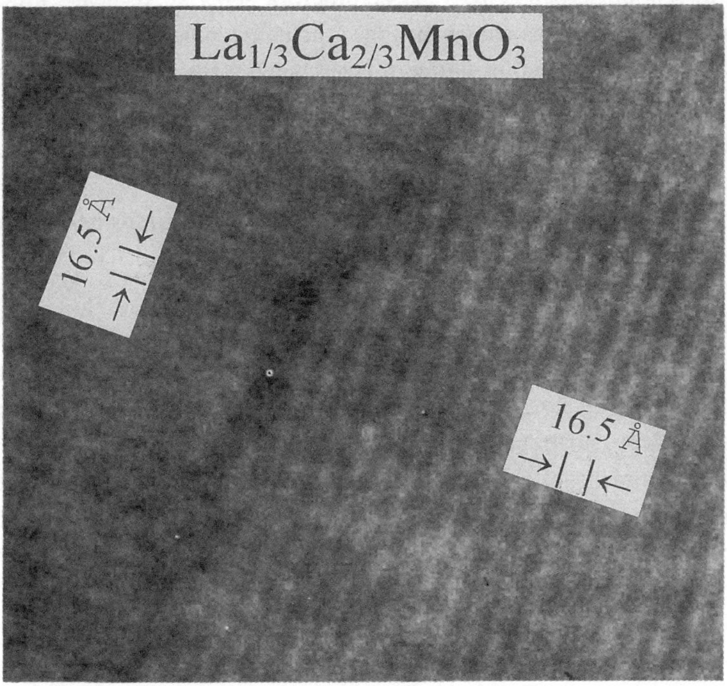

Fig. 9. High resolution lattice image of diagonal charge stripes of two adjacent charge ordered domains of $La_{1/3}Ca_{2/3}MnO_3$. Perpendicular stripes are seen from these neighboring domains.

The superlattice reflections shown in Fig. 7 suggest that the doped charge carriers localize and order along the diagonal direction of the Mn-O square lattice in the a-b plane forming the so-called diagonal charge-ordered stripes with a periodic spacing of 3a ≈16.5 Å. High resolution lattice images at low temperatures of the charge ordering (Fig. 2) show the presence of ~16.5 Å diagonal charge stripes. The instrumental resolution at low temperature is limited to ~5 Å in this case due to the mechanical and thermal instability of the specimen stage. Therefore, only the coarse charge modulation of ~16.5 Å and not the true atomic image of the modulation is seen. It is clear from Fig. 2 that the charge-ordered stripes are very regular without the presence of noticeable kinks or breaks. This implies that the phase slips of the charge ordering are energetically unfavorable. A schematic drawing showing ~16.5 Å charge stripes is depicted in Fig. 6(c). The appearance of stripes in only one diagonal direction in a given area also indicates that the two sets of perpendicular charge-ordering superlattice peaks around each fundamental Bragg spots as shown in Fig. 7 actually come from different areas of the sample. A high resolution lattice image showing the perpendicular stripes in two neighboring charge-ordered domians is shown in Fig. 9. The domain boundaries are generally parallel to the Mn-O-Mn edge of the square lattice.

[6] Charge ordering for $La_{1/2}Ca_{1/2}MnO_3$: discommensurations and competition between two ground states

As exhibited in Fig. 1, there exists a sharp phase boundary between ferromagnetic-metallic and charge-ordered-antiferromagnetic ground states at x=1/2 concentration in $La_{1-x}Ca_xMnO_3$, and the system at this x=1/2 concentration undergoes two phase transitions: one ferromagnetic transition and then a simultaneous commensurate charge ordering and antiferromagnetic transition at a lower temperature [23]. As shown in Fig. 10, this second transition is associated with a sharp drop of χ and a large increase of ρ. Furthermore, this transition is strongly-first-order, i.e., there exists a large thermal hysteresis as shown in Fig. 10.

The diffractional evidence of the charge ordering in $La_{0.5}Ca_{0.5}MnO_3$ is shown in Fig. 11 which represents a [001] zone-axis electron diffraction pattern obtained at 95 K. In Fig. 11(a), sharp superlattice spots are evident in addition to the fundamental Bragg reflections. Furthermore, the wave vector q of the superlattice modulation was found to be nearly commensurate at this temperature, and can be written as $q=(2\pi/a)(1/2-\varepsilon,0,0)$. with incommensurability parameter $\varepsilon=0.013$. In the tetragonal cell notation, the superlattice peaks occur near $(\pi/2,\pi/2)$. Results obtained from selected-area electron diffraction experiments show that ε could vary slightly from grain to grain, and area to area even within each grain. Commensurate charge ordering with $\varepsilon=0$ was often observed. We have also determined that the two perpendicular sets of superlattice spots around each fundamental Bragg reflection actually originate from different areas of the sample. We will elaborate this point later when we discuss the real space results obtained from dark-field images. In $La_{0.5}Ca_{0.5}MnO_3$, there are just as many $Mn^{4+}(3d^3)$ ions as $Mn^{3+}(3d^4)$ ions. We show in Fig. 6(b) a schematic picture of Mn^{4+}-Mn^{3+} ordering with wave vector of $q=(2\pi/a)(1/2,0,0)$.

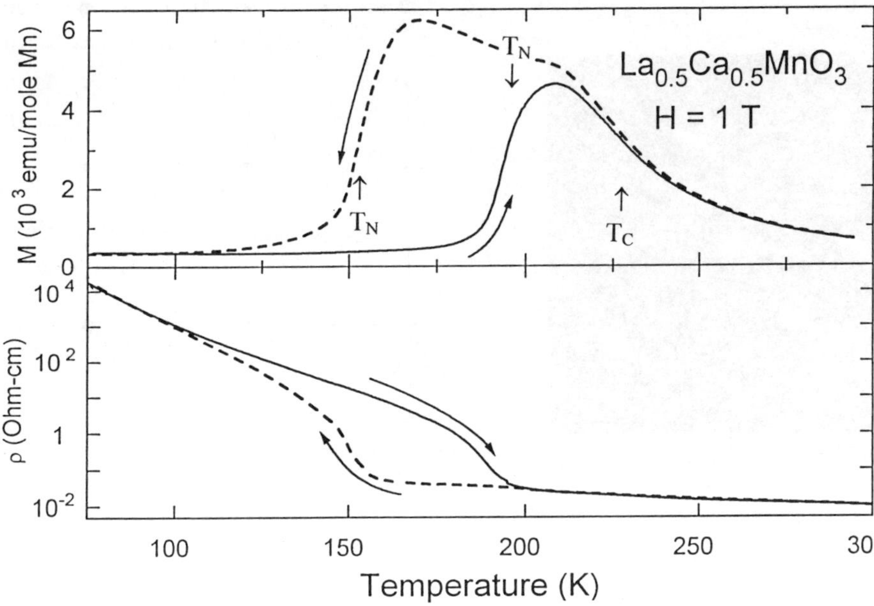

Fig. 10. Magnetization and logarithmic resistivity in 1 tesla field of $La_{0.5}Ca_{0.5}MnO_3$. The Curie temperature is ~225 K, and the simultaneous antiferromagnetic and charge ordering transition is ~190 or ~150 K, depending on thermal history.

Charge-ordered domains (CODs) can also observed by the dark-field imaging technique commonly used in electron microscopy. In this imaging mode, only superlattice spots from the charge ordering were used to form the dark-field images, and the contrast in these superlattice images is therefore directly associated with the charge ordering. Shown in Figs. 12(a) and (b) are a pair of complementary dark-field image for x=1/2 obtained from a pair of superlattice spots perpendicular to each other. The most striking feature of the dark-field images is the presence of CODs of a few thousand angstrongs in size. The complementary contrast shown in Figs. 12(a) and (b) is again proof that the two perpendicular sets of superlattice spots actually come from different, but twin-related CODs of the sample. We note that the COD structure can only be observed from the superlattice reflections (not

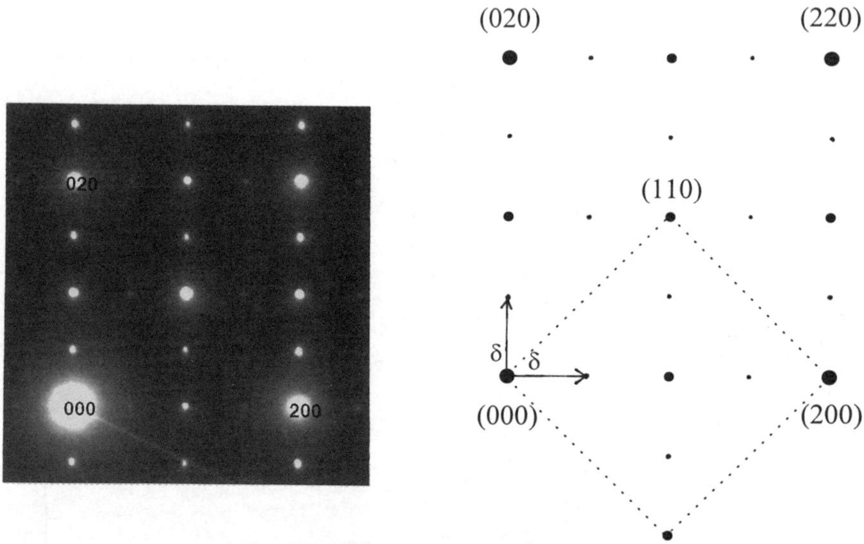

(a) (b)

Fig. 11. (a) [001] zone axis electron diffraction pattern obtained at 95 K for $La_{0.5}Ca_{0.5}MnO_3$. Kinematically forbidden (100) and (010) spots also appear as a result of multiple scattering. (b) A schematic representation for the first quadrant of the diffraction pattern (a), showing the location of the superlattice peaks of (1/2,0,0) and (0,1/2,0). The unit cell corresponding to the cubic perovskite structure with lattice parameter of $a_p \approx 3.9$ Å is also indicated by dash lines.

from the main Bragg peaks) and, therefore, they are directly associated with the charge ordering of doped holes. The boundaries of the CODs are more or less parallel to the Mn-O-Mn direction of the simple perovskite structure, which is 45° of the charge-ordering wave vector. This orientation becomes the natural choice of the COD boundaries in view of the twining structure of the CODs. The configuration and shape of CODs were found to remain essentially the same after repeated thermal cycle, suggesting the strong pinning effect of the COD boundaries to impurities and/or other defect structures in the sample. In this class of materials, extended defects such as dislocations are rarely observed and the defects, pinning the COD boundaries, are most likely impurities, point defects or small clusters of point defects.

Careful examinations of dark field images shown in Fig. 12 reveal the presence of finely spaced dark lines within the CODs. Experimentally, the dark line density was found to vary from area to area in the sample and to be inversely proportional to the incommensurability, ε, of the charge ordering wave vector $q=(1/2-\varepsilon,0,0)$ of the area. We identify these dark lines inside the CODs as discommensurations. Discommensurations were often seen in an incommensurate phase as domain walls which separate the small commensurate domains, and the modulation superlattice phase varies rapidly at discommensurations. In a commensurate phase, discommensurations generally are not expected to occur. However, residual discommensurations can sometimes remain in the commensurate phase because of the strong pinning by impurities and/or defects. The discommensurations are generally oriented in a direction perpendicular to the modulation wave vector when the discommensuration density is high (i.e., large incommensurabilities).

One interesting observation of the charge ordering in $La_{1-x}Ca_xMnO_3$ is the variation of ordering wave vector from area to area by as much as ~10 %. Dark-field images of the discommensurations inside the CODs reveal that this is due to the spatial fluctuations of the discommensuration density. The underlying reasons for the local density fluctuations of discommensurations remain unclear. We note, however, the length scale of the fluctuations is typically a few thousand angstroms and electron beam microanalysis does not reveal any significant chemical inhomogeneity of this length scale. Furthermore, we did not detect any obvious defect that can be responsible for the fluctuations.

To establish the correlation between the presence of charge-ordering superlattice peaks and the second transition with large thermal hysteresis, we have followed the

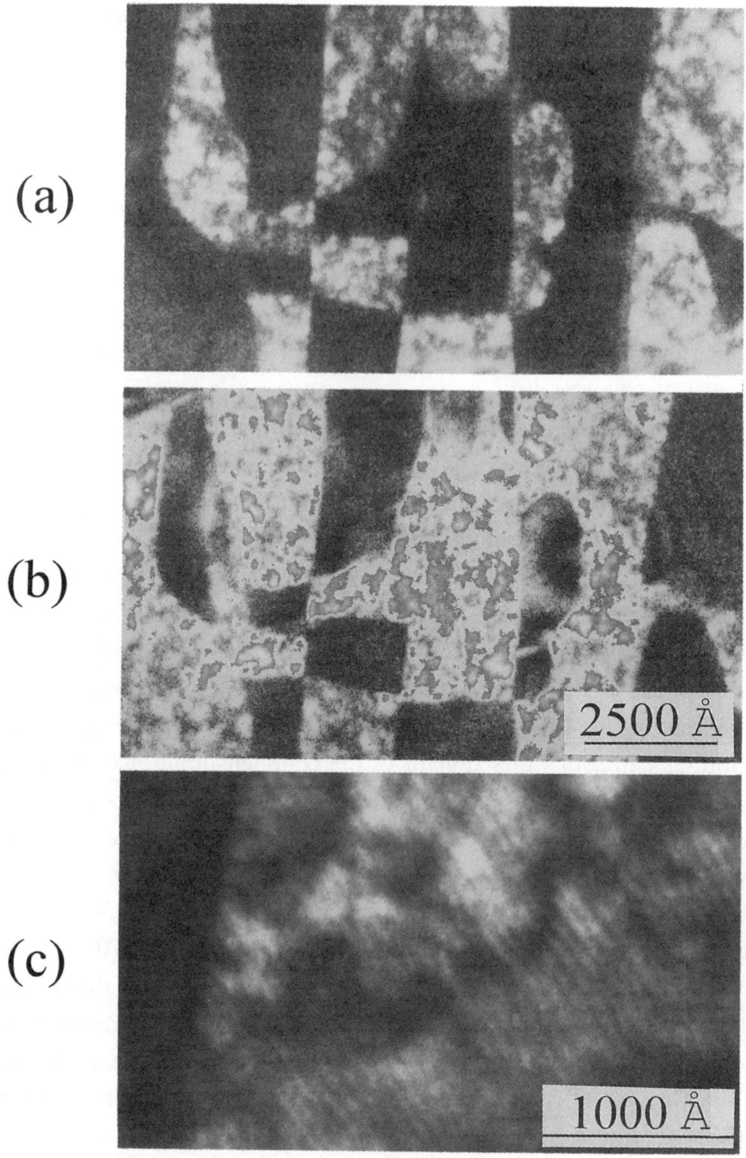

Fig. 12. (a,b) Complementary dark-field images obtained from superlattice peaks of $La_{0.5}Ca_{0.5}MnO_3$ that are perpendicular to each other. (c) The quasi-periodic discommensurations in the charge-ordered domains.

superlattice peaks and measured the charge-ordering wave vector as a function of temperature during cooling and warming. The results are shown in Fig. 13 in which the incommensurability of charge ordering ,ε, is plotted as a function of temperature. This Fig. 13 clearly demonstrates that the first-order transition is associated to an incommensurate-to-commensurate charge-ordering. transition. Upon cooling from room temperature, precursory incommensurate charge-ordering superlattice peaks first become visible below ~240 K, and ε decreases rapidly toward the ferromagnetic transition at ~225 K below which ε decreases rather slowly until ~150 K. Below ~150 K, ε decreases rapidly, and a commensurate charge-ordered state (ε=0) is observed below ~130 K. Due to spatial fluctuations of the charge-ordering wave vector, the charge-ordered state at lower temperatures is

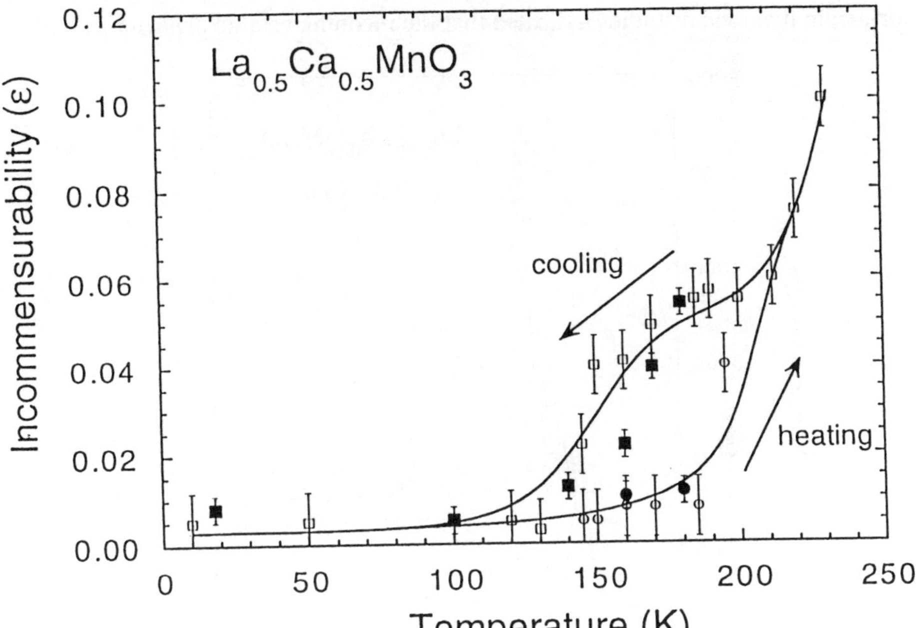

Fig. 13. Superlattice wave-vector incommensurability as a function of temperature upon cooling (squares) and warming (circles) from electron diffraction (open) and x-ray diffraction (closed) experiments on $La_{0.5}Ca_{0.5}MnO_3$. Lines are drawn as guides to the eyes.

not always commensurate, and nearly commensurate states are often observed. In that case, the incommensurate-commensurate transition described above will become an incommensurate-to-nearly commensurate transition. Upon warming, ε remains commensurate (ε=0) until temperature ≈150 K, above which the charge ordering becomes incommensurate, and increases rapidly and coincides with the cooling curve above ~210 K within the experimental uncertainties. Intensity measurements of superlattice peaks during cooling and warming also reveal similar hysteresis with weaker intensities in the ferromagnetic region. The commensurate-incommensurate charge-ordering temperatures at ~130 and ~180 K upon cooling and warming, respectively, correlate roughly with the first-order transition determined magnetization and ρ measurements (see Fig. 10). A similar temperature dependence of incommensurability with similar magnitude has also been observed in the nonmagnetic ferroelectric insulator of $Ba_2NaNb_5O_{15}$ [24]. First-principle theoretical studies indicated that the incommensurate behavior in

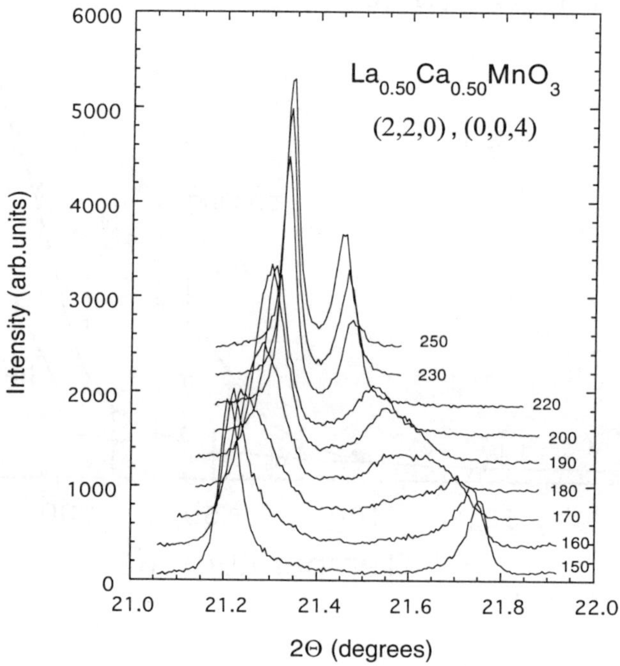

Fig. 14. Evolution of the (220) and (004) Bragg peaks as a function of temperature for $La_{0.5}Ca_{0.5}MnO_3$, based on synchrotron x-ray powder diffraction data with a resolution $\Delta 2\theta \approx 0.01$ Å.

the insulator was caused by the relaxation of highly unstable structural units exhibiting angstrom-size displacements with essentially zero energy change. Similar mechanisms may also be responsible for the incommensurate behavior observed in our case.

Charge-ordering phenomenon is mutually exclusive with ferromagnetism in double exchange mechanism which requires a charge carrier hopping from a Mn^{3+} ion across an intervening O^{2-} ion to an adjacent Mn^{4+}. It is very intriguing that incommensurate charge ordering do coexists with ferromagnetism in a narrow temperature window in $La_{0.5}Ca_{0.5}MnO_3$. Our recent preliminary experiments indicate that this coexistence occurs as a fine spatial mixture of the two competing phases.

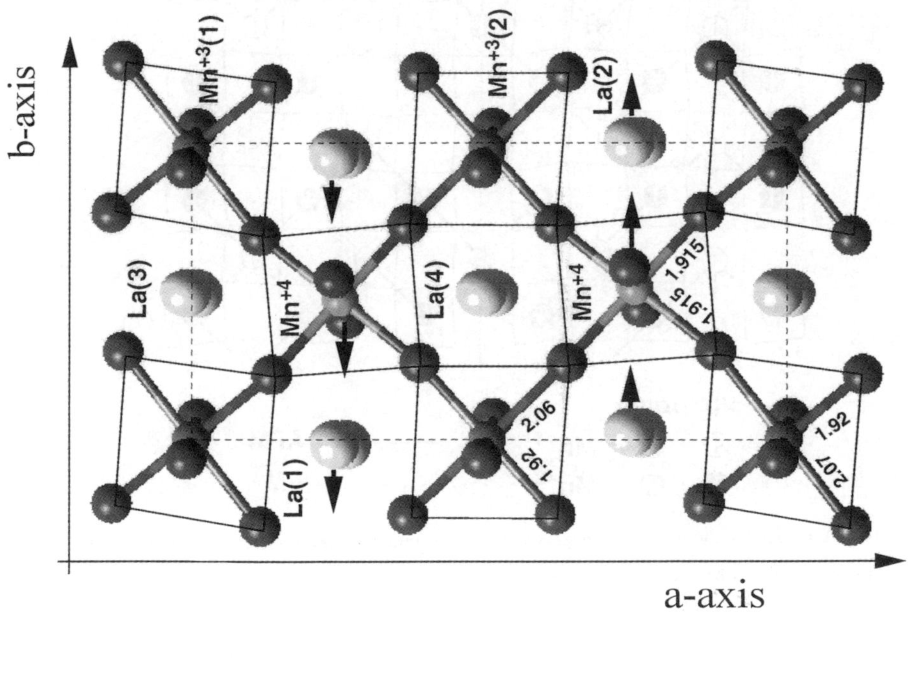

Fig. 15. Low temperature $La_{0.5}Ca_{0.5}MnO_3$ structure determined from the Rietveld refinement of neutron diffraction data. The a-b plane projection is shown. Arrows indicate the displacements of the $Mn^{4+}O_6$ octahedra and of the La(1) and La(2) atoms. Unlabeled atoms in the darkest gray shade are oxygen atoms.

The various conclusions from electron diffraction studies are further supported from synchrotron x-ray diffraction and neutron powder diffraction results [25]. First of all, superlattice peaks due to charge/orbital ordering can be clearly seen in synchrotron x-ray diffraction data, and the incommensurability ε obtained from synchrotron x-ray diffraction data shows the similar temperature dependence with that in electron diffraction results as shown in Fig. 13. A dramatic aspect of the structural change occurring in the incommensurate-ferromagnetic region is the unusual behavior of the x-ray peak widths [26]. Fig. 14 shows the evolution of the 220-004 Bragg reflections for x=0.5 in the temperature region 250-150 K (with cooling), where the largest change in lattice parameters takes place. The two Bragg reflections

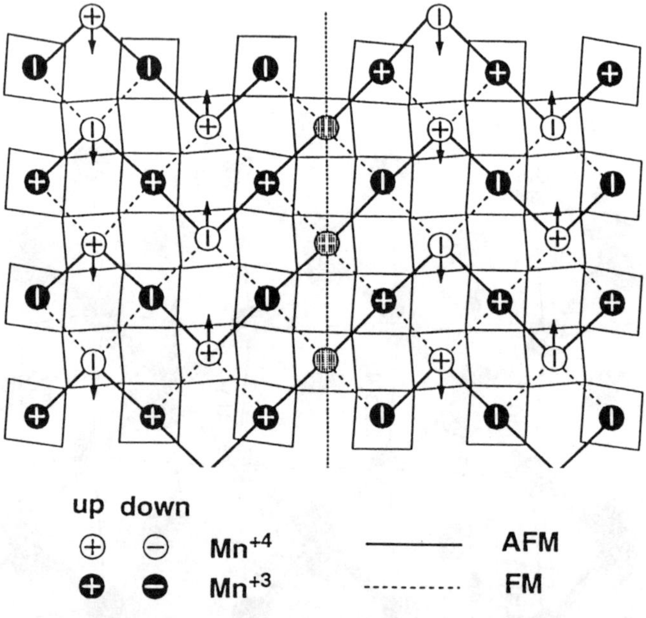

Fig. 16. A model for the structural/magnetic domain boundary (discommensuration) responsible for the incommensurability of the structural modulation and for the selective coherence length effects on the magnetic structure. The projection shows the a-b plane (orientation is the same as in Fig. 15). Filled and empty circles are Mn^{3+} and Mn^{4+}, respectively. Pluses (+) and minuses (-) indicate the directions of the spins (actually oriented in the same a-b plane). Arrows indicate the direction of the displacements giving rise to the structural modulation. Antiferromagnetic and ferromagnetic couplings are indicated with continuous and dashed lines, respectively. The discommensuration boundary is shown with a vertical dotted line.

are seen to be quite sharp at 250 and 150 K, but broaden dramatically in-between. In this intermediate region, intensity develops in a region of the pattern where no intensity is present at either 250 or 150 K. Our speculation, combined with electron diffraction results, is that the existence of a wide range of charge/ordering configurations with a large disorder or a high density of discommensurations is responsible for this Bragg peak broadening effect.

The low temperature superstructure of $La_{0.5}Ca_{0.5}MnO_3$ from Rietveld refinement of neutron powder diffraction pattern is shown in Fig. 15, showing the charge/orbital ordering. In this refined structure, Mn^{4+} environment is almost isotropic (all of the Mn-O bonds≈1.92 Å), and the environment with one Mn-O bond (~2.07 Å) significantly longer than the other two bonds (~1.92 Å) is consistent with the orbital ordering of Mn^{3+}. One of the intriguing aspect of this refinement result is the distortion pattern of Mn^{4+}, La(1) and La(2), presumably induced by the ordering of the long bond of Mn^{3+}. As shown with arrows in Fig. 15, Mn^{4+}, La(1) and La(2) distort in the direction transverse to the superlattice modulation vector.

A striking feature revealed by the neutron powder diffraction results is the presence of two different coherence lengths for the Mn^{3+} and Mn^{4+} magnetic sublattices. The magnetic Mn^{3+} reflections are significantly broadened while the nuclear peaks and the magnetic Bragg peaks associated with the Mn^{4+} sublattice are resolution-limited. The Mn^{3+} magnetic correlation length estimated from the peak width is ~310 Å . On the contrary, the correlation length of the Mn^{4+} magnetic sublattice is comparable to that of the nuclear structure ≥ 2000 Å. This seemingly striking results is understandable within the framework of the discommensurations discussed earlier. Since the Mn^{3+} propagation vector is probably coupled magnetoelastically rather strongly to one of the two inequivalent in-plane crystallographic axes (a or b), the magnetic twin boundaries are likely to be associated with the structural a-b twins, and also with the COD boundaries, observed by electron diffraction. On the other hand, the magnetic "spin flip" boundaries are likely to coincide with the much more finely spaced discommensurations, which are responsible for the incommensurability of the structural modulation. This hypothesis is confirmed by the fact that the relationship between the incommensurability ε and the domain size L ($L=2\pi/p\varepsilon$, with p=2 and $\varepsilon \approx 0.009$ at 20 K in synchrotron x-ray diffraction results [25]) would yield L≈350 Å, which is very similar to the magnetic correlation length for the Mn^{3+} sublattice. In Fig. 16, we depict a model for such a magnetic and structural discommensuration boundary. It can be seen that at this discommensuration boundary there is no loss of coherence for the charge ordering into Mn^{3+} and Mn^{4+} sublattices, but there exists a π phase shift of orbital ordering at this domain boundary. The spin structure of the Mn^{4+} magnetic sublattice is not perturbed through this discommensuration boundary, but the direction of the spins on the Mn^{3+} sublattices are reversed at the discommensuration. Evidently, discommensurations play important roles on the magnetic and electronic properties of $La_{0.5}Ca_{0.5}MnO_3$.

[7] Modulation wave vector for X=5/8 and other concentrations

The modulation wave vector of the charge ordering naturally depends upon the doped charge carrier concentration. For hole-carrier concentration of x, the ratio of Mn^{4+} ions to Mn^{3+} ions is x/(1-x) [27]. The periodic diagonal array of these two types of ions will result in a modulation wave vector δ which equals to 1-x for x>0.5, which is consistent with experimental results. Shown in Fig. 17 is a plot of δ as a function of x including a data point for x=1/2 and 2/3 (where $\delta \approx 1/2$ and 1/3, respectively). As expected, the charge ordering wave vector δ varies linearly with

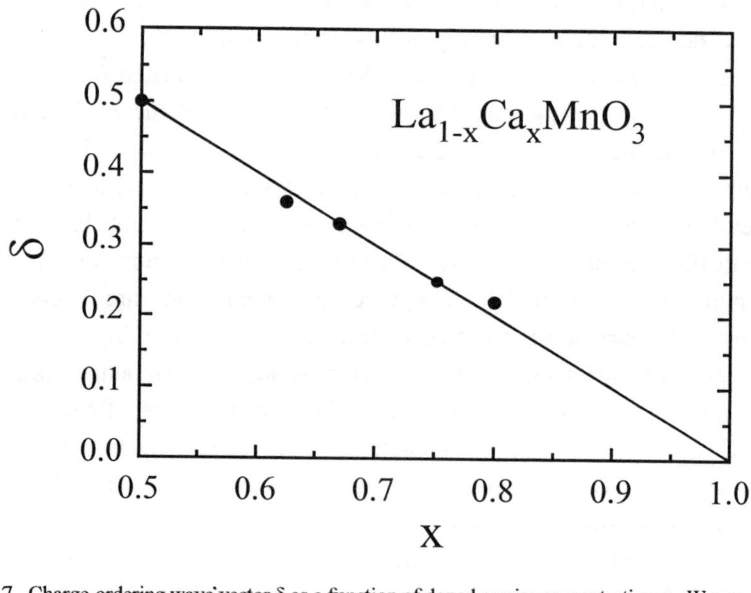

Fig. 17. Charge-ordering wave vector δ as a function of doped carrier concentration x. Wave vector δ closely follows the straight line of δ=1-x.

the doped carrier concentration x. Again, the charge-ordering wave vector varies typically ~10 % from area to area for samples with these compositions. The superlattice peaks of charge ordering in the, e.g., x=0.8 sample, however, are noticeably broader than other compositions, indicating a much shorter coherence length. This local fluctuation of wave vector is universally observed in the charge-ordered systems, including the two-dimensional transitional metal oxides such as the layered nickelates and the layered manganites.

At the x=5/8 concentration where the charge/orbital ordering transition temperature becomes maximum, x=1/2 type and x=2/3 type charge/orbital ordering configurations coexist with a fine mixture (~100 Å) of two types. As consistent with the level rule, the area ratio between the 1/2 type and the 2/3 type is roughly 3 to 1. It is challenging to understand why the charge/orbital transition temperature is highest when two types coexist with a certain ratio.

[8] Charge ordering in layered manganite

Charge ordering is also observed in perovskite-related layered manganite. Sr_2MnO_4, crystallizing in the K_2NiF_4 structure, is a two-dimensional antiferromagnetic insulator with Mn^{4+} (t_{2g}^3, S=3/2). χ shows a broad peak at ~280 K, typical of a two-dimensional antiferromagnet. Comparison with the results of high-temperature series expansion for S=3/2 yields an antiferromagnetic coupling constant $J/k_B \approx 73$ K. Earlier neutron diffraction showed that a three-dimensional antiferromagnetic order of the K_2NiF_4-type sets in at T_N=170 K [28], which can be seen as a small anomaly in our χ at T_N.

As soon as electrons are introduced via substituting La^{3+} for Sr^{2+}, two sets of peaks in x-ray diffraction pattern appear, corresponding to two different K_2NiF_4-related structures [29]. In other words, two phase region exists in $Sr_{2-x}La_xMnO_4$ for x>0. As shown in Fig. 18, with increasing La concentration, the (220) peak (in the orthorhombic notation: see below) at smaller 2θ angle, corresponding a larger tetragonal basal dimension a_p, becomes stronger while the (220) peak at larger 2θ, corresponding to a smaller a_p, becomes weaker and eventually disappears at x=0.38. This indicates that the phase with larger a_p grows at the expense of the phase with smaller a_p in the two-phase region. Surprisingly, the phase boundary for the two phases coincides x=3/8 within our experimental uncertainty (~0.02). This 3/8 feature in layered manganite reminds us the 3/8 anomaly in perovskite manganite showing the maximum Curie temperature (T_C) at x=3/8 in $La_{1-x}Ca_xMnO_3$. The variation of two sets of lattice parameters in $Sr_{2-x}La_xMnO_4$ are displayed in Fig. 19, clearly showing the two phase region for $0<x<3/8$.

Charge ordering transition occurs at ~235 K in $Sr_{1.5}La_{0.5}MnO_4$ where half of Mn ions are 3+, and the rest half are 4+ [29,30]. Upon cooling, χ sharply drops, and ρ and the charge activation energy derived from ρ increase rapidly at T_{co} (see Fig. 20). This behavior is very similar with that in, for example, perovskite $La_{1/3}Ca_{2/3}MnO_3$. Direct evidence for charge ordering is obtained from electron diffraction studies. Four sharp superlattice spots are observed around each fundamental Bragg

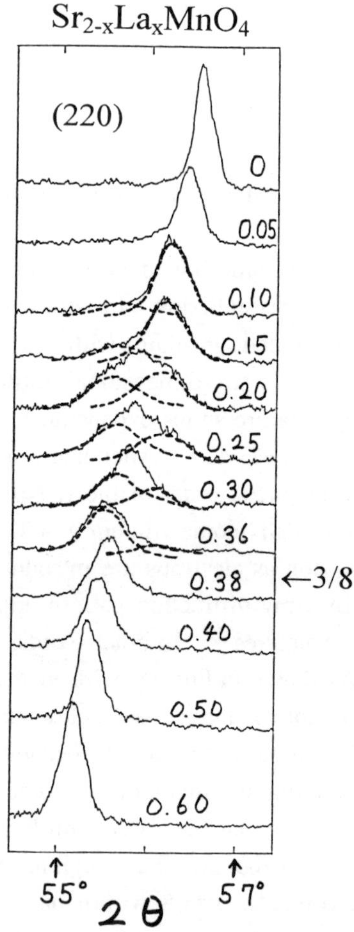

Fig. 18. Evolution of (220) peaks (orthorhombic notation) with x of $Sr_{2-x}La_xMnO_4$ in x-ray diffraction data. X-ray diffraction spectra for $0<x<0.38$ are composed of two peaks (shown with dashed lines), indicating the coexistence of two phases for $0<x<0.38$.

reflection in electron diffraction pattern at low temperatures, and no superlattice peaks can be seen at room temperature. The two different modulations, characterized by wave vectors q=2π/a(δ,0,0) and 2π/a(0,δ,0) arise from two different types of domains rotated by 90° to each other, as demonstrated by the selected area diffraction patterns in which only one type of modulation is observed. For comparison purpose we use orthorhombic unit cell of a≈5.5 Å in this case although the structure is tetragonal. The superlattice modulation is found to be commensurate, δ=1/2. However, we note that δ varies slightly from one grain to another, typically from 0.46 to 0.5. The superlattice reflections with δ=1/2, in general, are sharper than other values of δ, suggesting a longer coherence length for this preferred commensurate modulation in the case of x=0.5.

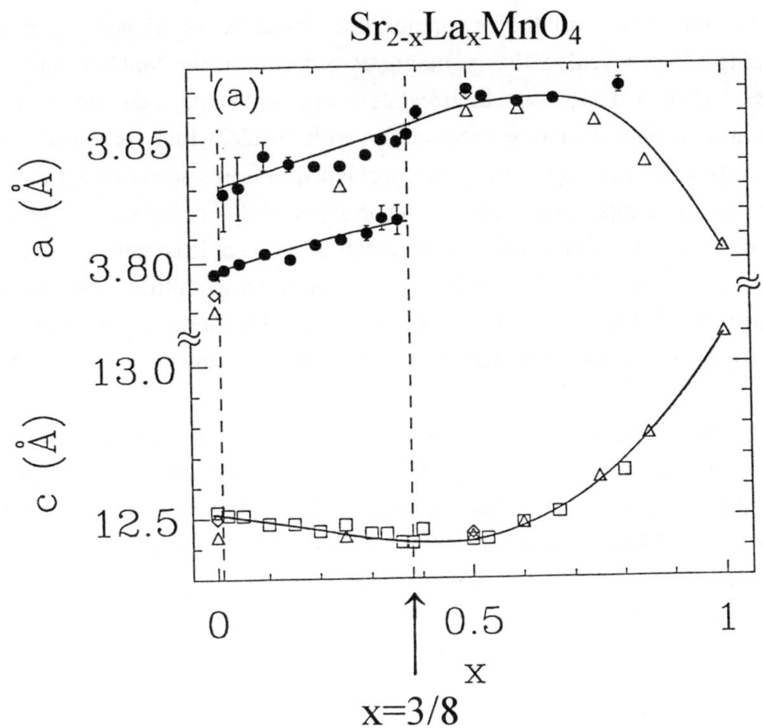

Fig. 19. Variation of lattice parameters in $Sr_{2-x}La_xMnO_4$ at room temperature. Squares are for c and closed circles a_p. Two phases with different a exist for 0<x<3/8.

Charge ordering is most pronounced at x=0.5, but is not limited to x=0.5 in $Sr_{2-x}La_xMnO_4$. A similar cusp in χ around 235 K also exists in samples with x=0.47 and 0.53. Much smaller cusp in χ can be detected near 235 K in specimens with x down to 0.05 in the two-phase region. Superlattice reflections have also been observed for, e.g., x=0.2 specimen. Actually, electron diffraction studies at low temperatures of the 0.2 sample show two kinds of diffraction patterns: one with superlattice spots quite similar to that shown in Fig. 11, and the other without superlattice reflections but showing a monoclinic distortion.

Two scenarios have been proposed for the charge ordering configuration in $Sr_{1.5}La_{0.5}MnO_4$. One resembles the charge/orbital ordering scheme in perovskite $La_{0.5}Ca_{0.5}MnO_3$ (see Fig. 6(b)) [31]. This scheme with orbital ordering in addition to charge ordering is consistent with the observed δ=1/2 charge modulation, and also appears to consistent with the magnetic superlattice peaks in neutron scattering experiments which reveals CE-type magnetic ordering in the MnO_2 layers. (Note that in Ref. 31, δ=1 peaks was identified as charge ordering peaks, but their recent experiment indicates a charge modulation with δ=1/2 which is consistent with electron diffraction results.) One of the problems with this scenario is the fact that the structure at temperatures above the charge/orbital ordering temperature is tetragonal so that the Jahn-Teller degeneracy is already, at least partially, lifted, and thus orbital ordering, at least partially, exists at high temperatures. Another puzzle is that large local structural distortions associated with the charge/orbital ordering in Fig. 6(b) have not been detected in the refinement of powder neutron diffraction data.

Another proposition is a bipolaronic charge ordering without orbital ordering. In this scheme, there is one row of Mn^{2+} for every three rows of Mn^{4+} along the diagonal direction in the Mn square lattice [29]. It has been known that Mn^{2+} with $3d^5$ high spin configuration is exceptionally stable. The energy of $Mn^{2+}(t_{2g}{}^3e_g{}^2)$-$Mn^{4+}(t_{2g}{}^3)$ configuration is lowered by Hund's rule exchange as compared with equal-electron-numbered $Mn^{3+}(t_{2g}{}^3e_g{}^1)$-$Mn^{3+}(t_{2g}{}^3e_g{}^1)$ configuration. It is beneficial to treat the $Mn^{4+}(t_{2g}{}^3)$ lattice as a background and to focus on doped electrons distributed among e_g orbitals on this square lattice. In one configuration, only one of the two empty e_g orbitals is occupied by a doped electron. Polarons with high spin Mn^{3+} character are formed by these e_g electrons through interaction with the surrounding lattice and background Mn^{4+} spins. In other configuration, two doped e_g electrons with parallel spins may become polaronic and mobile at high temperatures. They can be viewed as closely bound triplet bipolarons. While the total number of polarons and bipolarons is controlled by the La concentration, the ratio between them can vary. The formation of a bipolaron against two polarons, driven by Hund's rule on-site exchange, is hindered by the elastic strain energy which the large bipolaron has to pay to expand the underlying lattice. However, two neighboring bipolarons cost less strain energy than two separated ones, leading to bipolaron clustering. This mechanism can provide the driving force of phase separation into bipolaron-rich and polaron-rich phases in the range of $0<x<3/8$. The much larger ionic radius of Mn^{2+} than that of Mn^{3+} leads to larger in-plane square sizes for the bipolaron-rich phase than for the polaron-rich phase. Inspection of Fig. 19 also provides a further support for the Mn^{2+} formation, since the expansion of a_p above the two "end members" at x=0 (Mn^{4+}) and x=1 (Mn^{3+}) can be seen as due to Mn^{2+} formation. This proposition appears consistent with the electron diffraction results for x=0.2 sample. One phase with charge ordering can be attributed to the phase of bipolarons, which are of the isotropic Mn^{2+} character

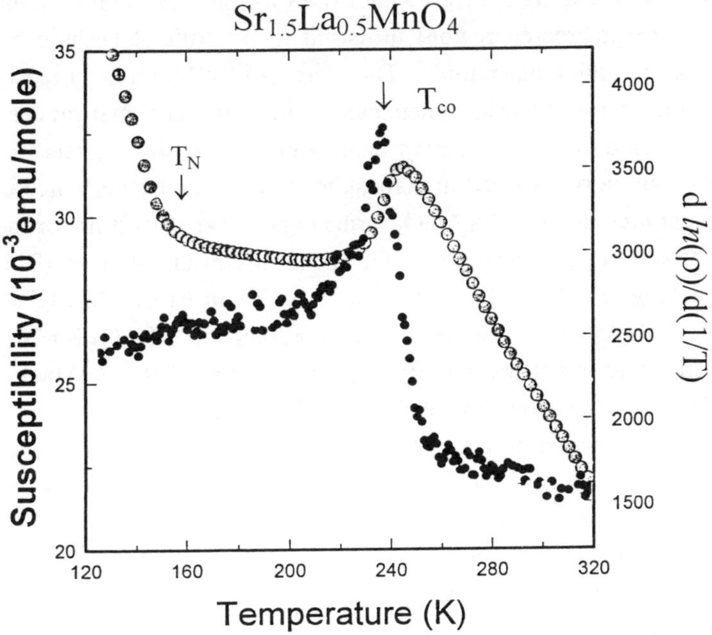

Fig. 20. Magnetic susceptibility and temperature derivative of resistivity (d *ln* (ρ)/d(1/T))as a function of temperature for $Sr_{1.5}La_{0.5}MnO_4$. Charge-ordering and spin-ordering temperatures are indicated.

and order below ~250 K, while the monoclinic phase to the polaron-rich phase, since Mn^{3+} with $t_{2g}{}^3e_g{}^1$ is susceptible to the Jahn-Teller distortion. Thus, this bipolaron scenario appears to explain the two phase issue, as well as the charge ordering modulation.

[9] Charge ordering vs. ferromagnetism in (Bi,Ca)MnO$_3$: origin of χ drop at T_{co}

In all charge-ordered systems, the magnetic susceptibility χ drops rapidly at T_{co}. Neutron scattering could unveil the origin of this χ drop at T_{co}, particularly the change of magnetic fluctuations at T_{co}, but it turns out that the growth of single crystals of charge-ordered compositions of, e.g., (La,Ca)MnO$_3$, necessary for neutron scattering experiments, is difficult. However, single crystals of the charge-ordered (Bi,Ca)MnO$_3$ system can be grown using self flux method. We note that floating-zone method has been successfully applied to grow single crystals of (La,Ca)MnO$_3$, but floating-zone growth of (La,Ca)MnO$_3$ for high Ca concentration region (i.e., charge-ordered region) turns out to be difficult because of high Ca vapor pressure at high temperatures. Therefore, (Bi,Ca)MnO$_3$ is an ideal system to investigate the charge ordering phenomenon in perovskite manganites by using neutron/x-ray scattering experiments which requires large size crystals.

$BiMnO_3$ can be synthesized in perovskite structure under high pressure [32], and under ambient pressure, $Bi_{1-x}Ca_xMnO_3$ forms in perovskite structure for the range of $0.3<x\leq 1$. Charge ordering occurs in the wide Ca concentration of (Bi,Ca)MnO$_3$. As shown in Fig. 21, T_{co} is found to peak at x=5/8 in $Bi_{1-x}Ca_xMnO_3$, identical to that in $La_{1-x}Ca_xMnO_3$. However, on contrary to (La,Ca)MnO$_3$, metallic ferromagnetic state is not observed in any composition of (Bi,Ca)MnO$_3$. Instead, charge ordering persists down to $x\approx 0.3$ in $Bi_{1-x}Ca_xMnO_3$.

Various measured physical properties for a $Bi_{0.18}Ca_{0.82}MnO_3$ crystal are summarized in Fig. 22(a). Similar with other manganites, there exists a sharp drop of χ, as well as a rapid increase of ρ, at T_{co} of ~210K [33]. The additional feature in χ at ~160 K originates from antiferromagnetic spin ordering. This antiferromagnetic transition is evident in the temperature dependence of magnetic (1,0,0) peak as shown in Fig. 22(b). It turns out that spins at low temperatures lie along c-axis, as found in (La,Ca)MnO$_3$ for a C-type antiferromagnet. Superlattice peaks in electron diffraction pattern are found with wave vector (~0.2,0,0) and (0,~0.2,0). The temperature dependence of these charge superlattice peaks is shown in Fig. 22(c) which also displays the temperature variation of lattice parameters determined from neutron diffraction study. This panel clearly indicates a significant structural transition at T_{co}.

Charge ordering drastically influences magnetic correlations in manganites. In double exchange picture, localized carriers cannot induce the double-exchange ferromagnetic coupling. This behavior is clearly demonstrated in inelastic neutron scattering results as shown in Fig. 22(d). There exist ferromagnetic correlations above T_{co} as evident from the finite ferromagnetic fluctuations in Fig. 22(d), as well as ferromagnetic Curie-Weiss temperature in χ. These ferromagnetic fluctuations decrease as charge ordering develops. In addition, antiferromagnetic fluctuations become dominant below T_{co}. Therefore, the drastic reduction of χ at T_{co}, a general phenomenon in various manganites and layered nickelates, is caused by a switch of spin fluctuations from ferromagnetic correlations to antiferromagnetic correlations. At low enough temperatures, the antiferromagnetic correlations induce a long-range antiferromagnetic ordering.

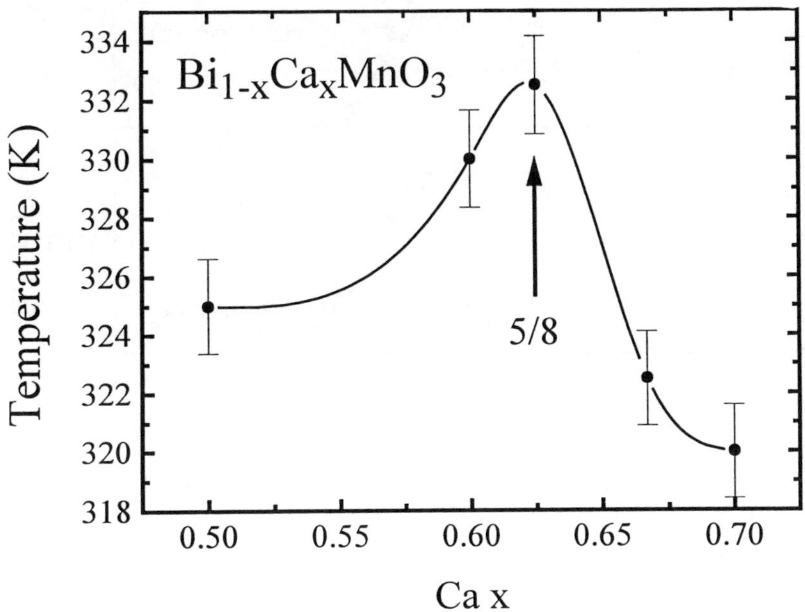

Fig. 21. Charge-ordering temperature as a function of Ca concentration in (Bi,Ca)MnO_3. T_{co} peaks at x=3/8. Lines are drawn as guides to the eyes.

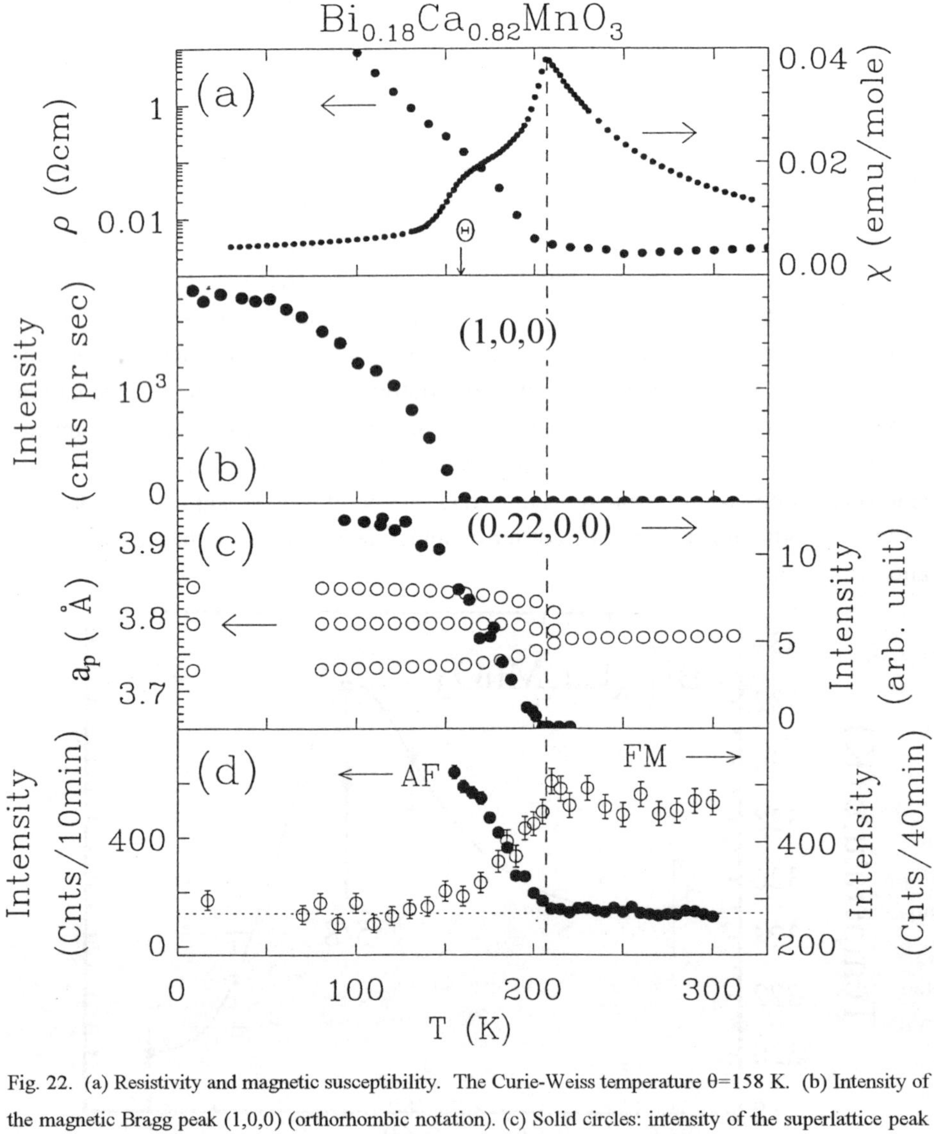

Fig. 22. (a) Resistivity and magnetic susceptibility. The Curie-Weiss temperature θ=158 K. (b) Intensity of the magnetic Bragg peak (1,0,0) (orthorhombic notation). (c) Solid circles: intensity of the superlattice peak (0.22,0,0). Open circles: lattice constant of the pseudocubic perovskite cell. (d) Inelastic neutron scattering intensity from paramagnetic spin fluctuations, probing antiferromagnetic correlations at **q**=(1,0,0) and $h\omega/\pi$=1.7 meV (solid circles) and probing ferromagnetic correlations near the forward direction at q=0.17 $Å^{-1}$ and $h\omega/\pi$=1 meV (open circles). The dotted line is background.

[10] Concluding remarks: dynamic charge stripe correlations

Charge ordering with stripe correlations is associated with a strong tendency toward electronic phase separation with stripe modulations when charge carriers are introduced into the Mott insulators by means of chemical doping. All of the parent compounds which show striped charge ordering when charge carriers are introduced chemically are antiferromagnetic Mott insulators. For example, $LaMn^{3+}O_3$, $BiMn^{3+}O_3$, $LaSrMn^{3+}O_4$ and $La_2Ni^{2+}O_4$ are antiferromagnets with half-filled band whose insulating behavior is due to electron correlations. The static correlation associated with the striped charge modulations in highly Ca substituted $LaMnO_3$ is basically limited by the charge ordering domain size and is long-ranged (~2000 Å), but the static stripe correlation length for $(La,Sr)_2NiO_4$ is intermediate-ranged (~500 Å) even at low temperatures. As discussed earlier, there exists a large electron-lattice coupling in mixed-valent manganites. It appears that this large electron-lattice coupling enhances the striped charge correlation, and also induce a large structural distortion associated with the charge modulation. This large lattice distortion associated with striped charge ordering results in strong superlattice peaks associated with the striped charge modulation, and these strong superlattice peaks enable for us to obtain the high-resolution real-space lattice images such as the one in Figs. 2 and 9.

It is interesting to note a possibility of dynamic stripe charge correlations in superconducting cuprates. Incommensurate inelastic neutron scattering peaks around antiferromagnetic instability point of (π,π) were discovered in metallic $(La,Sr)_2CuO_4$ in the early 90's, and were thought to be associated with Fermi surface nesting vectors [34,35]. However, it was found that they become spin-wave-like at high energies. Furthermore, static charge and spin peaks were observed in elastic neutron scattering experiments on insulating $(La,Nd)_{2-x}Sr_xCuO_4$ with x=1/8 [36]. The magnitude of the modulation wave vectors for x=1/8 is well correlated with that of inelastic peaks in metallic cuprates. These observations have led to the speculation that the inelastic peaks may be a result of fluctuating charge-spin stripe modulations [37]. Recent reports claim the existence of similar inelastic neutron peaks in Bi-based superconductors and superconducting $YBa_2Cu_3O_{7-\delta}$ [38,39]. The existence of dynamic charge/spin stripes in high Tc superconductors should have a significant impact on the way we attempt to understand layered cuprates. If dynamic stripes do exist in cuprates, then the general tendency from manganites through nickelates to cuprates is that charge/spin stripes become shorter-ranged or more dynamic, which appears to be associated with the decreasing electron-lattice coupling.

If dynamic charge stripes do exist in transition metal oxides, it may be speculated that the maximization of metallicity and the Curie temperature at the commensurate concentration of x=3/8 in $La_{1-x}Ca_xMnO_3$ is associated with the dynamic stripe correlations. No static charge/orbital ordering for x=3/8 is indicated in electron diffraction data. However, any existence of very-short-range or dynamic charge/orbital ordering cannot be ruled out. Future studies of charge/spin/lattice dynamics will elucidate this intriguing possibility.

Acknowledgments

This article is based on the results from close and friendly collaborations with enormous number of scientists and students. We sincerely thank our coworkers: H. Y. Hwang, S. Mori, P. G. Radaelli, W. Bao, C. M. Lopez, W. Ratcliff, A. P. Ramirez, P. Schiffer, B. Batlogg, M. Marezio, D. E. Cox, G. Aeppli, and A. S. Cooper. We also benefited greatly from discussions with P. B. Littlewood, B. I. Shraiman, A. Millers, Y. B. Kim, C. M. Varma, E. Abrahams, G. Kotliar.

References

[1] S. Jin, T. H. Tiefel, M. McCormack, R. A. Fastnacht, R. Ramesh and L. H. Chen, *Science* **264**, 413 (1994).

[2] C. Zener, *Phys. Rev.* **82**, 403 (1951).

[3] J. B. Goodenough, *Phys. Rev.* **100**, 564 (1955).

[4] Y. Tomioka, A. Asamitsu, Y. Moritomo, H. Kuwahara, and Y. Tokura, *Phys. Rev. Lett.* **74**, 5108 (1995).

[5] A. P. Ramirez, P. Schiffer, S-W. Cheong, C. H. Chen, W. Bao, T. T. M. Palstra, B. Zegarski, P. L. Gammel, and D. J. Bishop, *Phys. Rev. Lett.* **76**, 3188 (1996).

[6] C. H. Chen and S-W. Cheong, *Phys. Rev. Lett.* **76**, 4042 (1996).

[7] J. I. Pena, R. I. Merino, G. F. de la Fuente and V. M. Orera, *Adv. Mater.* **8**, 909 (1996).

[8] J. F. Dillon, Jr., *Magnetism* (ed. G. T. Rado and H. Suhl, Academic Press, New York and London, 1963), chp. 9.

[9] M. Seul and D. Andelman, *Science* **267**, 476 (1995).

[10] P. Matl *et al.*, to be published.

[11] T. T. M. Palstra, A. P. Ramirez, S-W. Cheong, B. R. Zegarski, P. Schiffer and J. Zaanen, *Phys. Rev.* **56**, 5104 (1997).

[12] C. Kittel, *Introduction to Solid State Physics* (John Wiley & Sons, Inc., 1986), chp. 21.

[13] J. P. Attfield and G. Ferey, *J. Solid State Chem.* **80**, 112 (1989).

[14] E. J. W. Verwey and P. W. Haaymann, *Physica* **8**, 979 (1941).

[15] J. Yoshida and S. Iida, *J. Phys. Soc. Jpn.* **47**, 1627 (1979).

[16] R. J. Cava , B. Batlogg, T. T. Palstra, J. J. Krajewski, W. F. Peck, Jr., A. P. Ramirez, and L. W. Rupp, Jr., *Phys. Rev.* B **43**, 1229 (1991).

[17] C. H. Chen, S-W. Cheong, and A. S. Cooper, *Phys. Rev. Lett.* **71**, 2461 (1993).

[18] S-W. Cheong, H. Y. Hwang, C. H. Chen, B. Batlogg, L. W. Rupp, Jr. and S. A. Carter, *Phys. Rev.* B **49**, 7088 (1994).

[19] J. M. Tranquada, D. J. Buttrey, V. Sachan, and J. E. Lorenzo, *Phys. Rev. Lett.* **73**, 1003 (1994).

[20] S. H. Lee and S-W. Cheong, *Phys. Rev. Lett.* **79**, 2514 (1997).

[21] A. J. Millis, to be published.

[22] H. Roder, Jun Zang, and A. R. Bishop, *Phys. Rev. Lett.* **76**, 1356 (1996).
[23] P. E. Schiffer, A. P. Ramirez, W. Bao, and S-W. Cheong, *Phys. Rev. Lett.* **75**, 3336 (1995).
[24] J. Schneck, J. C. Toledano, C. Joffrin, J. Aubree, B. Joukoff and A. Gabelotaud, *Phys. Rev.* B **25**, 1766 (1982).
[25] P. G. Radaelli, D. E. Cox, M. Marezio and S-W. Cheong, *Phys. Rev.* B **55**, 3015 (1997).
[26] P. G. Radaelli, D. E. Cox, M. Marezio, S-W. Cheong, P. E. Schiffer and A. P. Ramirez, *Phys. Rev. Lett.* **75**, 4488 (1995).
[27] C. H. Chen, S-W. Cheong and H. Hwang, *J. Appl. Phys.* **81**, 4326 (1997).
[28] J.-C. Bouloux, J.-L. Soubeyroux, G. Le Flem and P. Hagenguller, *J. Solid State Chem.* **38**, 34 (1981).
[29] W. Bao, S. A. Carter, C. H. Chen and S-W. Cheong, *Solid State Commun.* **98**, 55 (1996).
[30] Y. Moritomo, Y. Tomioka, A. Asamitsu, Y. Tokura and Y. Matsui, *Phys. Rev.* B **51**, 3297 (1995).
[31] B. J. Sternlieb, J. P. Hill, U. C. Wildgruber, G. M. Luke, B. Nachumi, Y. Moritomo and Y. Tokura, *Phys. Rev. Lett.* **76**, 2169 (1996).
[32] F. Sugawara, S. Iida, Y. Syono and S. Akimoto, *J. Phys. Soc. Jpn.* **20**, 1529 (1965).
[33] W. Bao, J. D. Axe, C. H. Chen, and S-W. Cheong, *Phys. Rev. Lett.* **78**, 543 (1997).
[34] S-W. Cheong, G. Aeppli, T. E. Mason, H. Mook, S. M. Hayden, P. C. Canfield, Z. Fisk, K. N. Clausen, and J. L. Martinez, *Phys. Rev. Lett.* **67**, 1791 (1991).
[35] Q. Si, Y. Zha, K. Levin and J. P. Lu, *Phys. Rev.* B **47**, 9055 (1993).
[36] J. M. Tranquada, B. J. Sternlieb, J. D. Axe, Y. Nakamura, and S. Uchida, *Nature* **375**, 561 (1995).
[37] M. I. Salkola, V. J. Emery and S. A. Kivelson, *Phys. Rev. Lett*, **77**, 155 (1996).
[38] P. Dai, H. A. Mook and F. Dogan, to be published.
[39] H. A. Mook *et al.*, to be published.

ELECTRONIC STRUCTURE OF PEROVSKITE-TYPE MANGANESE OXIDES

A. FUJIMORI, T. MIZOKAWA
Department of Physics, University of Tokyo, Hongo, Bunkyo-ku, Tokyo 113, Japan

T. SAITOH
Department of Physics, University of Colorado, Boulder, Colorado 80309-0390, U.S.A.

We give an overview of recent studies on the electronic structure of pure and hole-doped Mn oxides with the perovskite-type and related layered structures. Results of electron spectroscopy and band-structure calculations are presented and discussed.

1 Introduction

Transition-metal oxides of the perovskite-type structure with composition ABO_3 and their layered-structure relatives $A_{n+1}B_nO_{1+3n}$ (n = 1, 2, 3 .....) exhibit a variety of interesting physical properties including high-temperature superconductivity, metal-insulator transitions and magnetic phase transitions. Among them, manganites (B = Mn) have recently attracted particular attention because of the discovery of giant negative magnetoresistance in $La_{1-x}Sr_xMnO_3$ and $La_{1-x}Ba_xMnO_3$.[1,2] Their intriguing doping- and temperature-dependent metal-insulator transition and magnetic, orbital and charge ordering phenomena have been the subject of extensive research in these years. As a matter of fact, those properties have their origins in their unique electronic structures derived from the hybridized Mn 3d and oxygen 2p orbitals under the unique structural and chemical environment. Resulting intra-atomic exchange (Hund's rule coupling) and the orbital degrees of freedom of the Mn 3d electrons play essential roles. Furthermore, various types of structural distortions profoundly influence their electronic properties.

A large part of information about the electronic structure of the manganites has been deduced from photoemission spectroscopy, optical studies and band-structure calculations as described below. This article is organized as follows: In Sec. 2, we start with the global electronic structure of undoped manganites such as $LaMnO_3$ and $SrMnO_3$ on the basis of configuration-interaction (CI) analyses of photoemission spectra using the MnO_6 cluster model. The energy levels of the Mn 3d and O 2p states and the strength of the p-d hybridization are derived. Effects of Mn-Mn interaction in the periodic lattices are considered in Sec. 3 on the basis of *ab initio* and model-Hamiltonian band-structure calculations. The orbital degrees of freedom of the e_g sublevels of the Mn 3*d* states are shown to be coupled with lattice distortions of various types. In Sec. 4, experimental investigations of the doping and temperature dependence of the electronic structure are described in detail.

2 Local Electronic Structure of Undoped Compounds

2.1 Electronic Structure Parameters

The end members $R\mathrm{MnO}_3$, where R^{3+} is a trivalent rare-earth ion, and $A\mathrm{MnO}_3$, where A^{2+} is Sr^{2+} or Ca^{2+}, are antiferromagnetic (AFM) insulators with the partially filled *d* band, i.e., Mott insulators. In the ionic model, the Mn^{3+} ion in $R\mathrm{MnO}_3$ is in the high-spin ($S = 2$) state and has the electronic configuration of d^4 ($t_{2g\uparrow}^3 e_{g\uparrow}$); the Mn^{4+} ion in $A\mathrm{MnO}_3$ is also in the high-spin ($S = 3/2$) state with the d^3 ($t_{2g\uparrow}^3$) electronic configuration. In $R_{1-x}A_x\mathrm{MnO}_3$, the t_{2g} electrons behave as $S = 3/2$ local spins in the whole composition range $0 < x < 1$ while the e_g orbitals form a rather wide band, into which $1-x$ electrons per Mn ion are accommodated. The spins of the e_g electrons are aligned parallel to the t_{2g} spins through strong Hund's rule coupling (J_H ~1-2 eV). If we start from LaMnO_3, doped x holes per Mn ion go into the e_g band.

It has been known that the crystal-field splitting $10Dq$ of the *d* orbitals arises largely from mixing with neighboring *p* orbitals. It has also been known that Coulomb repulsion between *d* electrons is strong in 3*d* transition-metal oxides. The effect of the *p*-*d* hybridization in the presence of the strong *d*-*d* Coulomb repulsion can be properly taken into account by a CI treatment of the MnO_6 cluster model, as far as the local electronic structure is concerned. In the CI theory, the ground state of the MnO_6 cluster relevant to LaMnO_3 is given by[3]

$$\Psi = \alpha|d^4\rangle + \beta|d^5\underline{L}\rangle + ...,$$

where $\underline{L}$ denotes a hole in a ligand *p* molecular orbital.[3] Important parameters are the on-site *d*-*d* Coulomb energy U, $d^4 \to d^5\underline{L}$ charge-transfer energy Δ, and the *p*-*d* transfer integral ($pd\sigma$).

Those electronic structure parameters were determined by analysis of the Mn 2*p* core-level photoemission spectra by Bocquet *et al.* [4] Figure 1 shows the Mn 2*p* spectra of LaMnO_3 and SrMnO_3 as well as that of MnO (which has Mn^{2+} ions) and fits with cluster-model calculations; parameter values are listed in Table 1, from which one can see that Δ rapidly decreases with increasing Mn valence, reflecting the general chemical trend which has been established for a large number of 3*d* transition-metal compounds.[4] While Δ

Table 1: Charge-transfer energy Δ, on-site Coulomb energy U and p-d transfer integral $(pd\sigma)$ of Mn oxides in units of eV. Δ_{eff} and U_{eff} are "effective" charge-transfer and Coulomb energies defined with respect to the lowest multiplet term of each configuration.[3,4]

	Δ	U	$(pd\sigma)$	Δ_{eff}	U_{eff}
MnO	6.5	7.0	-1.1	8.8	11.6
$LaMnO_3$	4.5	7.8	-1.8	1.8	6.8
$SrMnO_3$	2.0	7.5	-1.5	-0.2	7.1

and U are defined with respect to the center of gravity of the multiplet of each configuration, Table 1 also show "effective" charge-transfer and Coulomb energies, Δ_{eff} and U_{eff}, defined with respect to the lowest multiplet term of each configuration.[5] The multiplet effect (or Hund's rule coupling) further reduces the charge-transfer energy, thereby increasing the degree of p-d covalency in the Mn^{3+} and Mn^{4+} oxides.[a] The parameter values indicate that $LaMnO_3$ and $SrMnO_3$ are charge-transfer type ($\Delta < U$) insulators, that is, the band gap is of the O $2p$-to-Mn $3d$ charge-transfer type. If Δ_{eff} is considered instead of Δ, $SrMnO_3$ is even a "negative-Δ" insulator.[6] Using those parameter values, Saitoh *et al.*[3] have calculated the band gap to be $E_{\text{gap}} = 1.7$ and 2.3 eV for $LaMnO_3$ and $SrMnO_3$, respectively, which is compared with the experimental value of ~ 1.1 eV for $LaMnO_3$.[7] The net d-electron number is calculated to be $n_d = 4.5$ and 3.8, respectively, indicating the strong p-d covalency. A hole doped into $LaMnO_3$, which has a spin antiparallel to the Mn^{3+} spin, have mostly (by ~ 75 %) oxygen p character, reflecting the charge-transfer nature of $LaMnO_3$.[3] This is analogous to the case of the high-T_C cuprates, where a doped hole enters oxygen p orbitals, is antiferromagnetically coupled with the Cu^{2+} spin, and forms a so-called Zhang-Rice singlet.[8] Somewhat different values for the Δ, U and $(pd\sigma)$ of $LaMnO_3$ have been obtained by Chainani *et al.*[9] using a similar approach. However, the net electron number n_d has been calculated to be quite similar to the above value.

2.2 Valence-Band Electronic Structure

Since the composition (x) dependence of the electronic structure of $R_{1-x}A_x MnO_3$ will be described in detail in Sec. 4.1, only results for $LaMnO_3$ and $SrMnO_3$ are presented here. Their ground states are basically well described by the classical ionic model, d^4 ($t_{2g\uparrow}^3 e_{g\uparrow}$) and d^3 ($t_{2g\uparrow}^3$) configurations,

[a] Note that for MnO, the same multiplet effect *increases* the charge-transfer energy thereby *reducing* the p-d covalency.

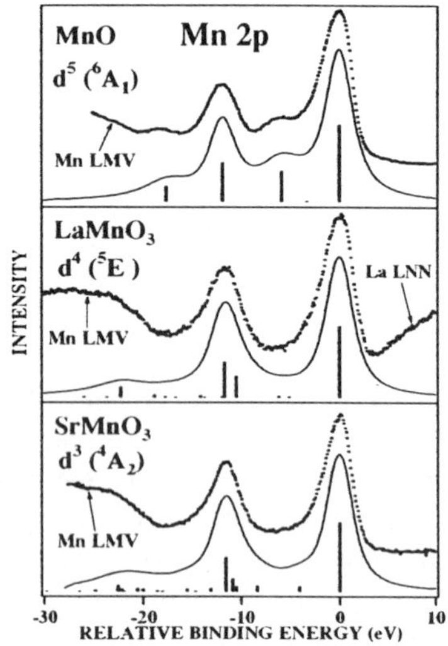

Figure 1: Mn $2p$ core-level photoemission spectra of MnO, $LaMnO_3$ and $SrMnO_3$. Experimental data (dots) and their cluster-model fits (vertical bars and solid curves).[4]

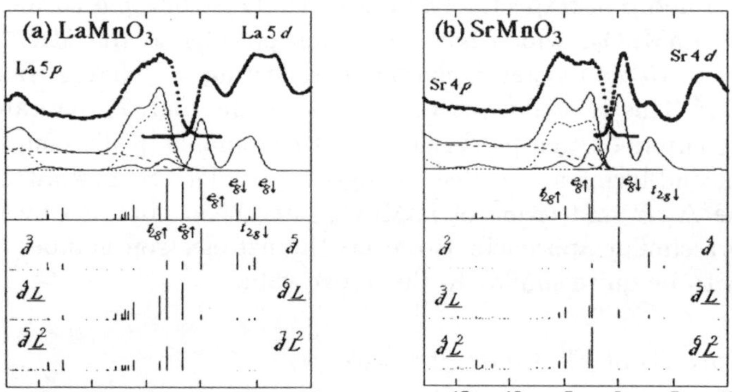

Figure 2: Cluster-model analysis of the valence-band photoemission and O $1s$ XAS spectra of $LaMnO_3$ and $SrMnO_3$.[3] Dots: Experimental spectra; Solid curves and vertical bars: Calculated spectra. Lower panels show decomposition of the spectra into final-state configurations.

respectively, although there is considerably strong p-d covalency. On the other hand, one would expect that the photoemission spectra may be fundamentally modified from the ionic model due to the charge-transfer nature of the band gaps since the highest occupied states in the charge-transfer insulators are oxygen p-like rather than Mn d-like.

In fact, as shown in Fig. 2,[3] the CI cluster-model calculation for $LaMnO_3$ has shown that $d^4 \to d^3$ spectral weight appears at binding energies as high as $\sim 10-15$ eV while $d^4 \to d^4\underline{L}$ spectral weight appears at lower binding energies. Nevertheless, the calculation has shown that the highest occupied states in $LaMnO_3$ have $e_{g\uparrow}$-symmetry and that the highest occupied states in $SrMnO_3$ have $t_{2g\uparrow}$-symmetry (Fig. 2), well corresponding to the energy levels predicted by the simplest ionic model. Indeed, the $d^4 \to d^3$ and $d^4 \to d^4\underline{L}$ components are strongly hybridized with each other; there is a significant amount of $d^4 \to d^3$ spectral weight in the $e_{g\uparrow}$-emission feature. The extra $e_{g\uparrow}$ spectral weight for $LaMnO_3$ compared to $SrMnO_3$ appears as a weak tail on the low binding energy side of the O $2p$ and $t_{2g\uparrow}$ bands.

3 Spin, Charge and Orbital Ordering

3.1 Undoped Compounds

The parent insulator RMnO$_3$ shows a layered AFM structure or so-called A-type AFM ordering as shown in Fig. 3(a).[10] Since one of the doubly-degenerate e_g orbitals is occupied in each of the Mn^{3+} (d^4) ions, RMn^{3+}O$_3$ shows interesting interplay between spin and orbital ordering. The orbital ordering is coupled with Jahn-Teller (JT) distortions:[11] $LaMnO_3$ shows a JT distortion as shown in Fig. 3(b).[12] In 1950's, Kanamori and Goodenough [13,14,15] have proposed that $3x^2 - r^2/3y^2 - r^2$-type orbital ordering accompanied by the JT distortion makes the superexchange coupling within the c-plane ferromagnetic (FM) and that between the c-planes AFM, resulting in the A-type AFM ordering.[10,13]

Recently, the interplay between spin and orbital ordering and the JT distortion in RMnO$_3$ has been studied using local-spin-density-approximation (LSDA),[16,17,18] LDA + U [19] and model Hartree-Fock (HF) [20,21] band-structure calculations. The *ab initio* LSDA calculations with the JT distortion correctly predict $LaMnO_3$ to be an A-type AFM insulator with the $3x^2 - r^2/3y^2 - r^2$-type orbital ordering. [16] The density of states of $LaMnO_3$ displayed in Fig. 4 shows that the e_g band just below the Fermi level (E_F) has a considerable amount of O $2p$ character. In the LSDA calculation,[16] the magnitude of the direct gap is calculated to be ~ 0.5 eV, which is smaller than the experimental value of ~ 1.1 eV.[7]

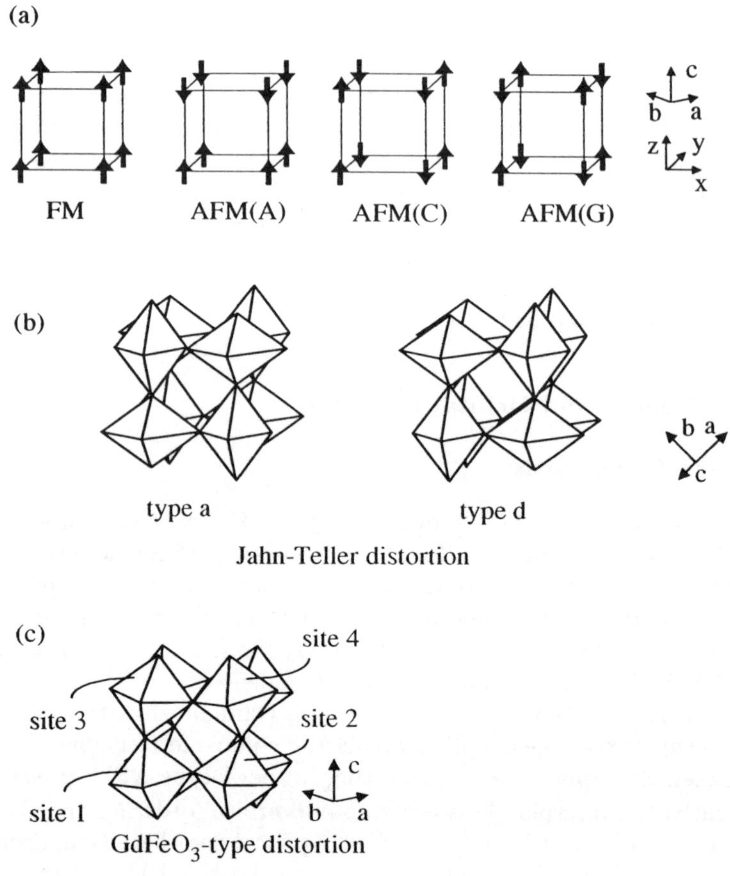

Figure 3: (a) Typical magnetic structures of the perovskite-type oxides. (b) Two types of Jahn-Teller distortions. (c) $GdFeO_3$-type lattice distortion.

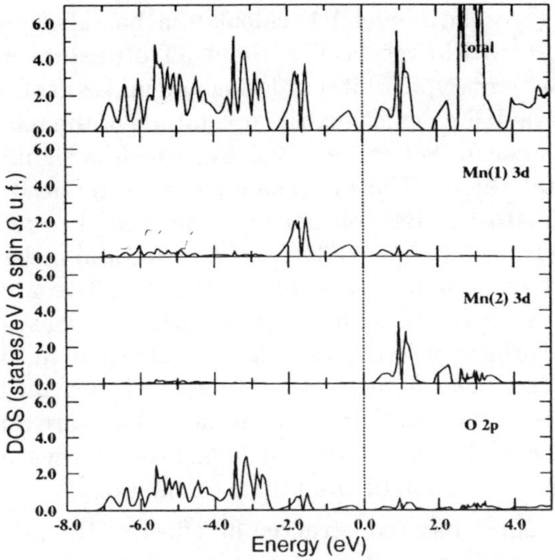

Figure 4: The density of states of A-type AFM $LaMnO_3$ obtained by Hamada *et al.*[16] using *ab initio* LSDA calculation.

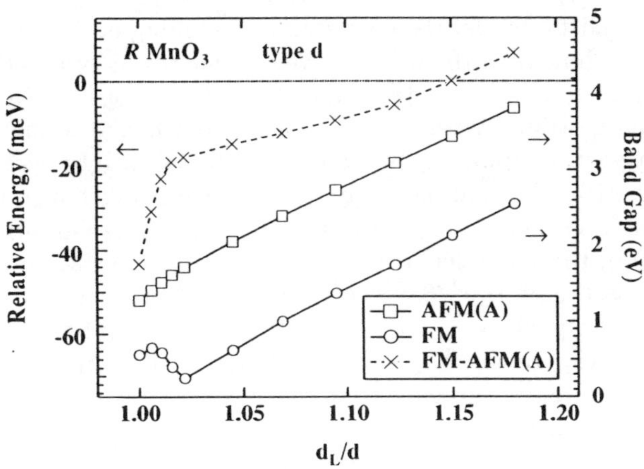

Figure 5: Energy difference between the FM and A-type AFM states and their band gaps for $LaMnO_3$ as functions of the strength of the JT distortion (d_L/d, where d_L and d denote the longer and averaged Mn-O bond lengths, respectively.) calculated using the HF approximation.

As shown in Fig. 5, the model HF calculation has also predicted that, while $RMnO_3$ would be a FM insulator without JT distortion, the distortion causes the $3x^2-r^2/3y^2-r^2$-type orbital ordering and makes $RMnO_3$ an A-type antiferromagnet. In the HF calculation, the magnitude of the band gap for the A-type AFM state is calculated to be $\sim$ 3.5 eV, which is significantly larger than the experimental value. The discrepancy can be reduced by including self-energy corrections to the HF solution, e.g., in second order in U.[22] The A-type AFM state with the $3x^2-r^2/3y^2-r^2$-type orbital ordering has been obtained in an exact diagonalization study of the $2 \times 2 \times 2$ cluster.[23] The two-dimensional character of the spin dynamics, which is consistent with the $3x^2-r^2/3y^2-r^2$-type orbital ordering, and the critical exponent has been found in the neutron scattering measurement of $LaMnO_3$.[24] The exchange interaction parameter within the c-plane and that between the c-planes are estimated to be $\sim$ 13.4 meV and $\sim$ -4.8 meV, respectively, by fitting the spin-wave dispersion,[24] in agreement with the prediction of the LSDA calculation.[17]

Kugel and Khomskii [11] have constructed an effective Hamiltonian by perturbation calculations with respect to the transfer integrals and have given a systematic way to describe the spin and orbital ordering in the degenerate Hubbard model. In the absence of JT distortion, calculations on the effective Hamiltonian of the Kugel-Khomskii (KK) type [11,25,26] predict that the orbital ordering for the A-type AFM state is of the z^2-x^2/z^2-y^2-type. On the other hand, the realistic model HF calculations including both the Mn $3d$ and O $2p$ orbitals [20,21] show that the orbital ordering for the A-type AFM state is a mixture of the z^2-x^2/z^2-y^2-type and the $3x^2-r^2/3y^2-r^2$-type. Since $LaMnO_3$ is a charge-transfer type insulator, superexchange pathways due to the oxygen-to-oxygen hopping term, which are not taken into account in the KK-type effective Hamiltonian, strongly affect the superexchange interaction and stabilize the intermediate orbital ordering. For a large $GdFeO_3$-type distortion [Fig. 3 (c)], the HF calculations [21] predict that the FM state becomes insulating and has the $x^2-y^2/3z^2-r^2$-type orbital ordering, in which sites 1, 2, 3 and 4 are occupied by z^2-y^2, $3x^2-r^2$, $3y^2-r^2$ and z^2-x^2 orbitals, respectively. Actually, $BiMn^{+3}O_3$, which has no JT distortion, shows a FM and insulating behavior,[27] implying that the FM state in $BiMnO_3$ would be driven by the orbital ordering.

It is important to reveal the relationship between the FM state driven by orbital ordering and the FM state by double-exchange mechanism. The orbital degrees of freedom may play an important role also in the FM metallic state of $La_{1-x}Sr_xMnO_3$, where the spin is fully polarized. Recently, a field theoretical calculation using the slave fermion method [28] has shown that the anomalous behaviors of the FM and metallic $La_{1-x}Sr_xMnO_3$ such as the broad incoherent component found in the optical conductivity [29] and the reduction of the photoemission spectral weight at E_F [30,31] may be caused by orbital fluctuations.

3.2 *Charge Ordering in* $R_{0.5}A_{0.5}MnO_3$

Extensive studies on $R_{1-x}A_x\mathrm{MnO_3}$ as a function of the sizes of the R and A ions have revealed that the $\mathrm{GdFeO_3}$-type distortion, which becomes larger as the size of the R and A ions become smaller, stabilizes the charge-ordered (CO) state compared with the FM state. The CO state tends to be stable at $x = 0.5$, where the ratio $\mathrm{Mn^{3+}} : \mathrm{Mn^{4+}} = 1 : 1$. The least distorted $\mathrm{La_{0.5}Sr_{0.5}MnO_3}$ is a FM metal below the Curie temperature (T_C) and a paramagnetic metal above T_C without charge ordering.[32] Strongly distorted $\mathrm{Pr_{0.5}Ca_{0.5}MnO_3}$ ($T_{CO} \sim 230$ K) [33,34] and $\mathrm{Nd_{0.5}Ca_{0.5}MnO_3}$ ($T_{CO} \sim 220$ K) [35] become CO insulators below the charge-ordering temperature (T_{CO}) and do not show ferromagnetism. $\mathrm{La_{0.5}Ca_{0.5}MnO_3}$ ($T_{CO} \sim 160$ K) [36,37,38] and $\mathrm{Nd_{0.5}Sr_{0.5}MnO_3}$ ($T_{CO} \sim 150$ K),[39,40] which are located between the two extremes, undergo a FM metal to CO insulator transition on cooling.

Most of the CO states in these compounds have CE-type AFM order [10,33] which is well explained by the model proposed by Goodenough.[13,15] In this model, $\mathrm{Mn^{3+}}$ and $\mathrm{Mn^{4+}}$ are arranged like a checkerboard as shown in Fig. 6(a) and the $\mathrm{Mn^{3+}}$ sites have a JT distortion. Along the c-axis, the same in-plane arrangement of $\mathrm{Mn^{3+}}$ and $\mathrm{Mn^{4+}}$ is stacked and the neighboring planes are antiferromagnetically coupled. Since the exchange coupling between $\mathrm{Mn^{3+}}$ and $\mathrm{Mn^{4+}}$ depends on which type of e_g orbital is occupied at the $\mathrm{Mn^{3+}}$ site, the orbital degrees of freedom at the $\mathrm{Mn^{3+}}$ site is also important in the doped $R\mathrm{MnO_3}$. The charge ordering and the JT distortion in the CE-type AFM state have been observed by x-ray and neutron diffraction measurements.[33,37,38] The CE-type AFM CO state in the three-dimensional manganites probably has the $3x^2 - r^2/3y^2 - r^2$-type orbital ordering at the $\mathrm{Mn^{3+}}$ site. However, the orbital ordering has not been observed directly. On the other hand, a neutron diffraction study [41] on $\mathrm{Pr_{0.5}Sr_{0.5}MnO_3}$ ($T_N \sim 130$ K [42]) showed that A-type AFM order sets in below T_N. Although the neutron diffraction study has failed to observe a clear sign of charge ordering, short range charge and/or orbital ordering would play an important role to stabilize the A-type AFM state.

The spin, charge and orbital ordering in $R_{0.5}A_{0.5}\mathrm{MnO_3}$ has been studied theoretically by LDA+U [43] and model HF [44] calculations. In the LDA+U calculation without the JT distortion,[43] it has been shown that $R_{0.5}A_{0.5}\mathrm{MnO_3}$ can have a CO state with the CE-type AFM ordering. In this calculation, however, its stability relative to the FM and other AFM CO states has not

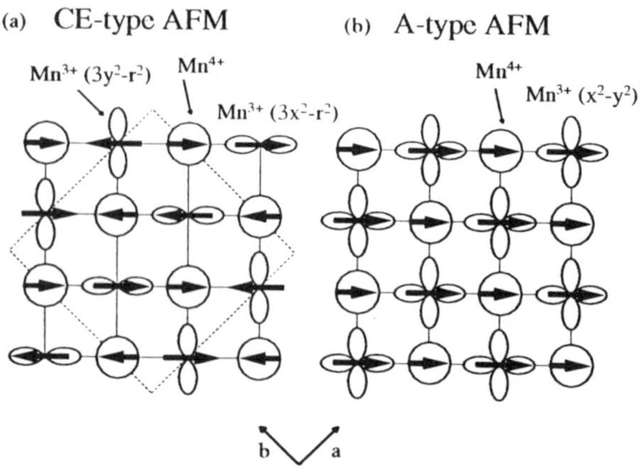

Figure 6: Spin, charge and orbital ordering for the CE-type AFM CO state (a) and for the A-type AFM CO state (b). The broken line shows the unit cell for the CE-type AFM CO order.

been studied. In the model HF calculation,[44] the CE-type AFM CO solution with a band gap of $\sim$ 0.6 eV is obtained but is higher in energy than the FM state without the JT distortion. In the CE-type AFM solution, the Mn^{3+} sites are accompanied by the $3x^2 - r^2/3y^2 - r^2$-type orbital ordering even without the JT distortion. This is because the e_g orbital of the Mn^{3+} site tends to point the neighboring Mn^{4+} sites in order to gain the kinetic exchange energy. The orbital ordering in $R_{0.5}A_{0.5}MnO_3$ is contrasted with that in $RMnO_3$, which is a mixture of the $z^2 - x^2/z^2 - y^2$-type and the $3x^2 - r^2/3y^2 - r^2$-type without the JT distortion. The calculated magnetic moments at the Mn^{3+} and Mn^{4+} sites are 3.44 and 3.34 μ_B, respectively, in the LDA+U calculation and 3.79 and 3.65 μ_B, respectively, in the model HF calculation. The difference in the magnetic moments between the Mn^{3+} and Mn^{4+} sites is small because $RMnO_3$ is a charge-transfer type insulator and therefore the net number of $3d$ electrons at the Mn^{4+} sites is close to that at the Mn^{3+} sites. The observed magnetic moments are 2.74 and 2.56 μ_B for $Pr_{0.5}Ca_{0.5}MnO_3$ [33] and 3.0 and 2.8 μ_B for $Nd_{0.5}Sr_{0.5}MnO_3$.[41] The model HF calculation also shows that the JT distortion that follows the $3x^2 - r^2/3y^2 - r^2$-type orbital arrangement stabilizes the CE-type AFM state compared with the FM metallic state as shown in Fig. 7(a). The difference in the magnetic moments increases with the JT distortion. These calculations give a scenario that, as the $GdFeO_3$-type distortion increases, the JT distortion becomes larger and consequently the

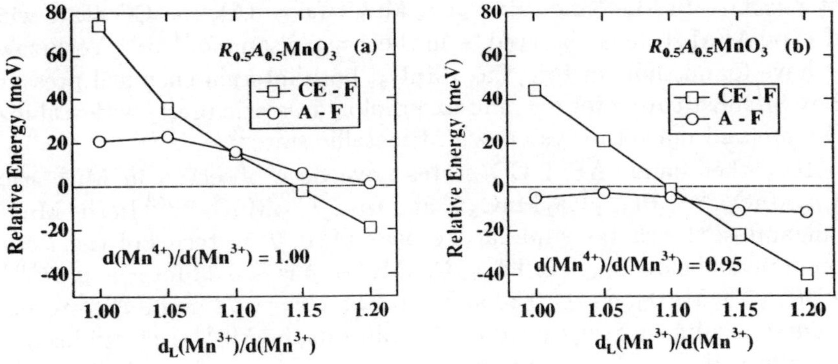

Figure 7: Energies per formula unit of the CE-type and A-type AFM states relative to the FM state calculated using the HF approximation as functions of the JT distortion (d_L/d) for the MnO_3 perovskite-type lattice with (b) and without (a) the in-plane breathing-type distortion.

CO state with the CE-type AFM ordering becomes favored.

In the A-type AFM CO state, the orbital ordering shown in Fig. 6(b) is expected to be favored. The orbital ordering couples with the in-plane breathing-type lattice distortion, in which the Mn^{3+} ion is expanded and the Mn^{4+} ion is contracted within the c-plane (without changing the Mn-O bond distance along the c-axis). This breathing-type distortion makes the MnO_6 octahedra at the Mn^{3+} sites expanded within the c-plane and favors the orbital ordering in which the $x^2 - y^2$-type orbital is occupied at the Mn^{3+} sites. The model HF calculation [44] shows that, without the JT distortion and with the breathing-type distortion, the A-type AFM CO state with the $x^2 - y^2$-type orbital ordering is the lowest in energy. This state may be related to the A-type AFM state found in $Pr_{0.5}Sr_{0.5}MnO_3$.

3.3 $R_{1-x}A_xMnO_3$ $(x \neq 0.5)$

A neutron diffraction measurement of lightly doped $La_{1-x}Sr_xMnO_3$ ($x < 0.17$) shows that the Mn^{3+}-rich sample undergoes a FM metal-to-canted AFM insulator transition on cooling.[45] Charge ordering in the Mn^{3+}-rich $La_{1-x}Sr_xMnO_3$ ($x = 0.10$ and 0.15) has been studied by neutron scattering measurements.[46] At $x = 0.15$, the system becomes FM below 240 K and the doped holes form isolated polarons ordered in real space below $T_{CO} \sim 190$ K. In the CO states at $x = 0.10$ and 0.15, the periodicity along the c axis is quadrupled, indicating that the charge ordering tends to be locked into the commensurate arrangement at $x = 1/8$. In Mn^{3+}-rich $Pr_{1-x}Ca_xMnO_3$ ($x = 0.3$), the CO state with the CE-type AFM state is observed as in the $x = 0.5$ sample.[33,47,48] Yoshizawa *et al.*[49] have found that, in $Pr_{0.7}Ca_{0.3}MnO_3$, both internal chemical pressure caused by Sr-substituting for Ca, and external hydrostatic pressure destabilize the CO state and make the system a FM metallic state.[49]

On the other hand, AFM CO states have been observed in Mn^{4+}-rich $La_{1-x}Ca_xMnO_3$,[50,51] $Pr_{1-x}Ca_xMnO_3$[33] and $Bi_{1-x}Ca_xMnO_3$.[52,53] In the Mn^{4+}-rich manganites, the charge superlattice peak (δ, 0, 0) in terms of the $Pbnm$ notation appears below T_{CO}. In $La_{1-x}Ca_xMnO_3$, δ is ~ 0.33 for $x = 0.67$[50,51] and, in $Bi_{1-x}Ca_xMnO_3$, δ is ~ 0.30 and ~ 0.22 for $x = 0.74$ and 0.82,[52] respectively. These results suggest that the CO states in the Mn^{4+}-rich oxides have domain walls as those in $La_{2-x}Sr_xNiO_4$.[54] $Bi_{1-x}Ca_xMnO_3$ of $x = 0.82$ has T_{CO} ~ 210 K and becomes C-type AFM below ~ 160 K probably accompanied by $3z^2 - r^2$-type orbital ordering at the Mn^{3+} sites.[52] The C-type AFM state was also observed in $La_{1-x}Ca_xMnO_3$ with $x \sim 0.7$[10] and $Pr_{1-x}Ca_xMnO_3$ with x ~ 0.7.[33] The $3z^2 - r^2$-type orbital ordering makes the superexchange coupling between the Mn^{3+} and Mn^{4+} sites along the c-axis FM and that within the c-plane AFM, causing the C-type AFM arrangement.[13,33]

3.4 Layered Manganites

In addition to the three-dimensional perovskites, an AFM CO state of the CE type has been observed in the layered perovskite $La_{0.5}Sr_{1.5}MnO_4$[55,56,57] which does not have the $GdFeO_3$-type distortion. In the two-dimensional manganites, the orbital ordering of the $3x^2 - r^2/3y^2 - r^2$ type at the Mn^{3+} site has been confirmed by an x-ray diffraction measurement,[58] which is the first direct observation of the orbital ordering. The model HF calculation for $La_{0.5}Sr_{1.5}MnO_4$[44] has shown that, without the JT distortion, the FM state is lower in energy than the CE-type AFM state which has the $3x^2 - r^2/3y^2 - r^2$-type orbital ordering and that the JT distortion makes the CE-type AFM CO state lower in energy than the FM state.

4 Low Energy Electronic Structure of Doped Compounds

4.1 Doping Dependence

The valence of the Mn ion can be widely controlled from 3+ (d^4) to 4+ (d^3) for most of the $R_{1-x}A_xMnO_3$ (113) compounds and the $La_{1-x}Sr_{1+x}MnO_4$ (214) compounds while the range of $0.3 \leq x \leq 0.5$ is available for $La_{2-2x}Sr_{1+2x}Mn_2O_7$ (327). Photoemission and x-ray absorption (XAS) studies of $La_{1-x}Sr_xMnO_3$

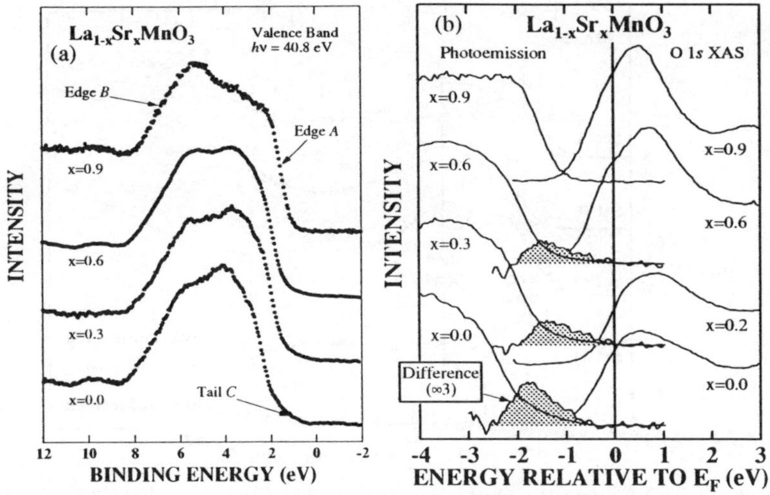

Figure 8: (a) Photoemission spectra of $La_{1-x}Sr_xMnO_3$ for various compositions. (b) Photoemission spectra and O 1 s XAS spectra of $La_{1-x}Sr_xMnO_3$ near E_F.[3] The photon energies for the XAS spectra are measured relative to the O 1s binding energy (530−531 eV) of each sample.

have been made by Chainani *et al*,[9] Abbate *et al*,[59] and Saitoh *et al.*[3] Figure 8(a) shows the ultraviolet photoemission (UPS) spectra of $La_{1-x}Sr_xMnO_3$.[3] The leading edge is shifted towards E_F with hole doping, while the spectral weight at the weak tail (tail C) of the leading edge (edge A) near E_F, a weak tail which is attributed to occupied $e_{g\uparrow}$ states, continuously loses its intensity and for x=0.9 the tail almost disappears. Combined UPS and O 1s XAS spectra near E_F are displayed in Fig. 8(b).[3] In order to highlight the changes of the low energy features with x, the appropriately shifted x=0.9 spectrum has been subtracted from the other spectra. The difference spectra thus represents the emission of $e_{g\uparrow}$ electrons; for $LaMnO_3$ the occupied $e_{g\uparrow}$ band is centered $\sim$ 1.7 eV below E_F. The intensity of the $e_{g\uparrow}$ emission decreases with Sr-substitution. The energy shift of the $e_{g\uparrow}$ emission is parallel to that of the leading edge and is very slow in the metallic phase. The same behavior of the depletion and shift of the $e_{g\uparrow}$ weight has been observed for $La_{1-x}Ca_xMnO_3$ by J. H. Park *et al.*[31] These authors observed a linear Fermi-level shift up to the Ca end, which behavior is different from that reported by Saitoh *et al.* However, the amount of the shift of 0.7−0.8 eV from d^4 to d^3 is similar to each other.

It should be noted that the intensity near and at E_F is unusually weak and

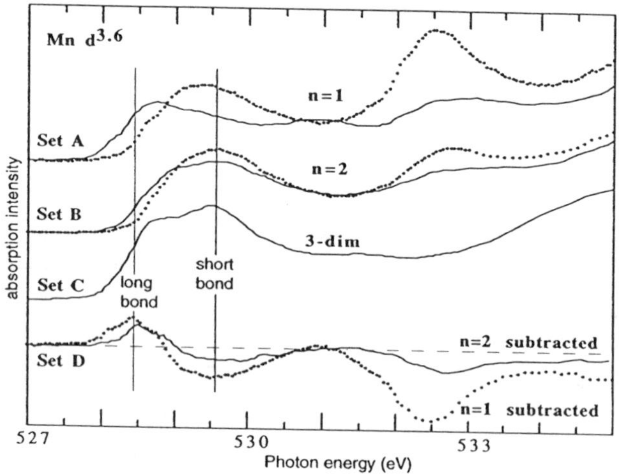

Figure 9: O 1s XAS spectra of the 214 (n=1), 327 (n=2) and 113 (3-dim) compounds with x=0.4. The solid curves and the dotted curves are grazing and normal incidence measurements, respectively.[62]

therefore that the photoemission studies with conventional energy resolution (150−350 meV) were not able to observe a clear Fermi edge [see Fig. 8(a)]. The $e_{g\uparrow}$ intensity is enhanced to some extent in the spectra taken at higher incident photon energies $h\nu$=500−1300 eV.[9,3,31] Nevertheless, in all the spectra taken with various photon energies, the intensity at E_F is *always* very weak.

As for the O 1s XAS spectra [Fig. 8(b)], the relatively broad structure centered at 0.5 eV for $LaMnO_3$ is due to Mn 3d-derived states. This broad structure has been resolved into two features in a recent report.[31] With hole doping into $LaMnO_3$, the broad structure grows into an intense peak, corresponding to the increase in the number of unoccupied states. In addition, a shoulder structure appears on the lower energy side of the peak, but finally almost disappears at the Sr end. A result consistent with this has been recorded in the fluorescence yield mode.[60] Very similar doping dependence has been observed in $La_{1-x}Ca_xMnO_3$.[31] They have interpreted the first feature of the x=0 spectrum as hole-induced spectral weight due to excess oxygens and the second feature as the $t_{2g}^3e_g^2$ ($^6A_1$) final state. Ca doping develops the first feature into an intense peak whose position shifts towards lower photon energies. Also the shoulder structure has been observed.

Pellegrin *et al.*[61] have measured the soft x-ray magnetic circular dichroism (SXMCD) spectra of $La_{1-x}Sr_xMnO_3$ (x=0.18 and 0.4). They have also observed the two features in the first structure, and assigned the lower energy

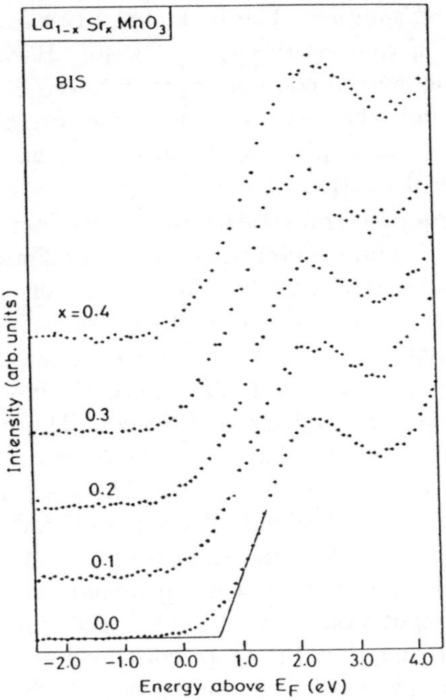

Figure 10: BIS spectra of $La_{1-x}Sr_xMnO_3$ near E_F.[9]

feature to the $e_{g\uparrow}$ states and the higher energy one to the $t_{2g\downarrow}$ states. Because the MCD signal of each feature shows little change with doping, the origin of each feature would not change with doping. Considering the peak and shoulder positions and their doping dependence, the two features which they assigned to the $e_{g\uparrow}$ and $t_{2g\downarrow}$ states in the x=0.4 spectrum should be identical to the two structures observed by Abbate *et al.*,[59] J. H. Park *et al.*[31] and recently by C.-H. Park *et al.*[62] In fact, the interpretation by Pellegrin *et al.* is consistent with that of $CaMnO_3$ by J. H. Park *et al.*

On the other hand, C.-H. Park *et al.*[62] have recently proposed another interpretation of those two features. They have investigated the O 1*s* XAS spectra of the 214, 327 and the 113 compounds with the same hole doping concentration of x=0.4 as shown in Fig. 9.[62] From the polarization dependence of the layered compounds, they have determined the character of the unoccupied orbitals of those samples: The first of the two features is the $d_{3z^2-r^2\uparrow}$ state and the second of them is the $d_{x^2-y^2\uparrow}$ state. Because of the similarity with the spectra of the layered compound, they inferred that the two features located at 528.7 eV and 529.5 eV have the same origin as those of the layered compound, namely, the splitting of those two peaks was attributed to a dynamical Jahn-Teller distortion.

Although XAS is a powerful tool to investigate unoccupied states, it has shortcomings that one cannot determine the exact Fermi-level position and that the spectral line shapes may be distorted by the core-hole potential. These shortcomings are overcome by inverse-photoemission spectroscopy (IPES) or bremsstrahlung isochromat spectroscopy (BIS) although the resolution and the signal-to-noise ratio are lower than XAS. Figure 10 shows the BIS spectra of $La_{1-x}Sr_xMnO_3$ by Chainani, Mathew and Sarma.[9] The structure at about 2.4 eV is due to Mn $3d$ states hybridized with O $2p$ states and corresponds to the first structure in the O $1s$ XAS spectra. The steep rise at higher energy is the tail of the intense La $4f$ signal. They found that the peak position of the first structure did not shift within the experimental uncertainty while the intensity of the edge located around 1 eV systematically increased with hole doping.[9] This extra weight would correspond to the shoulder structure in the XAS spectra. Considering that the O $1s$ core-level shift due to hole doping is basically monotonic towards lower binding energies,[3,31] the energy shifts of the XAS and BIS spectra are mutually consistent. Here we note that inverse photoemission in ultraviolet region (UV-BIS) has better resolution and a larger photoionization cross section for the Mn $3d$ states than x-ray BIS mentioned above, but there have been only few UV-BIS works on the current systems so far.[63,64]

Now we go back to the UPS and O $1s$ XAS spectra shown in Fig. 8(b). The figure clearly demonstrates spectral weight transfer from the unoccupied to the occupied states in going from x=0.9 to x=0.0 with *electron* doping. Starting from $SrMnO_3$ (d^3), introduced electrons will occupy the $e_{g\uparrow}$ orbitals and their spectral weight appears in the band gap of $SrMnO_3$. In the O $1s$ XAS spectra, at the same time, the intensity of the large peak due to the $e_{g\uparrow}$ band just above E_F decreases.[b] These behaviors are quite different from what would be expected from the rigid-band model, which would predict an upward shift of the Fermi-level without changes in the spectral line shape, and hence a sharp Fermi cutoff in the doped samples.

---

[b]It should be noted, however, that the decrease in the intensity of the unoccupied $e_{g\uparrow}$ states is not entirely due to the decrease in the number of the unoccupied states but is also due to the decrease in covalency between Mn $3d$ and O $2p$ associated with the increase in $\Delta_{\rm eff}$.

In analogy with the O 1s XAS spectra of the high-T_C cuprates, one would expect that $e_{g\uparrow}$–like "in-gap" spectral weight will appear in the unoccupied states within the bandgap of $LaMnO_3$ with hole doping. Although this is not obvious in the O 1s XAS spectra of $La_{1-x}Sr_xMnO_3$, the shoulder structure at the absorption edge would correspond to the in-gap spectral weight. In the case of $La_{1-x}Sr_xFeO_3$, the in-gap spectral weight corresponds to the $t_{2g\uparrow}^3 e_{g\uparrow}^2$ (d^5 high-spin) final state, which is strongly stabilized due to the strong Hund's rule coupling in the d^5 high-spin configuration.[5] Thus the in-gap spectral weight is located well below the bottom of the conduction band of $LaFeO_3$ so that the peak is clearly visible. In $La_{1-x}Sr_xMnO_3$, the in-gap spectral weight would be located near the bottom of the conduction band of $LaMnO_3$ because the bottom of the conduction band in $LaMnO_3$ ($t_{2g\uparrow}^3 e_{g\uparrow}^2$) has already been stabilized by Hund's rule coupling. Along this line, the two features in the $LaMnO_3$ spectrum are considered as the $t_{2g\uparrow}^3 e_{g\uparrow}^2$ and $t_{2g\uparrow}^3 t_{2g\downarrow}^1 e_{g\uparrow}^1$ final states, respectively. Upon hole doping, the shoulder structure grows and at the d^3 end, the shoulder and the peak structures are considered as the $t_{2g\uparrow}^3 e_{g\uparrow}^1$ and $t_{2g\uparrow}^3 t_{2g\downarrow}^1$, respectively.

4.2 Temperature Dependence

The very small spectral weight in the photoemission spectra of $La_{1-x}A_xMnO_3$ near E_F has recently been investigated using high-resolution photoemission measurements at various temperatures. Figure 11 shows high-resolution UPS spectra of $La_{0.6}Sr_{0.4}MnO_3$ near E_F taken below $T_C \sim 350$ K [32] by Sarma *et al.*[30] At the lowest temperature, both He I ($h\nu$=21.2 eV) and He II ($h\nu$=40.8 eV) spectra exhibit a clear Fermi cutoff. With increasing temperature, the intensity from E_F to ~ 0.8 eV below it decreases systematically. Since the Mn 3d-to-O 2p relative cross section increases with photon energy,[65] the higher E_F intensity in the He II spectra indicates that the Mn 3d contribution to the near-E_F emission is larger than the O 2p contribution. The decrease in the near E_F intensity has also been reported for $La_{0.65}Ca_{0.35}MnO_3$ by McIlroy *et al.*, but the Fermi cut-off was not clearly observed in their spectra due to low resolution.[66] J. H. Park *et al.*[31] have also studied the temperature-dependent spectra of $La_{0.67}Ca_{0.33}MnO_3$ (T_C=260 K) and $La_{0.7}Pb_{0.3}MnO_3$ (T_C=330 K) and found a systematic depletion with increasing temperature in the both samples, similar to the case of $La_{0.6}Sr_{0.4}MnO_3$. Since they used 110 eV incident photon energy, the near-E_F intensity and its depletion is clearer than that reported by Sarma *et al.*

The systematic decrease in the intensity at E_F with temperature is a very unusual phenomena since in conventional metals the Fermi edge is simply broadened with temperature and spectral weight integrated near E_F is con-

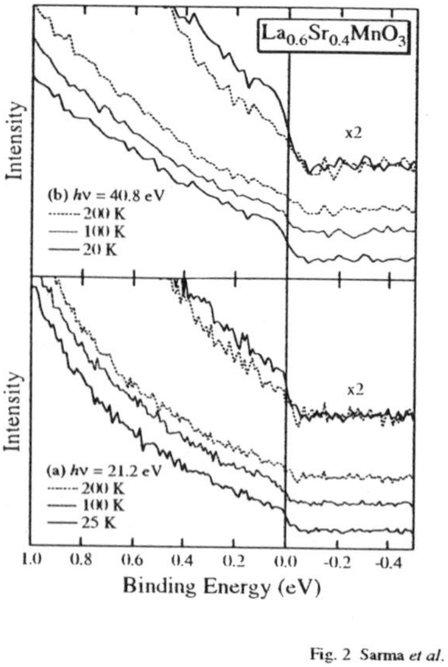

Figure 11: High-resolution photoemission spectra of $La_{0.6}Sr_{0.4}MnO_3$ near E_F.[30]

served. Furthermore, the spectral-weight depletion is not limited to the energy scale of T_C (~30 meV) but is spread over ~0.8 eV. In order to explain this unusual experimental result, Sarma *et al.* [30] have proposed that U increases and the $e_{g\uparrow}$ band width W decreases with temperature, resulting in an increase in U/W and hence increase in the incoherent spectral weight with temperature.[67]

Sarma *et al.* [30] and subsequently Saitoh *et al.* [68] have reported a strong temperature-dependent change of the whole valence-band spectra of $La_{1-x}Sr_xMnO_3$ as shown in Fig. 12. The intensity of the peak at ~3.5 eV increases and that at ~6 eV decreases with increasing temperature. For comparison, a similar compound $La_{0.6}Sr_{0.4}CoO_3$ has been studied and has not shown such a temperature-dependent change. The redistribution of spectral weight on the unusually large energy scale (~6 eV) is consistent with recent optical measurements.[69,70] The charge in the photoemission spectra with *decreasing* temperature is similar to the change induced by hole doping. Since both hole doping and decreasing temperature tend to delocalize the hole carriers in $La_{1-x}Sr_xMnO_3$, the observed changes probably reflect changes in the degree of carrier localization.

The compounds discussed so far are mostly typical double-exchange systems. On the other hand, there are another family of manganites which show

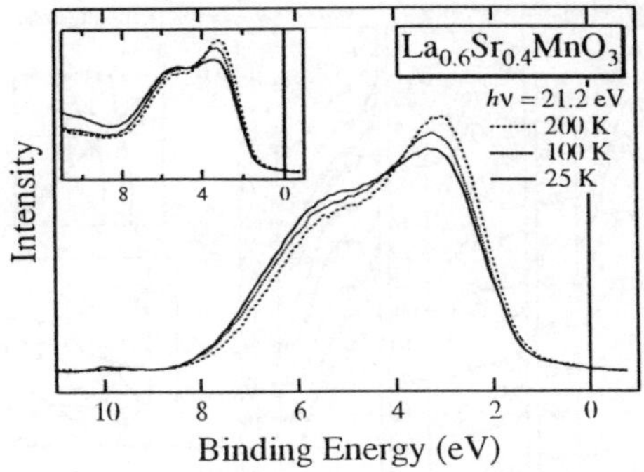

Figure 12: Temperature-dependent photoemission spectra of $La_{0.6}Sr_{0.4}MnO_3$ in the whole valence-band region.[30]

charge ordering as discussed in Sec. 3. Suga *et al.*[71] have investigated one of those charge-ordered manganites $Nd_{0.5}Sr_{0.5}MnO_3$, which has a Curie temperature of T_C ~250 K and a charge-ordering temperature of T_{CO} ~160 K.[72] The intensity near E_F monotonically decreases on cooling and correspondingly the Fermi-edge intensity also decreases, which is opposite to what has been observed for $La_{1-x}A_xMnO_3$. The Fermi-edge intensity is already very weak at the room temperature and almost disappears at 202 K. At 96 K, no detectable Fermi-edge intensity is observed.

4.3 Dispersive Energy Bands

Angle-resolved photoemission spectroscopy (ARPES) is the unique method to investigate dispersive energy bands of occupied states in solids. Unfortunately, it is technically difficult in many cases to perform ARPES measurements on three-dimensional compounds.[73] On the other hand, layered compounds are more suitable for ARPES measurements. Recently, Dessau *et al.*[74] have studied the electronic structure of $La_{1.2}Sr_{1.8}Mn_2O_7$ using high-resolution ARPES. In Fig. 13, one observes a clear dispersive band along the $(0,0)\rightarrow(\pi,0)\rightarrow(\pi,\pi)$ symmetry line [panels (b) and (c)]. Comparing the data with the LDA+U band structure calculations [panel (g)], the dispersive band is interpreted as the $e_{g\uparrow}$ band with dominant $d_{x^2-y^2}$ character.[62] Since the compound is metallic, one may expect a Fermi-surface (FS) crossing probably near the calculated Fermi surface. However, the spectra show no intensity at the FS crossing points. Instead, the peak reaches a minimum binding energy near the expected

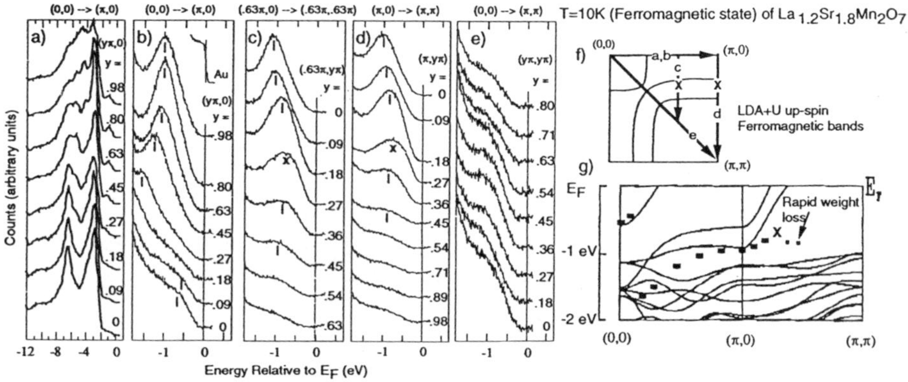

Figure 13: **Angle-resolved photoemission spectra of $La_{1.2}Sr_{1.8}Mn_2O_7$.[74] (a) Full valence-band spectra along the (0,0)→(π,0). (b)−(e) Near-E_F spectra along various directions, as indicated at the top of each panel and by the arrows along the Brillouin zone of panel (f). (g) The up-spin bandsin an LDA+U band-structure calculations vs. experimentally determined peak positions from panels (b) and (d).**

FS crossing point, beyond which the spectra exhibit a rapid weight loss and evidence of shifting away from E_F as if a gap were opened. This "pseudo-gap" behavior has been observed in the vicinity of all the expected FS crossing points explored in the measurements. The pseudo-gap can be considered due to very weak coherent weight of the spectral function and from this view point, Dessau *et al.* have explained the spectral line shapes using a strongly coupled electron-phonon model.

4.4 Comparison with Optical Spectra

Optical spectra of Mn oxides also show interesting doping and temperature dependence [29,69,70] which should be closely related to those of the photoemission spectra. Figure 14 shows the doping and temperature dependence of the optical conductivity spectra of $La_{1-x}Sr_xMnO_3$.[29] At x=0.1, the spectra show insulating behavior at every temperature while the spectra for x=0.175 and 0.3 develop Drude-like spectral weight with decreasing temperature. This spectral weight is transferred from higher energies with decreasing temperature, result-

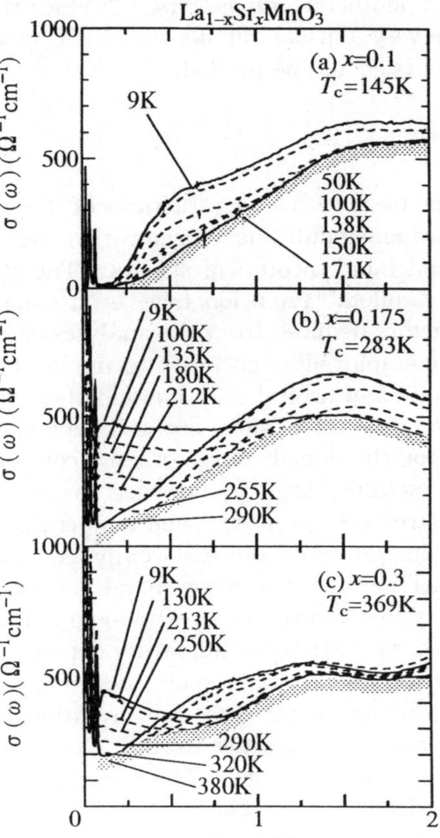

Figure 14: Optical conductivity spectra of $La_{1-x}Sr_xMnO_3$.[29]

ing in the spectral-weight transfer in a large energy scale of at least ~2 eV. This observation is in agreement with the spectral-weight transfer to near E_F in the photoemission spectra. It has recently been found that the temperature-dependent change in the optical conductivity spectra is actually spread over 3–4 eV.[69,70] This may be consistent with the temperature-dependent change in the whole valence-band photoemission spectra.[30,68] Since the changes on such a large energy-scale are very unusual, further investigation with photoemission and optical measurements should be needed.

5 Conclusion

We have shown that the basic electronic structures of the perovskite-type and related layered-structure manganites have been rather well understood by the recent spectroscopic and band-theoretical studies. The strong Mn 3*d*-O 2*p* hybridization and *d*-*d* Coulomb repulsion have been confirmed by the electronic structure parameters deduced from the analyses of spectroscopic data. If these parameters are employed as an input to model-Hamiltonian studies, the complicated magnetic and orbital ordering and their coupling to the JT-type and $GdFeO_3$-type lattice distortions are well explained for the undoped compounds as well as for the doped charge-ordered compounds. The nature of the doping-induced metallic states is, however, far from well understood. In particular, their spectroscopic properties and temperature dependence from low to high energy scales are quite unusual compared not only to ordinary metals but also to other metallic transition-metal oxides: (i) the density of states at E_F is unusually low and shows a pseudo-gap behavior on the energy scale of ~1 eV; (ii) it shows a strong temperature dependence in the metallic phase; and (iii) the entire valence band shows a strong temperature dependence. As a possible mechanism for these unconventional behaviors, strong electron-phonon interaction,[75] instabilities towards charge ordering and/or orbital ordering, etc., have been proposed. The understanding of the spectroscopic properties would lead to a fundamental understanding of the unusual magnetotransport properties of the manganites. Fortunately, there exit a variety of materials in the manganite series. Systematic and detailed studies of these materials would give more insight into their unique physical properties.

Acknowledgment

We would like to acknowledge fruitful collaboration and discussions with M. Takano, Y. Takeda, M. Abbate, D. D. Sarma, S. Suga, D. S. Dessau, C.-H. Park, Y. Tokura, H. Kuwahara, Y. Moritomo, T. Katsufuji, N. Hamada and

H. Yoshizawa. The present work is supported by a Grant-in-Aid for Scientific Research from the Ministry of Education, Science and Culture. T.S. acknowledges the hospitality of D.S. Dessau at the University of Colorado, and funding through the Office of Naval Research Young Investigator Grant N00014-96-1-0768.

References

1. R. von Helmolt, J. Wecker, B. Holzapfel, L. Schultz, and K. Samwer, *Phys. Rev. Lett.* **71**, 2331 (1993).
2. Y. Tokura, A. Urushibara, Y. Moritomo, T. Arima, A. Asamitsu, G. Kido, and N. Furukawa, *J. Phys. Soc. Jpn.* **63**, 3931 (1994).
3. T. Saitoh, A. E. Bocquet, T. Mizokawa, H. Namatame, A. Fujimori, M. Abbate, Y. Takeda, and M. Takano, *Phys. Rev.* B **51**, 13942 (1995).
4. A. E. Bocquet, T. Mizokawa, T. Saitoh, H. Namatame, and A. Fujimori, *Phys. Rev.* B **46**, 3771 (1992).
5. A. Fujimori, A. E. Bocquet, T. Saitoh, and T. Mizokawa, *J. Electron Spectrosc. Relat. Phenom.* **62**, 141 (1993).
6. T. Mizokawa, H. Namatame, A. Fujimori, K. Akeyama, H. Kondoh, H. Kuroda, and N. Kosugi, *Phys. Rev. Lett.* **67**, 1683 (1991).
7. T. Arima and Y. Tokura, *J. Phys. Soc. Jpn.* **64**, 2488 (1995).
8. F. C. Zhang and T. M. Rice, *Phys. Rev.* B **37**, 3759 (1988).
9. A. Chainani, M. Mathew, and D. D. Sarma, *Phys. Rev.* B **47**, 15397 (1993).
10. E. O. Wollan and W. C. Koehler, *Phys. Rev.* **100**, 545 (1955);
11. K. I. Kugel and D. I. Khomskii, *Sov. Phys. Usp.* 25, 231 (1982).
12. J. B. A. A. Elemans, B. van Laar, K. R. van der Veen, and B. O. Loopstra, *J. Solid State Chem.* **3**, 238 (1971).
13. J. B. Goodenough, *Phys. Rev.* **100**, 564 (1955).
14. J. Kanamori, *J. Phys. Chem. Solids*, **10**, 87 (1959).
15. J. B. Goodenough, A. Wold, R. J. Arnott, and N. Menyuk, *Phys. Rev.* **124**, 373 (1961).
16. N. Hamada, H. Sawada, and K. Terakura, in *Spectroscopy of Mott Insulators and Correlated Metals*, edited by A. Fujimori and Y. Tokura (Springer-Verlag, Berlin, 1995).
17. I. V. Solovyev, N. Hamada, and K. Terakura, *Phys. Rev. Lett.* **76**, 4825 (1996).
18. W. E. Picket and D. J. Singh, *Phys. Rev.* B **53**, 1146 (1996).
19. S. Satpathy, Z. S. Popović and F. R. Vukajlović *Phys. Rev. Lett.* **76**, 960 (1996).

20. T. Mizokawa and A. Fujimori, *Phys. Rev.* B **51**, 12880 (1995).
21. T. Mizokawa and A. Fujimori, *Phys. Rev.* B **54**, 5368 (1996).
22. T. Mizokawa, T. and A. Fujimori, *Phys. Rev.* B **53**, R4201 (1996).
23. W. Koshibae, Y. Kawamura, S. Ishihara, S. Okamoto, J. Inoue, and S. Maekawa, *J. Phys. Soc. Jpn.* **66**, 957 (1997).
24. K. Hirota, N. Kaneko, A. Nishizawa, and Y. Endoh, *J. Phys. Soc. Jpn.* **65**, 3736 (1996).
25. S. Ishihara, J. Inoue, and S. Maekawa, *Phys. Rev.* B **55**, 8280 (1997).
26. R. Maezono, S. Ishihara, and N. Nagaosa, unpublished.
27. H. Chiba, T. Atou, and Y. Syono, *J. Solid State Chem.*, **132**, 139 (1997).
28. S. Ishihara, M. Yamanaka, and N. Nagaosa, *Phys. Rev.* B **56**, 686 (1997).
29. Y. Okimoto, T. Katsufuji, T. Ishikawa, A. Urushibara, T. Arima, and Y. Tokura, *Phys. Rev. Lett.* **75**, 109 (1995); Y. Okimoto, T. Katsufuji, T. Ishikawa, T. Arima, and Y. Tokura, *Phys. Rev.* B **55**, 4206 (1997).
30. D.D. Sarma, N. Shanthi, S.R. Krishnakumar, T. Saitoh, T. Mizokawa, A. Sekiyama, K. Kobayashi, A. Fujimori, E. Weschke, R. Meier, G. Kaindl, Y. Takeda and M. Takano, *Phys. Rev.* B **53**, 6873 (1996).
31. J.-H. Park, C. T. Chen, S.-W. Cheong, W. Bao, G. Meigs, V. Chakarian, and Y. U. Idzerda, *Phys. Rev. Lett.* **76**, 4215 (1996).
32. A. Urushibara, Y. Moritomo, T. Arima, A. Asamitsu, G. Kido, and Y. Tokura, *Phys. Rev.* B **51**, 14103 (1995).
33. Z. Jirak, S. Krupicka, Z. Simsa, M. Dlouha, and S. Vratislav, *J. Mag. Mag. Mat.* **53**, 153 (1985).
34. Y. Tomioka, A. Asamitsu, Y. Moritomo, H. Kuwahara, and Y. Tokura, *Phys. Rev. Lett.* **74**, 5108 (1995).
35. T. Vogt, A. K. Cheetham, R. Mathendiran, A. K. Raychaudhuri, R. Mahesh, and C. N. R. Rao, *Phys. Rev.* B **54**, 15303 (1996).
36. P. G. Radaelli, D. E. Cox, M. Marezio, S-W. Cheong, P. E. Schiffer, and A. P. Ramirez, *Phys. Rev. Lett.* **75**, 4488 (1995).
37. P. G. Radaelli, D. E. Cox, M. Marezio, S-W. Cheong, *Phys. Rev.* B **55**, 3015 (1997).
38. C. H. Chen and S.-W. Cheong, *Phys. Rev. Lett.* **76**, 4042 (1996).
39. Y. Tokura, H. Kuwahara, Y. Moritomo, Y. Tomioka, and A. Asamitsu, *Phys. Rev. Lett.* **74**, 5108 (1995).
40. Y. Tokura, H. Kuwahara, Y. Moritomo, Y. Tomioka, and A. Asamitsu, *Phys. Rev. Lett.* **76**, 3184 (1996).
41. H. Kawano, R. Kajimoto, H. Yoshizawa, Y. Tomioka, H. Kuwahara, and Y. Tokura, *Phys. Rev. Lett.* **78**, 4253 (1997).
42. Y. Tomioka, A. Asamitsu, Y. Moritomo, H. Kuwahara, and Y. Tokura, *Phys. Rev. Lett.* **74**, 5108 (1995).

43. V. I. Anisimov, I. S. Elfimov. M. A. Korotin, and K. Terakura, *Phys. Rev.* B **55**, 15494 (1997).
44. T. Mizokawa and A. Fujimori, *Phys. Rev.* B **56**, R493 (1997).
45. H. Kawano, R. Kajimoto, K. Kubota, and H. Yoshizawa, *Phys. Rev.* B **53**, R14709 (1996).
46. Y. Yamada, O. Hino, S. Nohdo, R. Kanao, T. Inami, and S. Katano, *Phys. Rev. Lett.* **76**, 904 (1996).
47. H. Yoshizawa, H. Kawano, Y. Tomioka, and Y. Tokura, *Phys. Rev.* B **52**, R13141 (1995).
48. J. M. De Teresa, M. R. Ibarra, C. Marquina, P. A. Algarabel, and S. Oseroff, *Phys. Rev.* B **54**, R12689 (1996).
49. H. Yoshizawa, R. Kajimoto, H. Kawano, Y. Tomioka, Y. Tokura, *Phys. Rev.* B **55**, 2729 (1997).
50. A. P. Ramirez, P. Schiffer, S.-W. Cheong, C. H. Chen, W. Bao, T. T. M. Palstra, P. L. Gammel, D. J. Bishop, and B. Zagarski, *Phys. Rev. Lett.* **76**, 3188 (1996).
51. C. H. Chen, S.-W. Cheong, and H. Y. Hwang, *J. Appl. Phys.* **81**, 4326 (1997).
52. W. Bao, J. D. Axe, C. H. Chen, and S.-W. Cheong, *Phys. Rev. Lett.* **78**, 543 (1997).
53. Y. Murakami, D. Shindo, H. Chiba, M. Kikuchi, and Y. Syono, *Phys. Rev.* B **55**, 15043 (1997).
54. C. H. Chen, S.-W. Cheong, and A. S. Cooper, *Phys. Rev. Lett.* **71**, 2461 (1993); S. M. Hayden, G. H. Lander, J. Zarestky, P. J. Brown, C. Stassis, P. Metcalf, and J. M. Honig, *Phys. Rev. Lett.* **68**, 1061 (1992); V. Sachan, D. J. Buttrey, J. M. Tranquada, J. E. Lorenzo, and G. Shirane, *Phys. Rev. Lett.* **51**, 12742 (1993).
55. Y. Moritomo, Y. Tomioka, A. Asamitsu, Y. Tokura, and Y. Matsui, *Phys. Rev.* B **51**, 3297 (1995).
56. B. J. Sternlieb, J. P. Hill, U. C. Wildgruber, G. M. Luke, B. Nachumi, Y. Moritomo, and Y. Tokura, *Phys. Rev. Lett.* **76**, 2169 (1996).
57. W. Bao, C. H. Chen, S. A. Carter, and S.-W. Cheong, *Solid State Commun.* **98**, 155 (1996).
58. Y. Murakami *et al.*, unpublished.
59. M. Abbate, F. M. F. de Groot, J. C. Fuggle, A. Fujimori, O. Strebel, F. Lopez, M. Domke, G. Kaindl, G. A. Sawatzky, M. Takano, Y. Takeda, H. Eisaki, and S. Uchida, *Phys. Rev.* B **46**, 4511 (1992).
60. D. D. Sarma, O. Rader, T. Kachel, A. Chainani, M. Mathew, K. Holldack, W. Gudat, and W. Eberhardt, *Phys. Rev.* B **49**, 14238 (1994).

61. E. Pellegrin, L. H. Tjeng, F. M. F. de Groot, R. Hesper, C. F. J. Flipse, J. D. O'Mahony, Y. Moritomo, Y. Tokura, C. T. Chen, and G. A. Sawatzky, unpublished.
62. C.-H. Park, D. S. Dessau, T. Saitoh, Z.-X. Shen, Y. Moritomo, and Y. Tokura, unpublished.
63. S. Suga, S. Imada, T. Muro, T. Fukawa, T. Shishidou, Y. Tokura, Y. Moritomo, and T. Miyahara, *J. Electron Spectrosc. Relat. Phenom.* A **78**, 283 (1996).
64. D. N. McIlroy, C. Waldfried, Jindai Zhang, J.-W. Choi, F. Foong, S. H. Liou, and P. A. Dowben, *Phys. Rev.* B **54**, 17438 (1996).
65. J. J. Yeh and I. Lindau, *At. Data Nucl. Data Tables* **32**, 1 (1985).
66. D. N. McIlroy, J. Zhang, S.-H. Liou, and P.A. Dowben, *Physics Lett.* A **207**, 367 (1995).
67. A. Fujimori, I. Hase, H. Namatame, Y. Fujishima, Y. Tokura, H. Eisaki, S. Uchida, K. Takegahara, F. M. F. de Groot, *Phys. Rev. Lett.* **69**, 1796 (1992).
68. T. Saitoh, A. Sekiyama, K. Kobayashi, T. Mizokawa, A. Fujimori, D. D. Sarma, Y. Takeda, and M. Takano, *Phys. Rev.* B **56**, 8836 (1997).
69. S. G. Kaplan, M. Quijada, H. D. Drew, D. B. Tanner, G. C. Xiong, R. Ramesh, C. Kwon, and T. Venkatesan, *Phys. Rev. Lett.* **77**, 2081 (1996).
70. Y. Moritomo, A. Machida, K. Matsuda, M. Ichida, and A. Nakamura, *Phys. Rev.* B **56**, 5088 (1997).
71. S. Suga, T. Matsushita, M. Tsunekawa, T. Iwasaki, J. Samejima, Y. Kuwata, A. Kimura, A. Sekiyama, A. Fujimori, H. Ishii, T. Miyahara, H. Kuwahara, and Y. Tokura, *Physica* B, **237-238**, 413 (1997).
72. H. Kuwahara, Y. Tomioka, A. Asamitsu, Y. Moritomo, and Y. Tokura, *Science* **270**, 961 (1995).
73. There are a few ARPES studies on 113 compounds of thin films with conventional resolution [J. Zhang, D. N. McIlroy, P. A. Dowben, S. H. Liou, R. F. Sabirianov, and S. S. Jaswal, *Solid State Commun.* **97**, 39 (1996), C. Waldfried, D. N. McIlroy, S.-H. Liou, R. Sabiryanov, S. S. Jaswal, and P. A. Dowben, *J. Phys. Condens. Matter* **9**, 1031 (1997)].
74. D. S. Dessau, C.-H. Park, T. Saitoh, Z.-X. Shen, P. Villella, N. Hamada, Y. Moritomo, and Y. Tokura, unpublished.
75. A. J. Millis, P. B. Littlewood, and B. I. Shrainman, *Phys. Rev. Lett.* **74**, 5144 (1995).

SPIN-VALVE EFFECT IN MANGANESE-OXIDE PEROVSKITES: CMR

N. KUMAR
Raman Research Institute, Bangalore 560 080, and
Jawaharlal Nehru Centre for Advanced Scientific Research
Jakkur, Bangalore 560064, India

G.V. VIJAYAGOVINDAN
Department of Physics, Mahatma Gandhi University, Kottayam 686560, India

PRABHAKAR PRADHAN
Jawaharlal Nehru Centre for Advanced Scientific Research
Jakkur, Bangalore 560064, India

Abstract

A minimum model for manganese oxide perovskites is considered in the limit of infinite Hund's rule coupling when the spin of the mobile charge carrier is enslaved to the background ionic spins which are nominally coupled antiferromagnetically. This reduces to a spin-valve Hamiltonian describing the now effectively spinless electrons (holes in fact) whose intersite tunneling favours and is favoured by a ferromagnetic alignment of the background spins, taken to be classical and Ising-like. This coupling of the dynamics of the spinless electrons to the statistical mechanics of the background Ising spins describes several observed features of the phase diagram, in the limit of large one-electron bandwidth. Magnetic field and temperature dependence of the short range ferromagnetic order confining the electron in the paramagnetic matrix, and its relation to the thermal expansion anomaly observed above the curie temperature is also discussed. Mechanisms for the ionic-spin defined resistivity in the FM phase and for the Colossal Magneto-Resistance (CMR) near the PMI-to-FMM transition are also discussed.

1. Introduction

A class of manganese based transition-metal mixed-crystal oxides with the cubic perovskite-like structure is known to undergo an Insulator (I)-to-Metal (M) transition as the temperature is lowered through a critical value T_c at which, concomitantly, the material also undergoes a thermodynamic transition from the paramagnetic (PM) to the ferromagnetic (FM) phase.[1] The most striking feature of this PMI-to-FMM transition is the recently discovered negative magnetoresistance associated with it, namely that in the neighbourhood of the transition (Curie) temperature T_c, the resistivity drops dramatically, by several orders of magnitude on application of an external magnetic field of a few teslas $\sim$ the internal exchange field. This abnormally large negative magnetoresistance has been termed the Colossal Magneto-Resistance, or CMR for short.[2-5] The CMR is intrinsic and isotropic, and has been observed in polycrystalline films[2-4] as well as in single crystals.[5] The recent *oxide rush* derives from the very unusual magnetic and electronic properties, and their correlations, observed in these materials raising many fundamental questions, as also clearly from the great technological potential of CMR as an ultrasensitive magnetoresistive sensor even as lower operating fields and higher operating temperatures closer to the room temperature are being reported.

The class of CMR oxides has the general chemical formula $RE_{1-x}AE_xMnO_3$ where RE is a trivalent rare-earth (RE = $La^{3+}, Y^{3+}, Tb^{3+} \cdots$) and AE is a divalent alkaline-earth (AE = $Sr^{2+}, Ca^{2+}, Ba^{2+}$). These have relatively larger ionic radii and occupy the cube corners, i.e., the A-sites. The transition-metal ion Mn^{3+}/Mn^{4+} and its possible substitutions ($Cr^{3+}, Fe^{3+}, T_i^{4+}$) are all of relatively smaller radii and occupy the cube centres, the B-sites. The prototype is the manganese-oxide perovskite $LaMnO_3$, shown in Figure 1, and the prototypical CMR composition is $La_{1-x}Sr_xMnO_3$ with $x \approx 0.33$ and a resistance ratio $(\rho(0) - \rho(H))/\rho(H)$ reported[2] to be as high as $10^5\%$ for a field $H \sim 6T$. The CMR should be distinguished from GMR (Giant Magnetoresistance)[6,7] – GMR is a relatively small effect, typically with $\Delta\rho(H)/\rho(H) \sim$ 5-200%; CMR is observed in intrinsic materials, the mixed manganite perovskites while GMR is associated with atomically engineered or sputtered metallic superlattices of magnetic layers alternating with non-magnetic layers with the ultrathin layer

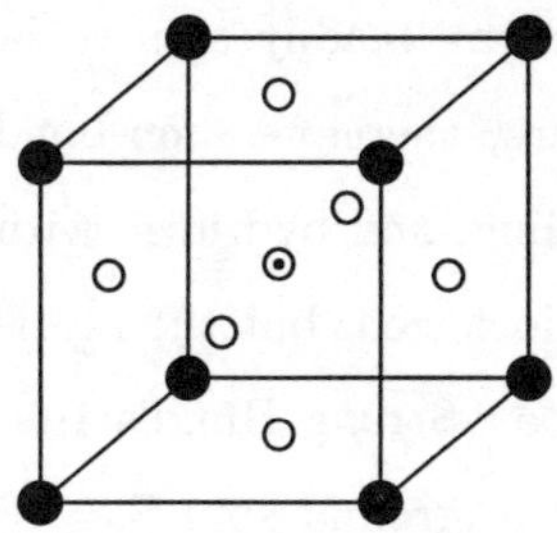

Figure 1: Conventional unit cell of cubic perovskite $LaMnO_3$.

thickness $\sim$ 100 Å, e.g., Co-Cu, Fe-Cr, Fe-Ni, etc.; CMR is observed in the critical neighbourhood of a PMI-to-FMM transition and almost certainly involves *spin-defined tunneling* while GMR is a non-critical effect involving *spin-defined scattering*; and finally CMR is non-saturable up to several teslas while GMR saturates at a relatively low field $\sim 0.1\,T$. The two may, however, have an underlying commonality, in that both involve spin-polarized currents, more so if we consider GMR with current flow perpendicular to the multilayer.[8] Essential to a proper understanding of CMR, its magnitude and its causal correlation with the PMI-to-FMM transition, is the fact of entrainment of the background ionic (t_{2g}^3)-spin orientations by the valence (e_g^1)-electron translation which favours and is in turn favoured by the ionic spins thus oriented. This background spin-defined tunneling is referred to as the spin-valve effect.[9] It is inherent to the electronic structure of these manganites having large Hund's rule coupling. In order to fix ideas, the main structural feature of these manganese-oxide perovskites may be isolated as[1] the $Mn^{3+} - O^{2-} - Mn^{3+}$ unit with the $Mn^{3+}(3d^4)$ at the cube centre, crystal-field split to a ground-state configuration $t_{2g}^3 e_g^1$.

The t_{2g}-orbitals (xy, yz, zx) are readily seen to be so directed as to poorly hybridize with O 2p and are, therefore, non-bonding, while the e_g-orbitals $(x^2 - y^2, 3z^2 - r^2)$ are bonding, and hydridize with the ligand O 2p orbitals. Thus, the t_{2g}^3 electrons are localized, but the e_g^1 electron may be itinerant or localized as the case may be. Strong Hund's rule coupling ($\gg$ crystal-field 10 Dq) then gives a local background spin S = 3/2 for the t_{2g}^3 state, while the spin 1/2 of the e_g^1-state must be aligned parallel to this local background (t_{2g}^3)-spin giving a total Mn^{3+}-ionic spin of 2. The two-fold orbital degeneracy of e_g^1 makes Mn^{3+} a Jahn-Teller (JT)-active ion, while $Mn^{4+} : t_{2g}^3 e_g^o$ is clearly JT-inactive. There is a single-ion JT-stabilization energy, denoted by E_{JT}. Now, the $Mn^{3+} - O^{2-} - Mn^{3+}$ provides the canonical 180°-superexchange[10] coupling $J_{AF}(> 0)$ between the Mn^{3+}-ionic spins via the diamagnetic O^{2-}, which is antiferromagnetic. Also, there is a charge transfer gap for the electron transfer from O 2p to Mn 3d state. This makes the end member $LaMnO_3$ a Charge Transfer (CT) insulator,[11] and, of course, an antiferromagnet. Divalent substitution (nominally hole doping x) in $La_{1-x}Sr_xMnO_3$ now forces a fraction x of the trivalent Mn^{3+} to become quadrivalent Mn^{4+}, introducing thereby the Zener pairs $Mn^{3+} - O^{2-} - Mn^{4+}$ and the Zener Double Exchange (DE).[12] The latter involves simultaneous virtual transfer of two electrons, one from O^{2-} to Mn^{4+} and the other from Mn^{3+} to O^{2-}, and energetically favours and is favoured by a parallel alignment of the local ionic spins in virtue of the strong Hunds' rule coupling. Hence DE favours ferromagnetism. Concomitantly, however, the Zener pair also allows for transport of charge by the real $Mn^{3+} - O^{2-} - Mn^{4+} \rightleftarrows Mn^{4+} - O^{2-} - Mn^{3+}$ processes (provided the local spins are parallel). Hence DE favours metallicity. Inasmuch as the Hund-rule coupling is the largest relevant energy scale in the problem, AFM order implies an insulating phase while FM order implies a metallic phase, consistent with the observed correlation between magnetic order and transport in the manganites $La_{1-x}Sr_xMnO_3$. Thus it should be clear that the strong Hund's rule coupling is all important and underlies the spin-valve effect.

The relevant energy scales in the problem, aside from the Hund-rule coupling which is the largest, are thus the super-exchange J_{AF} favouring antiferromagnetism, the one-electron e_g-bandwidth or the tunneling matrix element (-t) associated with $M_n^{3+} - O^{2-} - Mn^{4+} \rightleftarrows Mn^{4+} - O^{2-} - Mn^{3+}$ favouring ferromagnetism via DE, and the Zeeman energy Δ of the local ionic (t_{2g}^3) spins in an external magnetic field. In addition, there is the single-ion Jahn-Teller (JT) stabilization energy E_{JT} associated with the lattice distortion around the Mn^{3+} ion having the orbitally degenerate e_g^1 level with the possibility of a co-operative Jahn-Teller distortion of the whole crystal. Indeed, this JT effect is responsible for the cubic-tetragonal transition in $LaMnO_3^1$. These energy scales together with the band-filling (doping x) determine the equilibrium and the transport properties of these functional oxides. As for the relative magnitudes of these energy scales, systematics reveals that larger average A-cationic radius gives larger one-electron e_{g-} bandwidth that favours ferromagnetism and metallicity over antiferromagnetism and insulation. This sensitive dependence of the e_g-bandwidth on the A-cation radius derives from the fact that the Mn-O-Mn bond-angle decreases from the ideal 180° as the average cation radius decreases, as, e.g., in going from $La_{0.5}Sr_{0.5}MnO_3(< r_A >= 1.26\AA)$ through $Nd_{0.5}Sr_{0.5}MnO_3(< r_A >= 1.26\AA$ to $Nd_{0.5}Ca_{0.5}MnO_3(< r_A >= 1.17\AA)$ and $Y_{0.5}Ca_{0.5}MnO_3(< r_A >= 1.13\AA)$. The J_{AF} and E_{JT} are, of course, relatively insensitive to $< r_A >$. This leads to a rich phase diagram that has, in addition to magnetic orderings, the fascinating features of charge ordering (CO)[13–15]. This charge ordering can be of two different origins as can be readily understood from the following physical considerations. For not too large a one-electron bandwidth, when the super-exchange J_{AF} dominates the double-exchange (DE), the system becomes antiferromagnetic with the localized charge carriers (immobilized due to the unfavourable spin-valve setting) ordered in a charge-crystal so as to lower the configurational energy. This is thus an AFM induced CO. A sufficiently strong magnetic field (> the spin-flop field) can, however, alter the free-energy balance in favour of an FM metal stabilized by the Zeeman energy term and the communal entropy of delocalizatin. This melting of the AFM-charge crystal to an FM-charge liquid is a first-order transition. This is what happens in the case of $Nd_{0.5}Sr_{0.5}MnO_3$. On the other hand, in the limit of small one-electron bandwidth, the single ion E_{JT} dominates giving small lattice-polarons that again charge-order so as to lower the configurational energy. Co-operative Jahn-Teller distortion is also possible. This charge ordered insulating state may, of course, be paramagnetic or antiferromagnetic depending on the temperature of ordering. This JT-CO state is clearly distinguished from the AFM-CO state by the fact that it can not be melted into a charge-liquid state by the application of an external magnetic field, exceeding even the spin-flop field as high as $\sim$ 6T, as in the case of $Y_{0.5}Ca_{0.5}MnO_3$. The melting of the AFM-CO (charge crystal) to FMM (charge liquid) by an external magnetic field gives a CMR of up to 10 orders of magnitude, far greater than the CMR obtained in the case where the insulating phase is a PM without charge-ordering allowing conduction by uncorrelated hopping of the charge carriers while the FM is a dirty metal.

Both, the antiferromagnetic order and the lattice-polaronic distortion favour localization, in fact synergetically, while the e_g-bandwidth and the external magnetic field favour delocalization. A proper treatment of the CMR oxides must include both. In the following, however, we propose a reduced model and a model spin-valve Hamiltonian that describes qualitatively the main observed features of the CMR oxides in the limit of large one-electron e_g-bandwidth. The model enables us to discuss the PMI-to-FMM phase transition, the CMR, the resistivity of the FMM phase, and also the effect of introducing a weak lattice-polaronic effect to explain the thermal expansion anomaly observed above the Curie temperature.

2. Model and the Spin-Valve Model Hamiltonian

The recent small-angle neutron scattering (SANS) and volume thermal-expansion measurements of Teresa *et al.*,[16] have provided fairly unambiguously the following picture for cation doped manganese-oxide perovskites $(La-Y/Tb)_{2/3}Ca_{1/3}MnO_3$ that reasonably admits generalization to other CMR oxides. Thus there is a short-ranged ferromagnetic polarization (magnetic polaron) confining the charge-carrier in the PM matrix within which the carrier is free to move around. The size of this magnetic polaron (~12 Å) increases with the increasing external magnetic field and decreasing temperature. The magnetic polaron is also accompanied by lattice distortion – the small polaron. At the Curie temperature, these magnetically polarized regions begin to overlap, and presumably percolate making the carriers itinerant, and hence a (dirty) FM metal. Such a short-range magnetic polarization around the carrier (magnetic polaron), dressed with a lattice polarization (small polaron), was indeed proposed earlier to explain the anomalous transport in Eu-chalcogenides and magnetic semiconductors.[17] Recently,[18] a microscopic ferromagnetic Kondo-lattice model in the limit of infinite spin and infinite dimensions has been advanced to treat the purely magnetic-polaronic effects and CMR. More recently,[19] its large-coupling approximation (the Double-Exchange model), but incorporating the JT-lattice distortion, was treated following the claimed inadequacy[20] of the purely magnetic models without coupling to the lattice to quantitatively explain many aspects of the experimental data. In the following we propose and treat a minimum model that captures the essential physics of the problem and is qualitatively consistent with the main experimental features.

Our starting point is the realization that in the limit of large Hund's rule coupling, the e_g^1-electron spin is enslaved to the local ionic (t_{2g}^{3+}) spin (=3/2). In the limit of infinite Hund's coupling, therefore, this enslaving becomes a rigid constraint and we can consider the electron to be spinless, but with a modified tunneling matrix element $t_{ij} \rightarrow t_{ij}\cos(\theta_{ij}/2)$. Here we have assumed the local ionic spins to be classical. θ_{ij} is the angle between the spin vectors located at the sites i and j. We simplify matters further by taking these spins to be Ising-like ($\eta_i = \pm 1$). Thus, the reduced Hamiltonian is given by

$$H_{sv} = -\frac{t}{2}\sum_{<i,j>}(1+\eta_i\eta_j)c_i^+c_j + J_{AF}\sum_{<i,j>}\eta_i\eta_j - \Delta\sum_i \eta_i \ , \qquad (1)$$

where c_i^+, c_j are the spinless Fermion creation, annihilation operators, Δ is the Zeeman coupling to the local spin η_i, and $< i, j >$ implies summation over nearest neighbours. Also we take $t > 0$. H_{sv} is the minimum Hamiltonian that now governs the electron dynamics through the spin-valve effect operating through $(1 + \eta_i\eta_j)$ due to the strong (infinite) Hund's rule coupling to the background antiferromagnetically coupled local spins. Clearly, the Ising spins provide a statistical mechanical environment for the electron dynamics. We will now study this Hamiltonian, first numerically and then semi-phenomenologically. The effective loss of the spin degrees of freedom for the e_g-band electrons is an essential simplifying feature of this model, and is physically well motivated.

3. Numerical Study of H_{sv} at $T = 0$: Polaronic Short-Range Order

First, we present the results of a quick numerical (Monte Carlo) calculation for the ground-state of the H_{sv} for a finite 2-dimensional 16×16 square lattice with periodic boundary conditions. Figure 2(a) shows the case of a single electron on the lattice in the presence of an external magnetic field for certain choices of parameters given in the figure. One can clearly see the background AFM order except around the carrier where there is a short-ranged FM order polarized parallel to the external magnetic field. The size of this *magnetic*

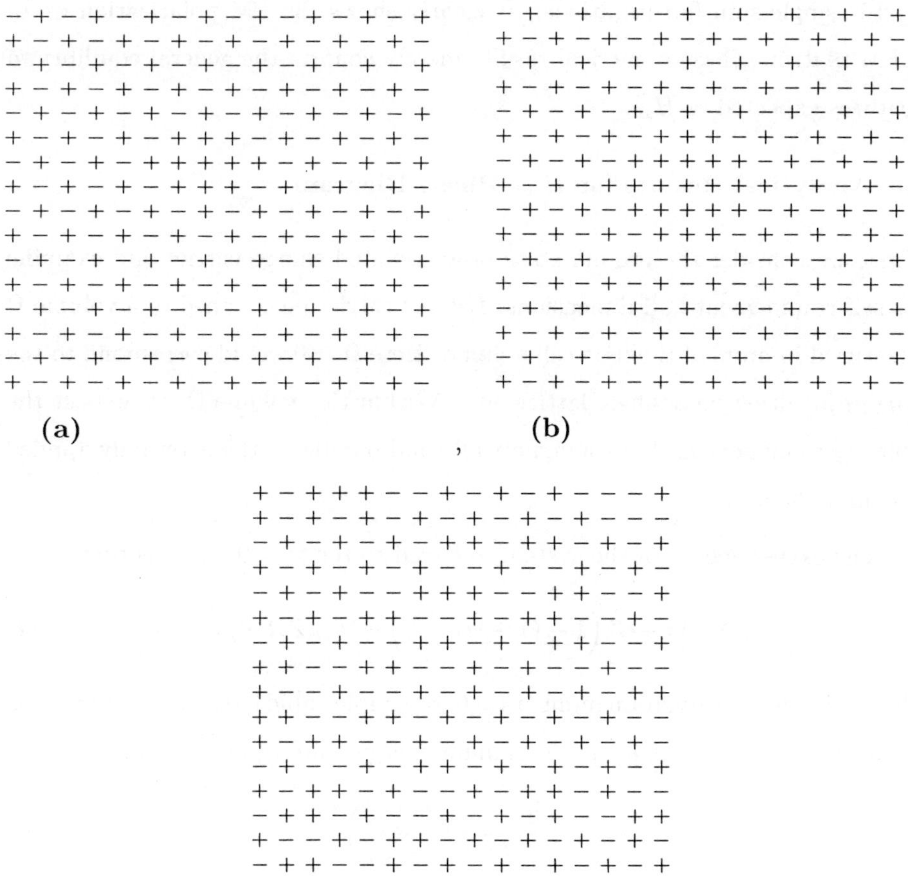

(a) (b) (c)

Figure 2: (a) Ferromagnetically polarized cluster in the antiferromagnetic background for a single carrier on a 2-dimensional square lattice at T = 0 K; (b) Same as in (a) but with non-zero magnetic field δ = -0.05. Notice the increased cluster size; (c) Ground state spin configuration for finite band-filling $n = 1/3$ for $t = -1$, $j = 0.4$ and $\delta = 0$.

polaron increases as we increase the magnetic field, Fig. 2(b). Figure 2(c) is the case for a finite band-filling $n = 1/3$. The ground state was obtained after 5×10^3 single-spin flip iterations. It clearly shows the FM polarization of an AFM matrix. These numerical results merely confirm the general equilibrium features expected of H_{sv}.

4. Analytical Results for H_{sv}: Phase Diagram

First, we consider the ground state of an isolated charge carrier and examine its self-confinement in 3-dimensions. Let the carrier be confined to a volume Ω measured in units of the unit-cell volume. Here $\Omega = 0$ would correspond to the carrier localized on a single lattice site. Within this volume Ω, we expect the spins η_i to be polarized ferromagnetically and parallel to the externally applied magnetic field.

The excess energy of the system, referred to the $\Omega = 0$ state, is then

$$\Delta E = -tZ\left(1 - (1+\Omega)^{-3/2}\right) + J_{AF}Z\Omega - \Delta\Omega. \tag{2}$$

Here Z is the co-ordination number (= 6 for simple cubic lattice). Minimization with respect to Ω gives for the minimum energy state (ground state)

$$\Omega_{min} = \left(\frac{2}{3}\,\frac{tZ}{(ZJ_{AF} - \Delta)}\right)^{3/5} - 1 \tag{3}$$

with the proviso that for $t < 3/2(J_{AF} - \Delta/Z)$, Ω will shrink to zero, while for $t > 3/2(J_{AF} - \Delta/Z)$, Ω will expand to Ω_{min}. Thus, Ω_{min} is the size (volume) of the FM polaronic cluster.

Figure 3 shows the variation of the polaron size with the magnetic field for different values of the antiferromagnetic exchange (J_{AF}) and the one-electron bandwidth ($2Zt$). We have introduced dimensionless quantities $j = J_{AF}/Zt$, $\delta = \Delta/Zt$, with Zt = half the one-electron bandwidth. Increase of the FM cluster

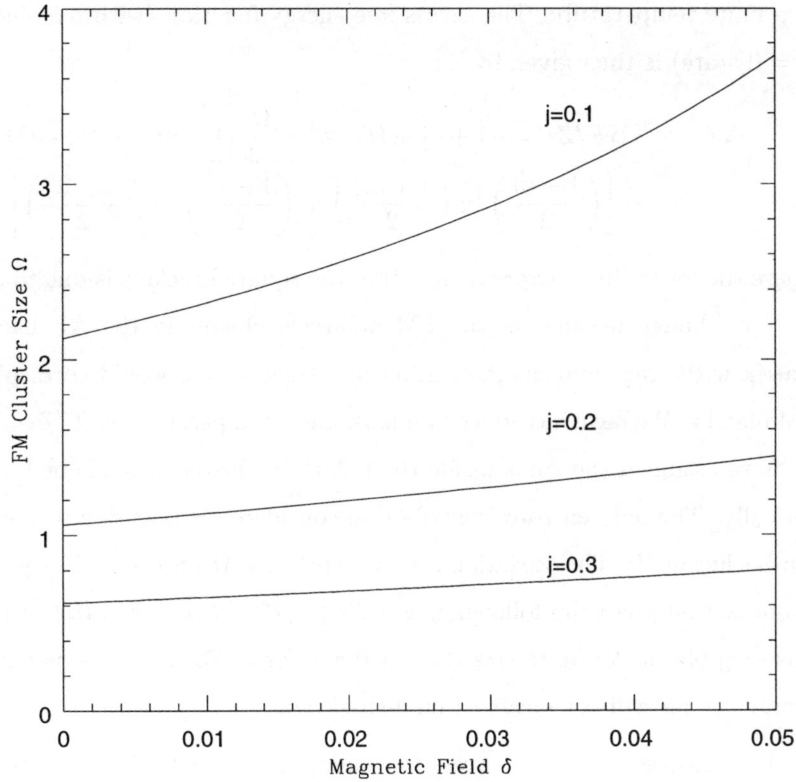

Figure 3: Ferromagnetically polarized (magnetic polaron) volume Ω for an isolated carrier at zero temperature as function of external magnetic field δ for different values of antiferromagnetic coupling j. All couplings normalized to half one-electron bandwidth.

size (Ω) with increasing magnetic field (δ) and decreasing antiferromagnetic coupling (j) is clearly seen.

Next, we consider the same problem of an isolated charge carrier but now at a finite temperature. The excess free-energy function (with reference to the $\Omega = 0$ state) is then given by

$$\Delta f \equiv \Delta F/Zt = -1 + (1+\Omega)^{-2/3} + \frac{\Omega j}{2}(1+m^2) - \delta\Omega - \Omega\tau \cdot$$
$$\cdot \left[\left(\frac{1+m}{2}\right) ln\left(\frac{1+m)}{2}\right) + \left(\frac{1-m}{2}\right) ln\left(\frac{1-m)}{2}\right)\right], \quad (4)$$

where the logarithmic expression within the square brackets is due to the spin-entropy change because of the FM polarized cluster in the AF background matrix with staggered magnetization m. Thus $m = 0$ would correspond to a PM matrix. We have introduced dimensionless temperature $k_BT/Zt = \tau$. Here we have assumed the spins inside the polarized cluster fully aligned ferromagnetically. The only entropy contribution considered is that due to the spins in the background. The variational parameters are Ω and m . Straightforward minimization gives the following: For $Zt > 3/2(ZJ_{AF} + k_BTln2)$ a polarized cluster (polaron) of finite size $\Omega_{min} \neq 0$ will form. For reversed inequality, the charge carrier will get localized on a single site.

This expression for the excess free energy for a single charge carrier can be easily generalized to a finite band-filling of n carriers per site, but assuming one carrier per cluster. This enables us to discuss the percolative transition from the high-temperature PM (Insulator) to low temperature FM-(Metal) as the volume fraction ($\alpha = n\Omega$) associated with the polaron size (Ω) reaches the percolation threshold $\alpha \approx 0.2$ (in 3 dimensions).

Figure 4 plots the cluster size Ω of the isolated CM polaron as function of temperature (τ) for different values of the antiferromagnetic coupling (j) and magnetic field (δ). Again, the variation is consistent with what is expected from physical considerations discussed above.

Figure 5(a), (b) and (c) give the phase diagram in the temperature-bandwidth plane for a finite band-filling n. Larger bandwidth ($j < 2/3$) favours a percolative PM-to-FM transition, while smaller bandwidth ($j > 2/3$) favours PM-to-AFM transition.

Finally, one can treat the general and somewhat realistic problem of more than one carriers sharing the same FM cluster in a crude approximation by treating these clusters as degenerate (spinless) electron puddles occupying a volume fraction $\alpha = n\Omega$. The excess free-energy per unit volume can be readily shown to be

$$\begin{aligned}\Delta f &= -n + 2^{2/3}\left(\frac{3}{5}\right)\epsilon_F\frac{n}{\alpha^{2/3}} + \frac{\alpha j}{2}[(1+m^2)] \\ &\quad -S\alpha - \tau\alpha\left[\left(\frac{1+m}{2}\right)ln\left(\frac{1+m)}{2}\right) + \left(\frac{1-m}{2}\right)ln\left(\frac{1-m)}{2}\right)\right] \quad (5)\end{aligned}$$

where $\epsilon_F = E_F/Zt$, the Fermi energy for a virtual crystal with n conduction electrons per unit volume. Here the variational parameters are α and m. Again we ignored the entropy of electrons and of the polarized spins in the FM clusters. Here $\alpha \approx 0.2$ marks the percolation threshold for bulk FM and metallicity.

5. Transport, CMR at PMI-to-FMM Transition at T_c and Thermal Expansion Anomaly above T_c

The above treatment and the results obtained therefrom provide a reasonably clear background for discussing transport and thermal expansion anomalies in these CMR materials.

The colossal magnetoresistance is a direct consequence following from the percolative nature of the PMI-to-FMM transition. As the size of the FM polaronic clusters, and hence their volume fraction increases with increasing magnetic field it is clear that the effect of the magnetic field close to the percolative

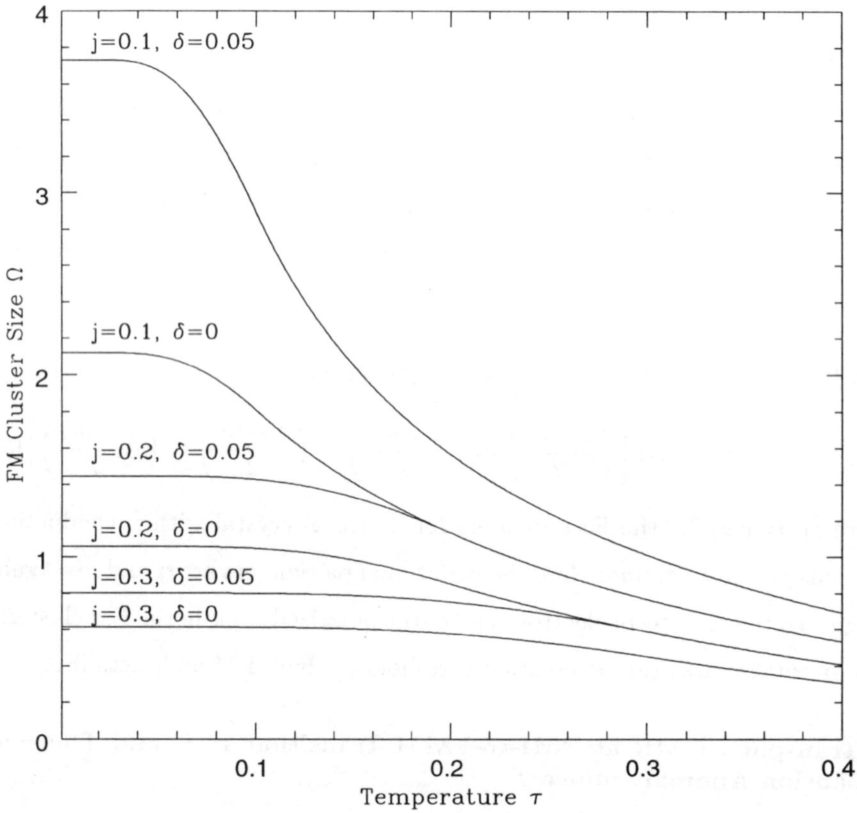

Figure 4: Isolated FM cluster size Ω as function of temperature (τ) for different value of AF coupling (j) and magnetic field (δ).

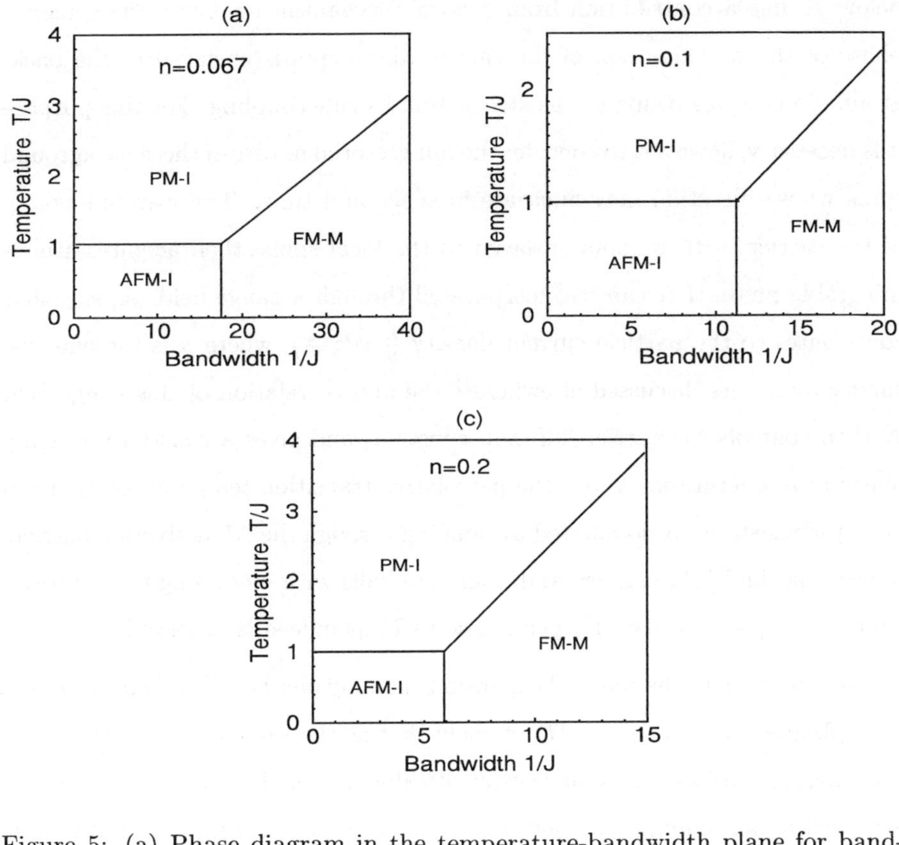

Figure 5: (a) Phase diagram in the temperature-bandwidth plane for band-filling $n = 0.067$, (b) same for band-filling 0.1, (c) same for band-filling 0.2.

PMI-to-FMM transition will be to enhance the percolation probability from zero to a non-zero value, causing an insulator to-metal transition, and hence the CMR. It is greatly more pronounced if the phase above T_c is a CO insulator.

The temperature dependent resistivity in the dirty metallic (FMM) phase below T_c has a contribution from a novel mechanism involving the spinorial phase of the wavefunction of the carrier whose spin is enslaved to the background ionic spins owing to the strong Hund's rule coupling. For this purpose it is necessary, however, to consider the full vectorial nature of these background spins whose direction may fluctuate in space and time. The wavefunction ψ of the carrier, with its spin enslaved to the local spins, then acquires a non-integrable phase that can be incorporated through a gauge field, $\mathbf{A}$, say, that contributes to the particle current density $\frac{\hbar}{\mu}(\psi^*\psi\mathbf{A})$, where μ is the effective carrier mass. As discussed elsewhere[21] the autocorrelation of this gauge field $\mathbf{A}$ then controls the carrier diffusion constant, and gives a resistivity roughly linear in temperature. Above the percolative transition temperature T_c, however, the resistivity is dominated by hopping through the AF activation barriers separating the FMM clusters and, therefore, falls with increasing temperature. Thus, we expect a resistivity peak close to T_c as indeed is observed.

Next we turn to the thermal expansion anomaly observed[16] in $(La{-}Y/Tb)_{2/3}$, $Ca_{1/3}MnO_3$ for $T_c < T < 1.8T_c$, namely that as the sample was cooled in the PM phase, its volume remained systematically greater than what was normally expected by an amount that showed a sudden drop to zero at T_c. Within our polaronic picture we can ascribe this to the degeneracy pressure of the electrons confined to the FM polarized clusters of volume Ω. The point is that the degeneracy pressure is exerted on the misaligned spins at the FM cluster boundary that confines the electrons by the spin-valve effect. As these spins are localized on lattice sites, the positive internal pressure acts on the lattice giving an added expansion (ΔV). The effect is, of course, in a mean field sense, proportional to the volume fraction $\alpha = n\Omega$ occupied by these electronic puddles, and hence increases as the temperature is lowered towards T_c. At T_c, however, the electronic puddles percolate throughout the sample and the pressure is no longer borne by the internal confining boundaries. Hence the sudden fall of ΔV at T_c. It is to be pointed out here that a JT distortion corresponds to traceless strain tensor and hence may not directly contribute to the observed ΔV.

6. Concluding Remarks

We have discussed the phase diagram, transport and the anomalous thermal expansion of the manganese-oxide based perovskites in terms of a minimum model that incorporates the spin-valve effect associated with the strong Hund's rule coupling, in the limit of large one-electron bandwidth. Our model shows short-ranged ferromagnetic polarization (polaron) associated with the charge carrier in the background paramagnetic/antiferromagnetic matrix. The size of these polarons increases with decreasing temperature and increasing magnetic field. The PM-to-FM transition is of a cluster percolative nature. In the FM phase it is a dirty metal with resistivity arising partly from the spinorial dephasing. In the PM phase the transport is by thermal hopping involving activation energy of the order of the AF exchange energy, and the distances under hopping of the order of the mean distance between the FM polarized cluster boundaries (separated by AF regions). The PM phase is insulating. Application of magnetic field close to the PM-FM transition enhances the percolation probability by increasing the polarized cluster size (and, therefore, its volume fraction). This accounts for the large magnetoresistance – the CMR, a field induced metal-to-insulator transition of the percolative type. An interesting point of this percolative picture of PMI-FMM transition to be noted is the following. Inasmuch as at the percolation threshold there is ideally one and only one percolating path with a single critical bond somewhere along the percolation path (whose cutting would break the continuity), one would expect a universal sample resistance corresponding to the Sharvin contact resistance between the electronic puddles (the FM polarized metallic clusters) so critically connected. Magnetic tuning of the percolation can perhaps reveal this universality.

As for the lattice-polaronic effects, except for the case where we have a cooperative JT effect, the main effect of the single-ion JT distortion should be to cut-off the tunneling matrix element-t by the Holstein overlap factor. This certainly would co-operate with the J_{AF} and enhance the spin-valve effect favouring localization. But the JT distortion, being a traceless strain, is unlikely to cause the observed thermal expansion anomaly above T_c. The explanation proposed by us, based on the degeneracy pressure of the electrons confined in the FM polarized clusters, is quite robust.

The present work calls for generalization so as to include the co-operative JT effect, which can be done, e.g. by treating the single-ion $e_g^1 - JT$ distortion as an Ising pseudo-spin carried by the e_g^1 electrons and by introducing interactions of elastic origin among these pseudo-spins. Such an Ising fluid can then be treated for pseudo-spin- and charge-ordering.

Finally, it seems quite apt to comment on the possible relationship between the manganese oxide cubic perovskites showing CMR and the copper-oxide layered perovskites showing high-T_c superconductivity (HTSC). CMR manganites involve spin-polarized tunneling intrinsically because of the strong Hund's rule coupling giving the spin-valve effect. The HTSC cuprates seem to involve tunneling of the oxygen holes involving a copper-spin flip process – there is no Hund's rule coupling here and hence non spin-valve effect. Thus CMR and HTSC seem mutually exclusive. It has been suggested, however, that these two classes of materials share a common feature relevant to transport, namely, that they exhibit dynamic phase segregation[27] at the I-M transition where the magnetic polarons, dressed by cooperative lattice distortion involving several lattice sites, exist and move diffusively.

Acknowledgement

One of us (NK) would like to thank Prof. C.N.R. Rao for many a fruitful discussion on the problem of CMR and CO in these oxides.

References

[1] For a review, see J.B.Goodenough and J.M.Longon, *Magnetic and Other Properties of Oxides and Related Compounds*, eds. K.-H. Hellwege and O. Madelung, Landolt-Börnstein, New Series, Group III, Vol. 4, Pt a (Springer-Verlag, Berlin, 1970).

[2] S. Jin, M. McCormack, T.H. Tiefel and R. Ramesh, *J. Appl. Phys.* **76**, 6929 (1994); S. Jin *et al. Science* **264**, 413 (1994).

[3] R. Von Helmolt, I. Wecker, B. Holzapfel, L. Schultz and K. Samwer, *Phys. Rev. Lett.* **71**, 2331 (1993).

[4] K. Chahara, T. Ohno, M. Kasai and Y. Kozono, *Appl. Phys. Lett.* **71**, 1996 (1993).

[5] A. Urushibara, Y. Moritomo, T. Arima, A. Asamitsu, G. Kido and Y. Tokura, *Phys. Rev. B* **51**, 14103 (1995).

[6] M.N. Baiblich, J.M. Broto, A. Fert, F. Nguyen van Dau, F. Petroff, P. Etienne, G. Creuzet, A. Friederich and J. Chazelas, *Phys. Rev. Lett.* **61**, 2472 (1988).

[7] S.S.P. Parkin, N. More and K.P. Roche, *Phys. Rev. Lett.* **64**, 2304 (1990).

[8] J.Z. Sun *et al.*, *Appl. Phys. Lett.* **69**, 3266 (1996).

[9] Y. Lu *et al. Phys. Rev. B* **54**, 8357 (1996); see also, H.Y. Hwang, S.-W. Cheong and B. Batlogg, *Appl. Phys. Lett.* **68**, 3494 (1996).

[10] P.W. Anderson and H.Hasegawa, *Phys. Rev.* **100**, 675 (1955); P.G. de Gennes, *Phys. Rev.* **118**, 141 (1960).

[11] T. Arima, Y. Tokura and J.B. Torrance, *Phys. Rev. B* **48**, 17006 (1993).

[12] C. Zener, *Phys. Rev.* **82**, 403 (1951).

[13] Y. Tokura, Y. Tomioka, H. Kuwahara, A. Asamitsu, Y. Moritomo and M. Kasai, *J. Appl. Phys.* **79**, 5288 (1996) and references cited therein.

[14] C.N.R. Rao, A.K. Cheetham and R. Mahesh, *Chem. Mater.* **8**, 2421 (1996).

[15] N. Kumar and C.N.R. Rao, *J. Solid State Chem.* **129**, 363 (1997).

[16] J.M. De Teresa, M.R. Ibarra, P.A. Algarabel, C. Ritter, C. Marquino, J. Blasco, J. Garcia, A. del Moral and Z. Arnold, *Nature* **386**, 256 (1997).

[17] T. Kasuya and A. Yanase, *Rev. Mod. Phys.* **40**, 684 (1968); S. von Molnar *et al.*, *Phys. Rev. Lett.* **51**, 706 (1983).

[18] Nobuo Furukawa, *J. Phys. Soc. Jpn.* **65**, 3214 (1994).

[19] H. Röder, Jun Zang and A.R. Bishop, *Phys. Rev. Lett.* **76**, 1356 (1996).

[20] A.J. Millis, P.B. Littlewood and B.I. Shraiman, **Phys. Rev. Lett. 74**, 5144 (1995).

[21] N. Kumar and A.M. Jayannavar, in *Metal-Insulator Transition Revisited*, eds. P.P. Edwards and C.N.R. Rao (Taylor and Francis, London, 1995), p.309; See also, Gen Tatara and H. Fukyama, *Phys. Rev. Lett.* **78**, 3773 (1997).

Theory of Colossal Magnetoresistance: Some Questions

T V Ramakrishnan
Department of Physics
Indian Institute of Science
Bangalore 560 012, India

Abstract

Some of the phenomena observed in doped rare earth manganites and the questions they raise are discussed. The nature of the paramagnet to ferromagnet transition and the origin of giant magnetoresistance, the unusually large T^2 term in the low temperature resistivity, the unique occurrence of residual resistivities several orders of magnitude larger than the Mott maximum, the singular effectiveness of magnetic fields in determining phase stability and transport and the characteristic electron excitation spectrum reflected in optical conductivity, are some of the themes touched upon. The extreme sensitivity of relative phase stability to apparently small changes in ionic radii/composition is unusual. We argue that available theories and ideas are probably not adequate.

The discovery of colossal magnetoresistance in mixed valent manganites (1,2) has led to a resurgence of interest in this family of compounds. Many spectabular and unusual new phenomena have been discovered, raising basic questions about the nature of electronic states and phases. Many of the observed effects are due to the strong connection between magnetism and the insulator to metal transition special to these systems. While a general set of ideas exists, the theoretical explanations seem inadequate for a number of observed effects.

I begin by outlining some of the basic phenomena, and the idea of double exchange. Some recently developed models and theories are briefly described. These furnish us with the basic ingredients. I focus then on a number of unanswered questions mentioned in the abstract.

1. Introduction

1.1. Setting

The stoichiometric manganites, eg., $LaMnO_3$ are Mott insulators, namely they are insulators both above and below the Neel temperature of about 150K. The Mott-Hubbard U, or the net energy cost of adding another electron to the highest occupied e_g orbital is rather large, of order 5 eV or so. From several properties of $LaMnO_3$ it is clear that other interactions are important (3). For example, there is e_g orbital ordering and lattice distortion (with respect to a perfect perovskite structure) in $LaMnO_3$. Both these clearly point to the importance of the Jahn-Teller effect i.e. the removal of the double degenracy of the e_g orbital by a static lattice distortion. The Jahn Teller energy is about 0.5 eV. The facts that M_n^{3+} (with one e_g and three t_{2g} orbitals occupied) has high spin and that the Mn^{3+} spins (S=2) have **ferromagnetic** alignment in plane (with an antiferromagnetic interplanar arrangement) point to the strong Hund's rule tendency. The d electrons take up the hightest possible total spin configuration to minimize coulomb repulsion and super exchange processes can be ferromagnetic. The effective Hund's rule energy J_H is large, $\simeq$ 2 eV. The bandwidth is small, of order an eV or less.

Thus even at the insulating $LaMnO_3$ end, strong correlation, Hund's rule and Jahn Teller effect are important. On substituting trivalent La with a divalent ion such as Ca or Sr, one d electron is transferred out from Mn^{3+} to oxygen, so that $La_{1-x}Sr_xMnO_3$ for instance has a fraction $(1-x)$ of the Mn ions in the 3^+ state and fraction x in the 4^+ state. This mixed valent crystal is metallic because the e_g electron can hop from an Mn^{3+} site to a neighbouring Mn^{4+} site. Because of the strong Hund's rule coupling, the hopping is proportional to the probability amplitude that the initial and final states have the maximum spin S=2. Thus, the e_g electron kinetic energy favours parallel or ferromagnetic orientation of nearest neighbour spins. This is the double exchange of Zener (4), and of Anderson and Hasegawa (4). The ground state of $La_{1-x}Sr_xMnO_3$ is ecpected to be a ferromagnetic metal, because of the kinetic energy gained by the e_g electron moving parallel to the aligned, t_{2g} spins. On increasing the temperature, the 'resident' t_{2g} spins disorder because of entropic reasons, the e_g electron does not propagate well, and there is transition to a paramagnetic insulator phase. The moving e_g electron carries its Jahn-Teller distortion with it; this polaronic effect aids the transition or may even be the main cause.

The ideas mentioned, namely mixed valence of Mn, double exchange, and Jahn-Teller distoration related polaronic effects seem appropriate at least qualitatively. For example, $La_{1-x}Sr_xMnO_3$ shows a paramagnet to ferromagnet transition for $x > 0.17$, with a $T_c \sim 200$ to 300K and nearly coinciding with it an insulator to metal transition. There is considerable structural evidence for polaronic effects. The colossal magnetoresistance that is observed near the transition can be rationalized as being due to the fact that a magnetic field, by partially aligning moments makes it easier for electrons to move. This effect is maximum near the transition where the magnetic suscptibility is the largest.

It becomes clear, however, that many significant features of the transition are not well explained. The so called metallic state often has resistivities such that the electron mean free path is smaller than a lattice spacing. This is true even in the best prepared single crystalline or thin film samples, say near T_c. This is uncommon, though possible if e_g electrons hop largely inelastically. The size of the colossal magnetoresistance is quite unusual. A magnetic field of a few Teslas ($\sim$ energy of a few ^oK) makes a large difference to electronic states and transport at a temperature of a few hundred degrees, which is itself more than an order of magnitude smaller than the electron kinetic energy ($\sim$ 1 eV). Another curious feature is that the paramagnetic to ferromagnetic transition generally takes place below (often about 50K below) the insulator metal transition, defind as the temperature where the electrical resistivity has a maximum.

Near $x \simeq 1/2$, and /or when ionic sizes are not optimal, charge ordering (Mn^{3+}, Mn^{4+} ordering, nominally) is endemic. The interplay between charge ordering, magnetic order, lattice distortion and the metal insulator transition is a major new area where very little is understood except in broad qualitative terms. For example, $Pr_{1-x}Sr_xMnO_3$ has for $x = 0.5$ discontinuous transition from a metallic phase to a charge ordered insulating phase (5) at about 140K. A field of only 6-7 Tesla brings back the metallic phase at $T = 0$. In $(Nd\,La)_{0.5}\,Ca_{0.5}MnO_3$, (6), well developed charge, ordering correlations, on cooling lead not to an insulating state with long range charge order, but to a ferromagnetic metallic state.

1.2. Models and Theories

We describe briefly, modes which have been proposed to describe the electronic and magnetic behaviour of CMR oxides.

(i) Kondo Lattice Model

In a site based description, at each lattice site, there are 4 or 3 d electrons, for Mn^{3+} and Mn^{4+} respectively. It is conventional to regard the 3 t_{2g} d electrons on each site as spins (S=3/2), and moreover as classical spins. That is, the t_{2g} orbitals have always, only three electrons, in the highest spin state (Hund's rule). The fourth d electron, in an e_g orbital, can move from site to site. There is a strong tendency for the spin of this electron to be aligned with the S=3/2 spin resident on site. Suppose the energy differnce between parallel and antiparallel alignment is $J_H S$. One simple Hamiltonian which focusses on this effect is

$$H_{KL} = \sum_{i,j,o} t_{ij} a^{\dagger}_{i\sigma} a_{j\sigma} - J_H \sum_{i} a^{\dagger}_{i\alpha} (\vec{\sigma}_i)_{\alpha\beta} a_{i\beta} \cdot \vec{S}_i \tag{1}$$

This ferromagnetic Kondo lattice Hamiltonian describe **uncorrelated** electrons hopping from site i to site j with amplitude t_{ij}, and interacting ferromagnetically with the resident spin $\vec{S}_i$ at site i, the coupling strength being J_H. The Hamiltonian and its consequences have been discussed at length by Furarkawa (7).

Interestingly, even though the electrons are uncorrelated, the strong J_H leads to effective correlations. For example, at one electron per site, the ground state is seen to be an antiferromagnetic insulator, with a Hubbard like charge excitation gap of size J_H!

Most of the calculations done using this model use the single site self consistent dynamical mean field theory (or $d = \infty$ theory) (8). This method allows one to investigate strong coupling behaviour ($J_H \gg W$) selfconsistently, though spatial correlations are ignored A paramagnetic to ferromagnetic transition is obtained for $x \tilde{>} 0.2$, at realistic temperatures. In the paramagnetic state, the electron propagates incoherently, and the system is a strongly disordered metal. Within the approximation scheme used, an insulating paramagnetic (high temperature) state is not possible in contrast to observations of both insulating and barely 'metallic' states, as in $L_{a-x}Ca_x\,MnO_3$ and in $La_{1-x}Sr_x\,MnO_3$ respectively. Whether this is due to approximations used, or to the neglect of Jahn-Teller coupling, is not clear.

The large magnetoresistance is attributed in this model to the coherence in quasiparticle motion induced by the magnetic field. This increases the quasiparticle spectral density near the Fermi energy. In scaled units, the (ρ/ρ_o) plots as a function of $(\Delta\, M(M_s)\,)$ look similar to those found experimentally for small ΔM.

The approach describes in a consistent, single picture several features of CMR systems where the electrons are itinerant both above and below T_c. There are no well defined quasiparticles above T_c. However, it is probably inadequate in several respects discussed later. Some of these might well stem from the approximations used which are the best available currently for treating strong local correlations in quantum systems, but are single site approximations. The neglect of Jahn-Teller or orbital-lattice coupling, and of electron correlations are some other obvious shortcomings.

(ii) Projected Electron Hopping

Very early, Anderson and Hasegawa (4) realized that in the limit of strong Hund's rule coupling ($J_H = \infty$), the problem simplifies. The e_g electron hops from one site to the next to the extent that its spin state projects on to that of the resident spin $\vec{S}$ (represented by a classical vector) (See also ref.8) Thus, the hopping Hamiltonian can be written as

$$H = \sum_{i,j} \psi_i^\dagger \frac{(1 + \vec{\sigma}_i.\vec{S}_i)}{2} t_{ij} \frac{(1 + \vec{\sigma}_j.\vec{S}_j)}{2} \psi_j \qquad (2)$$

where ψ_i^+, ψ_j are two component spinors and $\frac{1+\vec{\sigma}_i\vec{S}_i}{2}$ is a projection operator, or as

$$H = \sum_{i,j} a_i^\dagger \tilde{t}_{ij} a_j \qquad (3a)$$

where

$$a_i^\dagger = cos\frac{\theta_i}{2}a_{i\uparrow}^+ + \sin\frac{\theta_i}{2}\, e^{i\phi_i} a_{i\downarrow}^+ \tag{3b}$$

and

$$\tilde{t}_{ij} = t_{ij}\left(cos\frac{\theta_i}{2}cos\frac{\theta_j}{2} + sin\frac{\theta_i}{2}sin\frac{\theta_j}{2}exp(i(\phi_j - \phi_i))\right) \tag{3c}$$

In Eq.(3) a spin $\vec{S}i$ is described by the angles (θ_i, ϕ_i). Notice that the large energy J_H has dropped out of the description; one has just 'projected' fermions hopping. Surprisingly little is known about the properties of this system, characterized by one base energy scale, $Zt = W$. At $T = 0$, this just describes a tight band of free electrons with bandwidth W, since the resident spins are fully aligned. As temperature increases, spins disorder and there is randomness (annealed, thermally distributed) in electron hopping. Does it lead to localization of electrons? How does this affect spin order in turn?

A more complete model, including the effect of e_g orbital degeneracy, and its lifting (Jahn-Teller splitting) by coupling to lattice ion displacement, has been developed by Millis and coworkers (9). The model is described by a Hamiltonian with orbital degrees of freedom μ , ν on each site:-

$$H = -\sum_{\langle ij\rangle_{\mu\nu\alpha}} t_{ij}^{\mu\nu}\, a_{i\mu\alpha}^\dagger\, a_{j\nu\alpha} - J_H \sum_{i\mu\alpha\beta} i\vec{S}.a_{i\mu\alpha}^\dagger \vec{\sigma}_{\alpha\beta}\, x\, a_{i\nu\beta} + g\sum_{iab\alpha} \vec{r}_i a_{i\mu\alpha}^\dagger \vec{\tau} a_{i\nu\alpha} + \frac{1}{2}k\sum_i \gamma_i^2 \tag{4}$$

The first two terms are just the Kondo lattice model. The third term describes how lattice displacement $\vec{r}_i$ mixes orbitals μ and ν, and the last term is the elastic energy cost. The dimensionless parameter which measures the strength of electron phonon coupling (or $J-T$ polaronic effects) is $\lambda = (g^2/kt)$, namely elctron phonon coupling in relation to the electron bandwidth.

The model again cannot be solved exactly. In the strong coupling regime of interest, namely $(J_H/Zt) > 1$, and $\lambda \sim 1$, the single site dynamical mean field theory is the method used. The on site electron self energy is determined selfconsistently for a given local lattice distortion. For large λ the energy is minimized with distortions such that the electron is locally trapped, with no spectral density at the Fermi energy. As a function of temperature, there is a paramagnet to ferromagnet phase transition. However, depending on λ, both the former and the latter can be quasi metallic or insulating. This covers, qualitatively many of the possibilities actually realized. There is however a static distortion over a broad temperature range. The qualitative behaviour of resitivity with temperature, of T_c, and of magnetoresistance are similar to what is seen for one electron per site. On moving away from $n = 1$, however, there are empty sites, not $J-T$ distorted, which always contribute states at the Fermi energy. This leads to much poorer qualitative features in relation to experiment.

In addition to the above shortcoming which the authors argue will change on inclusion of breathing mode distortions and short range coulomb interactions, the solution has several unrealistic elements. An obvious one is the single site nature of the approximations, so that spatial magnetic or $J-T$ (polaronic) correlations, as well as localization effects due to spin/ lattice disorder are absent. There is experimental (neutron scattering) evidence for the former, and electron localization is one of the open questions in this broad area. Another issue is that of dynamic Jahn Teller effect. That these are significant is clear from the strong isotopic mass dependence of T_c (see below). However, there are at present no systematic ways of including such effects in the strong correlation/ electron phonon coupling regime.

There is a considerable body of electronic structure calculations which have addressed insightfully questions of basic energy scales and ground states (10). Finite site exact diagonalization studies of say, the Kondo lattice Hamiltonian, are being done (11). The results suggest possibilities of two phase regions (hole poor AF, hole rich FM), charge ordered structures etc.

We now turn from a description of models and theories to several phenomena which suggest the need for new ideas.

2. Low Temperature Resistivity

The ferromagnetic, metallic state of the optimally doped $La_{2-x}(Sr,Ca)xMnO_3(x \simeq 0.3)$, in high quality single crystals as well as thin films, has a residual resistivity as small as 100 $\mu\Omega$ cm, corresponding to a mean free path of order 40 to 60 Å. The resistivity increases very rapidly with temperature (Fig.1, Ref. 12).

The increase can be fitted to the dependence

$$\rho(T) = \rho_o + AT^2 + BT^{9/2} \tag{5}$$

where the T^2 term is much the bigger one. If the ground state is fully magnetized, eg. with all spins pointing up, single spin wave processes (Fig.2a) have very little weight since there is an exponentially low density of final down spin electron states. The leading current relaxing process at low temperatures involves two spin waves (Fig.2b) and depends on temperature as $T^{9/2}(2)$. This is not the leading term, which goes as T^2 and is very large. The T^2 term if extrapolated, would lead to a close to the Mott maximum value resistivity for a temperature of order the ferromagnetic Curie temperature T_c! Such a large T^2 term is uncommon. At $T \simeq (T_c/2)$, the mean free path has decreased from about 60 Åto about 6 to 10 Å.

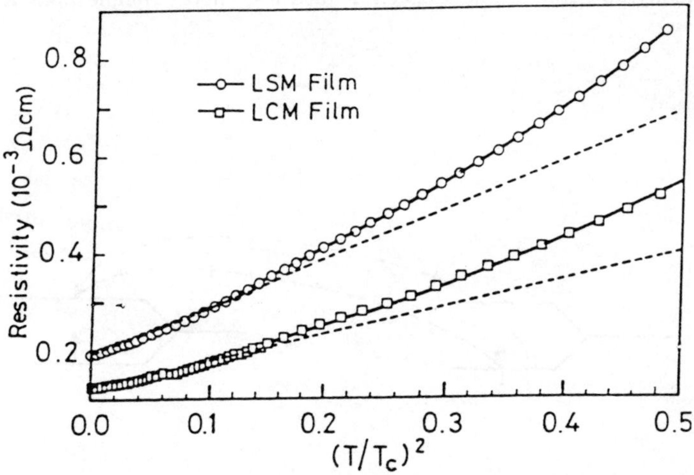

Fig.1 The low temperature resistivity of $La_{0.67}(Sr/Ca)_{0.33}MnO_3$ films. The solid lines are a fit to $R_0 + R_2T^2 + R_{9/2}T^{9/2}$. The dotted line is the best fit to a constant plus a quadratic term (From Ref. 4).

Fig.2 Spin wave scattering processes contributing to low temperature resistivity in a ferrmagnetic metal. (a) Single spin flip process, possible if both up and down spin electron densities are non zero (b) The leading, two spin wave, process for a fully polarized metal.

It is clear from the above that the picture of the low temperature phase as an effectively noninteracting, fully spin polarized electron system with spin waves as the elementary excitations is inadequate even when the ground state is very well ordered magnetically and spin waves are well defined low energy magnetic excitations.

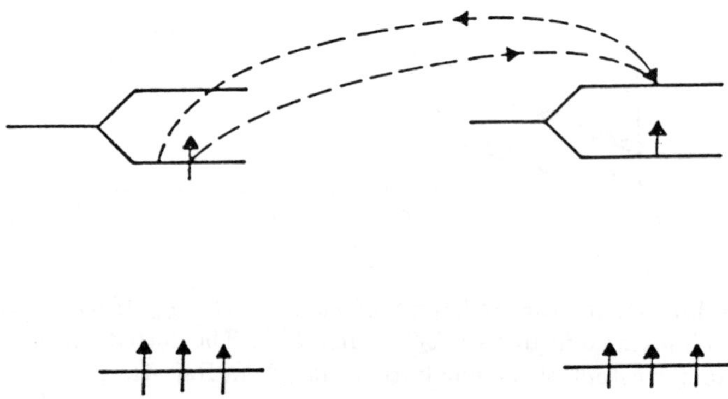

Fig.3 A possible nearest neighbour virtual hopping process that leads to a nearest neighbour attractive interaction.

One can certainly imagine interactions arising out of nearest neighbour hopping terms in higher order; one such is shown in a Fig.(3). It is a ferromagnetic nearest neighbour attraction, $-V_{ij}n_in_j$, lowering the energy of the state with one e_g electron on site i and another on site j, in the spin configuration shown, for example. Whether this interaction is strong enough to cause a T^2 resistivity of the observed size is not clear.

It needs to be pointed out that low temperature specific heat of metallic manganites shows an electronic linear term γ close to what is expected from a tight binding band; there is no strong enhancement of the linear specific heat while the T^2 coefficient is highly enhanced, approximately by the ratio (T_F/T_c). $[\Delta\rho(T) \simeq \rho_{Mott}(T/T_c)^2$ and not $\rho_{Mott}\,(T/T_F)^2$!]. This is in contrast to systems near the Mott transition where both γ and $\Delta\rho\,(T)$ are enhanced similarly.

3. Residual Resistivity

As mentioned above, in clean metallic manganites, the residual resistivity can be quite small, smaller by a factor of ten or more than the maximum metallic value of about 1 to 2 $m\Omega$ cm. However, many polycrystalline systems with nearly optimum size of the rare earth (or substituent) ions as well as systems with non optimal (generally smaller) ion size show extremely high values of residual resistivity. These values can be several orders of magnitude higher than the maximum metallic resistivity, and yet the resistivity seems to flatten out at $T \rightarrow 0$ without increasing to infinity as it should for an insulator. This is in contrast to the behaviour of all other strongly disordered and or correlated systems in which as a function or disorder or correlation, a maximum metallic resistivity (the Mott value corresponding to $k_F\ell \sim 1$) separates the disordered metal from the insulator. There are scale dependent localization effects near critical disorder due to which, at very low temperatures the conductivity goes smoothly to zero as the disorder approaches a critical value; however on the large, experimental scales of temperature and disorder in most disordered systems these quantum critical effects are invisible and the Mott dividing line is very real indeed.

One example of the spectacular violation of the Mott limit is the resistivity of $Re_{0.7}A_{0.3}MnO_3$ films measured by Coey et al (15), where the Lanthanide Re is either Nd or La, and the alkaline earth ion A is either Sr or Ba. The $Nd_{0.7}Sr_{0.3}MnO_3$ sample has a residual resistivity of $10^4\Omega$cm, and a ferromagnetic ordering temperature of about 100K, while $La_{0.7}Sr_{0.3}MnO_3$ has a resistivity of about $10^{-4}\Omega$cm and a T_c close to 300K. The two systems have the same carrier density. Nd^{3+} is much smaller than La^{3+}, and this can cause various structural and electronic effects. A common way of parametrizing the structure is through the composition averaged radius $(1-x)r_{Re} + xr_A = \langle r_A \rangle$. An exponential variation of ρ_o with $\langle r_A \rangle$ has been noted by Mahesh et al (17). For ceramic samples with $x = 0.3$, in which $\langle r_A \rangle$ is varied by mixing two different reare earth ions, keeping the total fraction the same (i.e. 0.7) per formula unit, Garcia-Munoz et al (18) obtain a dependence of residual resisitivity on $\langle r_A \rangle$ shown in Fig.(4). We note that the residual resisitivity can be as much as 10^6 times larger than the maximum metallic value!

This is certainly unique and quite astonishing, because the Mott criterion is well obeyed experimentally (within factors of two), the main reason being its geometrical nature. It is based on the equality of two lengths, namely the elastic scattering mean free path and the de Broglie wavelength of the electron. As such it is independent of the bandwidth. This is easily seen in a tight binding model on a lattice of spacing a. In three dimensions, the minimum metallic conductivity is $(e^2/\hbar)$ and does not depend on the bandwidth $|zt|$. (It depends weakly on band filling). It can also be shown, at least within perturbation theory, that quasiparticle renormalization (due to frequency dependent self energy) has no effect on the residual Boltzmann conductivity. Thus it is difficult to sustain the common belief that polaronci reduction in band-width due to decreasing $\langle r_A \rangle$ is the cause of the enormous residual resistivities observed. Moreover, the reduction in bandwidth with a few percent change in r_A is not normally expected to be very large.

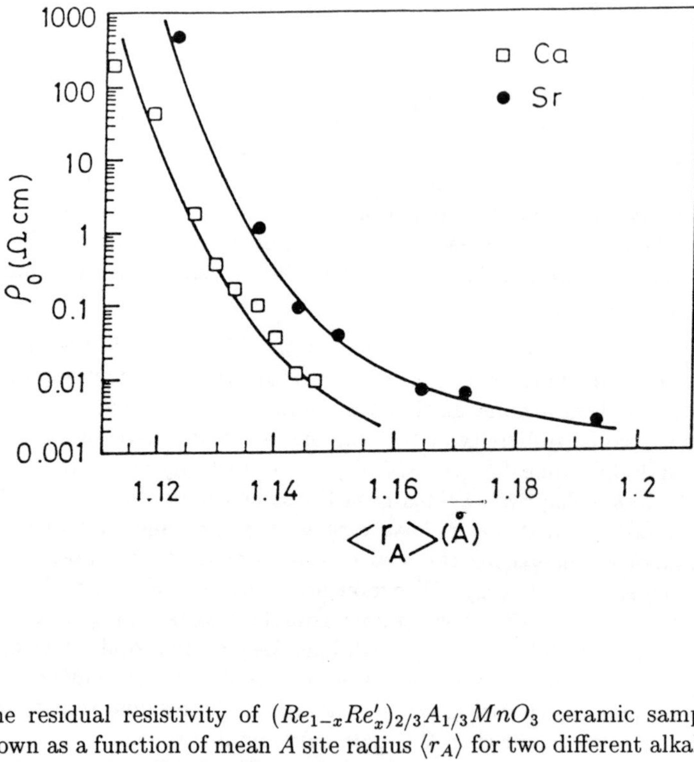

Fig.4 The residual resistivity of $(Re_{1-x}Re'_x)_{2/3}A_{1/3}MnO_3$ ceramic samples shown as a function of mean A site radius $\langle r_A \rangle$ for two different alkaline earth ions. $A = Ca, Sr$. (From Ref. 18)

If the effective bandwidth W_{eff} due to $J-T$ polaron formation is very small, one expects the following temperature dependence for the conductivity. At $T \gg W_{eff}$, one has localized polarons hopping incoherently from site to site. If the intersite hopping probability is p_i, the conductivity of this classical gas is $\sigma \sim (p_i/kT)$. If the intersite hopping is not activated, p_i does not depend on temperature. Thus the conductivity increases with decreasing temperature, and at $T \sim W_{eff}$, goes over to the tight binding Mott value ($e^2/\hbar a$). (Crudely, $p_i \propto W_{eff}$ so that the conductivity at $T \simeq W_{eff}$ is independent of the latter). This is not the observed behaviour. Other purely thermally activated hopping processes give exponentially increasing resistivity with decreasing temperature, the exponent depending on the actual process (variable range hopping etc.).

The following observations are suggestive as to the possible causes of the large residual resistivity.

The reduction in the mean rare earth ionic radius is accompanied by a bending of the Mn-O-Mn bond, from 170° to 155° as r_A decreases from 1.20 to 1.12 Å. This implies that there is an increasing antiferromagnetic component to the Mn-Mn spin coupling, which is reflected in the rather rapid decrease in the saturation magnetization as well as T_c with r_A (18). Thus there is magnetic frustration and the energetically favourable arrangement is not one in which all Mn spins are parallel to each other.

The consequences of possible magnetic frustration for transport can be phenomenologically explored via the dependence of resistivity on saturation magnetization. This is seen to be approximately exponential. (Fig.5). This relation, suggested by several authors, works over several decades (nearly six) in resistivity. The data of Garcia Munoz et al (18) can be reexpressed as

$$\rho(T=0) \;=\; \rho_{Mott}\, exp(\lambda(\delta M/M_s))$$

where $\rho_{Mott} \simeq 2m\Omega$ cm, δM is the deviation from saturation the maximum or intrinsic magnetization M_s, and λ is a large constant, being about 18 for Sr and about 71 for Ca (in the family $(Re_{1-y}Re'_y)_{2/3}\,A_{1/3}M_nO_3$ where A = Sr or Ca). The origin of these large factors in the exponent is not clear. The exponential dependence seems to be applicable not only to residual resistivity but also to $\rho(T)$ as a function of thermal equilibrium magnetization M(T) at temperature T. (19) In another experiment, Heffner et al (20) directly measured the total magnetic field in ferromagnetic $La_{2/3}Ca_{1/3}MnO_3$ by muon spin precession. They again found an exponential relation between $\rho\,(T)$ and average muon site internal magnetic field. A Gaussian correlation between $\rho\,(T)$ and M(T) has been found (21) in MBE grown thin films of $La_{0.7}Ca_{0.3}MnO_3$

Colossal magnetoresistance can also be described in terms of the dependence of resistivity on magnetization, as noted by Urushibara et al (22). The fact that magnetoresistance is largest near T_c is due to the large magnetic susceptibility there. At low temperatures and in very clean systems, the magnetization saturates to the maximum possible value, and a magnetic field has only a small effect on it and thus on resistivity. For small values of the magnetization, the resistivity decreases as M^2.

In addition to magnetic frustration, multiple grain structure and grain boundary effects could be connected with the high residual resistivity of polycrystalline ferromagnetic samples (23,24). For example, Gupta et al (24) found that the residual resistivity of $La_{2/3}Ca_{1/3}MnO_3$ films increases from about $10^{-2}\Omega$ cm to $2 \times 10^{-1}\Omega$ cm as the grain size decreases from 24 μm to 3μm. Mahesh et al (23) find for a comparable system that with a further reduction in grain size to 0.02 μm, the resistivity increases to 320 Ω cm! Gupta et al analyze their data in terms of a two medium model with low resistivity (ρ_1) grains of size L separated by high resistivity grain boundary layers of thickness L' and resistivity ρ_2. In a simple effective medium theory, the net resistivity is

$$\rho \cong \rho_1 + (L'/L)\rho_2 \tag{6}$$

This fits the data reasonably well; the fitting parameter $L'\rho_2$ is about $6\times10^{-5}\Omega\, cm^2$. (The very small grain data of Mahesh et al (23) suggests a stronger increase in ρ with decreasing grain size). For an interface of thickness $L \simeq 10 - 12\mathring{A}$, the grain boundary conductance works out to nearly 10^{-5} in units of (e^2/h). Such a small conductance for a barrier three atomic layers thick is unusual. Grain boundaries seem to have a very deleterious effect on electron transport. One possibility could be magnetic frustration induced by grain boundaries on crystallites adjacent to them.

An obvious concern about these results is whether the system is a multiphase mixture, or if it is single phase, the spatial scale of inhomogeneities. There is strong x-ray and neutron evidence in many cases that the systems are single phase, even though because of the small ionic radii, there is considerable distortion of the octahedra. Not much is known about the spatial scale of the disorder. Are the systems collections of small grains? What is the connection between grain size distribution and high residual resistivity? How does the former depend on $\langle r_A \rangle$ and effect magnetization in the systematic way one sees experimentally? How can electron transport through a collection of grains and grain-boundaries evade the Mott limit at $T = 0$?

The other possibility is homogeneous frozen in local disorder due to magnetic frustration, size mismatch, orbital or Jahn-Teller distortion randomness etc.. Again, it is not clear how such disorder can lead to very large residual resistivities.

It does seem from properties of a closely related system with small $\langle r_A \rangle$, namely single crystal $Pr_{1-x}Sr_{xMnO_3}$ with $x = 0.3$ and 0.35 (25) that competing interactions that lead to many magnetic phases also cause electron localization. A relatively small magnetic field turns this system into a ferromagnetic metal.

4. Fluctuation Effects

We collect here some observations which suggest clearly that spatial and temporal fluctuations in magnetic moments (direction, size) and lattice ion displacements play an important role in the properties of manganites.

4.1. Oxygen isotope effects on T_c

The ferromagnetic transition temperature T_c depends rather strongly on the oxygen isotope mass (26). Fig. (6) shows the isotopic exponent $(dT_c/T_c)/(\Delta m/m) = \alpha_o$ for different systems. For convenience, α_o is poltted as a function of the mean radius $\langle r_A \rangle$ for several $Re_{1-x}A_xMnO_3$ manganites. α_o decreases from a large value of nearly one for small $\langle r_A \rangle$ ($\sim 1.16\mathring{A}$) to about 0.1 for a large $\langle r_A \rangle \simeq 1.22\mathring{A}$. Clearly fluctuations in the oxygen ion position strongly affect T_c; T_c larger for the lighter O^{16} ion because its oscillation amplitude is larger. Such a dependence is absent in a theory where classical or static Jahn Teller distortion causes the ferromagnetic transition in combination with double exchange.

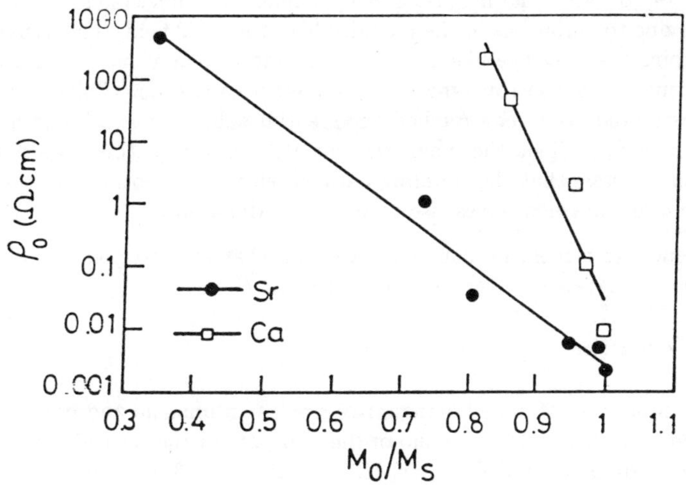

Fig.5 The residual resistivity ρ_o shown in Fig.4, plotted against T=0 saturation magnetization M for Sr and Ca ions. (From Ref.18).

However when quantum (zero point) lattice fluctuations are included in a polaronic theory, the effective e_g bandwidth depends strongly (exponentially) on the Debye frequency or the ion mass in the site bound limit. Since the ordering temperature is proportional to the bandwidth in the exchange model, the strong oxygen mass dependence of T_c is argued to point towards the existence of Jahn-Teller polarons (26). At present there are no estimates of the $(\Delta T_c/T_c)$ expected, and of its strong dependence of $\langle r_A \rangle$ [due to variation in the bond angle? Effective electron phonon coupling?].

4.2. μsr Broadening

The muon spin precession spectrum shows broadening which is commoner in magnetic glasses. For example, Heffner et al (20) found that in $La_{2/3}Sa_{1/3}MnO_3$, the muon spin orientation changes with time in a way unlike that of a decaying a particle precessing in a uniform magnetic field. It has instead of an exponential time dependence, a dependence of the form $G_2 \sim A_2 exp[-(\wedge t)^K]$ where K decreases rapidly from about unity above $T_c(\sim 260K)$ and close to it, to about 0.5 or so within a range of forty degrees or so (250 K ¡ T ¡ 300 K). Further, the characteristic rate $\wedge$ shows a rather sharp peak at about 260K. This implies that within the lifetime of the μ meson ($\sim 10^{-6}$sec), the magnetic field at muon site fluctuates. There could also be effects due to variations in the magnitude of the local field. The dynamics of coupled Mn spins is not Debye like; such slow relaxation behaviour is characteristic of glassy systems and not ferromagnets in the magnetically ordered phase. Another commonly postulated origin of stretched exponential behaviour is a broad distribution of internal fields. Since the magnetic transition is sharp, this does not seem likely. It seems possible that the coupling between electronic motion and thermally fluctuating spin arrangement gives rise to very slow dynamics.

There is evidence from inelastic neutron scattering that high frequency spin waves are anomalously broadened at temperatures close to T_c.

4.3. Optical Conductivity

The optical conductivity of doped manganites revelas an unusual and poorly understood excitation spectrum. Fig. (7) shows the real part of the optical conductivity $\sigma(\omega)$ for metallic single crystal $La_{1-x}Sr_xMnO_3$ with $x = 0.3$, as a function of temperature, in an energy range $\hbar\omega \sim< 2eV$ (16). At low temperatures, well below T_c, there is a roughly Lorentzian structure centered around $\omega = 0$ with the conductivity increasing again to a broad peak at around 1 eV. Fitting it with a Drude or Lorentz form consistent with the measured *dc* conductivity leads to a relaxation rate $(\hbar/\tau) \sim 0.02\,eV$, rather small, and a Drude number of 0.014 (i.e. $n_{eff} \sim 0.014n$ if m is assumed to have the free electron value and n is the nominal chemical formula carrier density). A more assumption free estimate of n_{eff} is obtained from the conductivity sum rule, in which the measured $\sigma(\omega)$ is integrated upto a cutoff ω_c. In this case, the cutoff is taken to be the minimum of $\sigma(\omega)$, i.e. one neglects the oscillator strength in the high frequency peak (this is not very large at low T). The corresponding carrier density is $\sim$ 0.03n in the free electron approximation.

Thus there is an enormous depletion (by a factor of 70 to 30, depending on the rather rough mode of analysis) in the ratio (n/m^*). This is not due to effective mass enhancement, since the specific heat γ shows no signs of it, being not far from the value inferred in a free electron picture. In the ferromagnetic ground state, with resident spins fully aligned, one expects the e_g electron to move coherently, more or less like a free particle. This does not seem to happen. Further, the sum rule weight decreases very rapidly with increasing temperature, even in the temperature range where the saturation magnetization does-not change much. In a tight binding model, the conductivity sum rule is related to the mean kinetic energy. This is anomalously small.

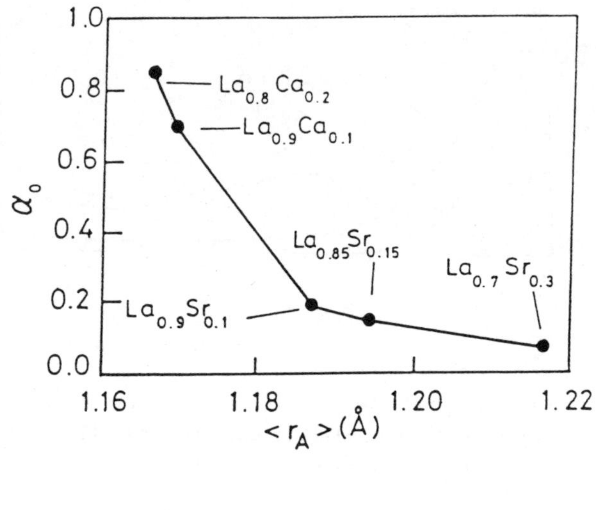

Fig.6 The dimensionless oxygen isotope mass dependence α_o of T_c $\alpha_o = d(\ell n T_c)/d(\ell n m_o)$ for several manganites, shown as a function of mean A site ion size $\langle r_a \rangle$. (From Ref. 26).

As temperature increases, there is a dramatic transfer of spectral weight from low frequencies ($\sim 0.5\,eV$) to high frequencies ($0.5\,eV < \hbar\omega < 2eV$), and above T_c the absorption is peaked at about 0.9 eV. This can perhaps be identified with a transition between 'bands' split by the Hund's rule coupling. There is, however no theoretical understanding of most of the unusual features mentioned above, especially the small low energy oscillator strength, and the complete transfer of spectral weight as a function of temperature. It is possible that the large temperature independent incoherent background that grows with ω, is also to be included in the sum rule. The nature of this background is not very clear.

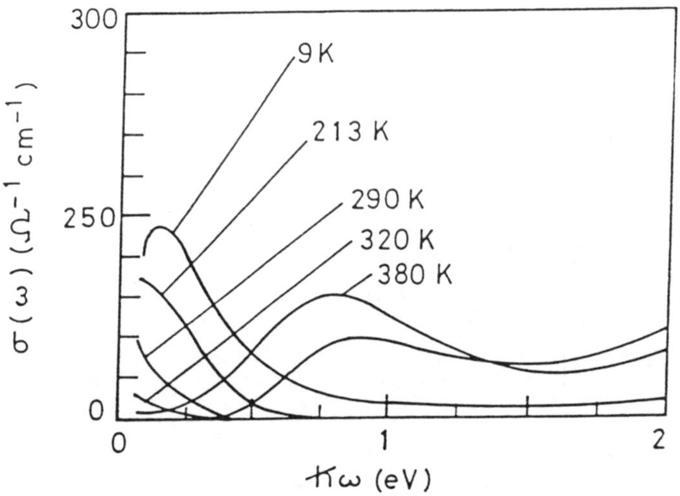

Fig.7 Real part of optical conductivity $\sigma(\omega)$ as a function of frequency $\hbar\omega$, for $La_{0.7}Sr_{0.3}MnO_3$ single crystals (From Ref. 16). The temperature independent background and phonon contributions have been removed.

5. Near the Transition

The paramagnet to ferromagnet transition has several unusual features.

5.1. Short Range Magnetic Order

Small angle neutron scattering experiments on $La_{2/3}Ca_{1/3}MnO_3$ (27) suggest that there is well defined short range magnetic order in this system over a wide range of temperatures above T_c. The magnetic correlation length ξ changes little, from about 12Åto about 18Åas the temperature decreases from 1.15 to 1.03 T_c. Between $1.03T_c$ and T_c ξ increases rapidly. This is several times the lattice spacing, and is different from what is expected in, say, nearest neighbour coupled Heisenberg spin systems above T_c. Also intriguing is the increase in ξ by a factor of two (from 12 to 25 Å) on applying a magneti field of 5T, at 1.1 T_c ! The existance of short range order is also indicated by a reduced paramagnetic susceptibility (with respect to the Curie Weiss value) observed over a wide temperature range of $T_c < T < 2T_c$ (28).

5.2. Hall Effect

Direct measurement of the number of free carriers near T_c would be very valuable. Two extreme possiblities are that the carriers are localized in the paramagnetic regime, or that there is no change in the free carrier density but only in carrier mobility. The most direct experiment would be Hall resistivity. However, the interpretation of results is difficult because of the large magnetoresistance, and the likelihood of an anomalous Hall term. Recently, Matl et al (28) argue on the basis of their results on high quality thin film samples of $La_{1-x}Ca_xMnO_3$ that the mobile carrier concentration changes strongly with temperature near T_c, and not the mobility. With an applied magnetic field as well, the main change, it is argued, is in the extended carrier density as if in the spin disordered regime, carriers are localized (See also ref.29).

Some of the basic questions raised by these results are:- why is there so much short range order above T_c? Why is it so sensitive to magnetic field? How does this short range order affect localization/delocalization of elctronic state? It is clear that single site theories cannot directly address these questions.

6. Charge Ordering

It is well known that near the equiatomic composition $Re_{1/2}A_{1/2}MnO_3$ where Re is a rare earth ion and A is a divalent alkaline earth, the Mn^{3+} and Mn^{4+} ions which are nearly equal in concentration, order in space.(5). Depending on the sizes of the Re and A ions, a rich variety of phases eg. paramagnetic insulator, ferromagnetic metal, charge ordered paramagnetic insulator, antiferromagnetic insulator has been observed (12). One of the earliest observations (5) is of the transition from a paramagnet to a ferromagnetic metal to a charge ordered antiferromagnetic insulator in $Pr_{1/2}Sr_{1/2}MnO_3$ which has a large $\langle r_A \rangle \simeq 1.24\mathring{A}$ and thus little lattice distortion. The magnetic field temperature phase diagram is shown in Fig.(8). A remarkable feature of the transition is the singular effectiveness of the magnetic field in suppressing charge ordering.

From the transition temperature $T_c \sim 150\,K$, one might naively conclude that the zero temperature energy difference between the two phases (ferromagnetic metal and antiferromagnetic insulator) is about $k_B T_c \sim 150\,K$, which is the balance between kinetic energy loss due to e_g carrier localization in the antiferromagnetic phase, and the coulomb or Madelung energy gain on charge ordering. (The Madelung energy is large, but differences in coulomb energy between a homogeneous metallic mixed valent fluid, namely the ferromagnet, and a valence or charge ordered solid may not be that large).

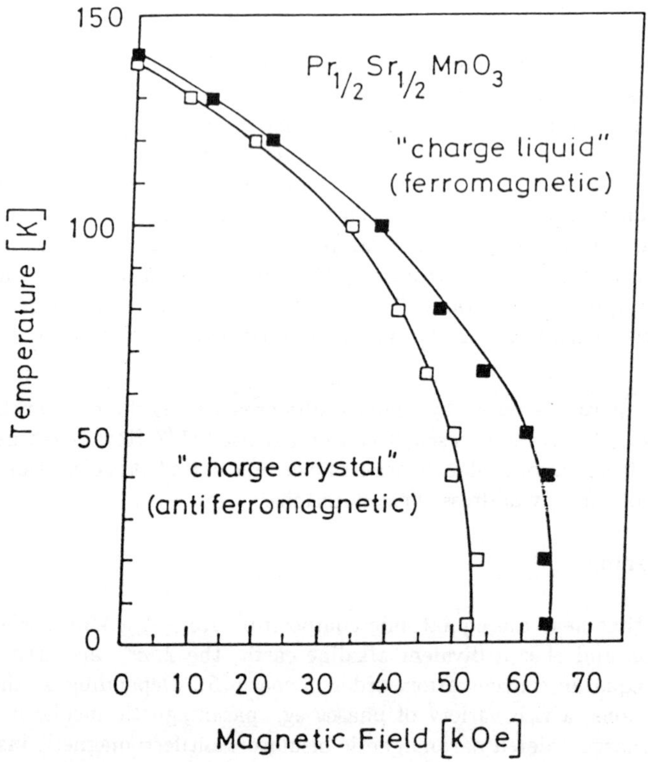

Fig.8 Temperature magnetic field phase diagram of $Pr_{1/2}Sr_{1/2}MnO_3$. (From Ref. 5). The full squares represent the transition on heating, and the empty squares the transition on cooling.

However, the magnetic field required to stabilize the ferromagnetic phase at T=0 is only 6T or an energy $3.5\mu_B H \simeq 15K$. This is a factor of ten less than $k_B T_c \simeq 150K$. From single particle tunnelling, the energy gap in the antiferromagnetic state is about 0.3 eV (30) which is more than an order of magnitude higher than $k_B T_c$!

Thus while the qualitative explanations for the possibility of various phases appear reasonable, different energy scales which ought to be comparable, differ by orders of magnitude.

In general, near equiatomic composition, depending on small changes in the ion size or $\langle r_A \rangle$, there is a variety of magnetic and charge ordered phases with dramatically different electrical properties (1). This extreme sensitivity of electronic behaviour to small local size change is unique. This is emphasized by recent results on the reentrant ferromagnetic metal transition (6).

7. Prospects

The active exploration of the properties of CMR oxides has just begun. The systems are very rich in unusual and spectacular phenomena. The basic ingredients for describing these systems all seem to be known (strong Hund's rule coupling, strong correlation, Jahn-Teller effect, and possibly Coulomb or Madelung energies). However, we seem unable to put these ingredients together and cook theories that have the full flavour of reality. In particular, there seems to be some localizing mechanism which is very sensitive to magnetization and magnetic frustration and to small differences in ion size. This has not been identified. In this contribution, I have tried to describe some phenomena from this perspective.

Acknowledgements:

I would like to thank Professors C N R Rao and A K Raychaudhuri for interesting me in this problem, and Mr. G V Pai for help in finalizing this manuscript.

References

1. C N R Rao, A K Cheetham and R Mahesh, *Chem. Mater.* **8** 2421 (1996).
2. J M D Coey, M Viret and S von Molnar, *Adv. Phys.* (to appear, 1997).
3. D I Khomskii and G A Sawatzky, *Solid State Comm.* **102** 87 (1997).
4. C Zener, *Phys. Rev.* **82**, 403 (1951).
   P W Anderson and H Hasegawa, *Phys. Rev.* **100**, 675 (1955).
5. Y Tomioka, A Asamitsu, Y Moritomo, H Kuwahara and Y Tokura, *Phys. Rev. Lett.* **74**, 5108 (1995).
6. A Arulraj, Amlan Biswas, A K Raychaudhuri, C N R Rao, A K Cheetham, P M Woodward, T Vogt and D E Cox (to be published).

7. N Furukawa, *J. Phys. Soc. Japan* **63**, 3214 (1994), *ibid* **64**, 2754 (1995), **64**, 3164 (1995).

8. E Müller-Hartmann and E Dagatto, *Phys. Rev. B***54**, R6819 (1996).

9. A J Millis, R Mueller and Bons I Shraiman, *Phys. Rev. B***54** 5389 (1996); *ibid* **54** 5405 (1996).

10. S Satpathy, Z S Popovic and F R Vukajlovic, *Phys. Rev. Lett.* **76**, 960 (1996).
W E Pickett and D J Singh, *Phys. Rev. B***53**, 1146 (1996).
D D Sarma, N Shanthi, S R Barman, N Hamada, H Sawada and K Terakura, *Phys. Rev. Lett.* **75** 1126 (1995).

11. S Yunoki, J Hu, A L Malvezzi, A Moreo, N Furukawa and E Dagotto, *cond-mat/* 9706014.

12. G J Snyder, R Hiskes, S Di Carolis, M R Beasley and T H Geballe, *Phys. Rev. B***53**, 1443 (1996).

13. K Kubo and N Ohata, *J. Phys. Soc. Japan* **33**, 21 (1972).

14. J J Hamilton, E L Keatley, H L Ju, A K Raychaudhuri, V N Smolyaninova and R L Greene, *Phys. Rev. B***54**, 14926 (1996).

15. J M D Coey, M Viret, L Ranno and K Oundajela, *Phys. Rev. Lett.* **75**, 3910 (1995).

16. Y Okimoto et al, *Phys. Rev. B***55**, 4206 (1997).

17. R Mahesh, R Mahendiran, A K Raychaudhuri and C N R Rao, *J. Solid State Chem.* **12**, 204 (1995).

18. J L Garcia-Muñoz et al, *Phys. Rev. B***55**, R668 (1997).

19. M F Hundley et al, *J. Appl. Phys.* **79** 4535 (1996).

20. R H Heffner et al, *Phys. Rev. Lett.* **77**, 1869 (1996).

21. J O' Donnell, M Onellion, M S Rzchowski, J N Eckstein and I Bozovic, *Phys. Rev. B***54**, R6841 (1996).

22. A Urushibara et al, *Phys. Rev. B***51**, 14103 (1995).

23. R Mahesh, R Mahendiran, A K Raychaudhuri and C N R Rao, *Appl. Phys. Lett.* **68**, 2291 (1995).

24. A Gupta et al, *Phys. Rev. B***54**, R15629 (1996).

25. Y Tomioka et al, *J. Phys. Soc. Jpn.* **64**, 3626 (1995)
H Yoshizawa et al, *Phys. Rev. B***52**, R13145 (1995).

26. G M Zhao, K Conder, H Keller and K A Müller, *Nature* **381**, 676 (1996).

27. J M De Teresa et al, *Naure* **386**, 256 (1997).

28. P Matl et al (to be published; N P Ong, private communication).

29. C M Varma, *Phys. Rev. B***54**, 7328 (1996).
J B Goodenough (*MRS Conference Proceedings, Boston* 1996).

30. A Biswas and A K Raychaudhuri, *J. Phys. Cond. Matt.* **8** L739 (1996).